**Inorganic Chemistry
in Focus III**

*Edited by
Gerd Meyer,
Dieter Naumann
and Lars Wesemann*

Related Titles

Meyer, G., Naumann, D., Wesemann, L. (Eds.)

Inorganic Chemistry Highlights

2002, ISBN 3-527-30265-4

Meyer, G., Naumann, D., Wesemann, L. (Eds.)

Inorganic Chemistry in Focus II

2005, ISBN 3-527-30811-3

Driess, M., Nöth, H. (Eds.)

Molecular Clusters of the Main Group Elements

2004, ISBN 3-527-30654-4

Balzani, V. (Ed.)

Electron Transfer in Chemistry
5 Volumes

2001, ISBN 3-527-29912-2

Inorganic Chemistry in Focus III

Edited by
Gerd Meyer, Dieter Naumann and Lars Wesemann

WILEY-VCH

WILEY-VCH Verlag GmbH & Co. KGaA

The Editors

Prof. Dr. Gerd Meyer
Institut für Anorganische Chemie
Universität zu Köln
Greinstr. 6
50939 Köln

Prof. Dr. Dieter Naumann
Institut für Anorganische Chemie
Universität zu Köln
Greinstr. 6
50939 Köln

Prof. Dr. Lars Wesemann
Institut für Anorganische Chemie
Universität Tübingen
Auf der Morgenstelle 18
72076 Tübingen

Library of Congress Card No.: applied for

British Library Cataloguing-in-Publication Data
A catalogue record for this book is available from the British Library.

Bibliographic information published by Die Deutsche Bibliothek
Die Deutsche Bibliothek lists this publication in the Deutsche Nationalbibliografie; detailed bibliographic data is available in the Internet at <http://dnb.ddb.de>

Typsetting K+V Fotosatz GmbH, Beerfelden
Printing Strauss GmbH, Mörlenbach
Bookbinding Litges & Dopf Buchbinderei GmbH, Heppenheim
Cover Design Adam Design, Weinheim

Printed in the Federal Republic of Germany
Printed on acid-free paper

ISBN-13: 978-3-527-31510-9
ISBN-10: 3-527-31510-1

Dedicated to Professor John D. Corbett on the occasion of his 80th birthday

In Praise of Synthesis

This book is about passion. A passion for chemistry. A passion for John D. Corbett, Distinguished Professor of Science and Humanities at Iowa State University of Science and Technology and Senior Chemist at the Ames Laboratory. A passion and admiration for John's way of conducting research in solid state chemistry and for the way he passes on his vast amount of accumulated knowledge to his students, postdoctoral associates and the community as a whole. John Corbett is a truly outstanding solid state inorganic chemist, an individual of immense and different talents, who has influenced not only his discipline but, in many ways, has led the renaissance in solid state chemistry over the past several decades.

"First comes the synthesis." What else? But it is as simple as that. It must have been sheer luck, both for him and for the scientific community, or perhaps, fate, that John D. Corbett, born in Yakima, Washington, on March 23rd, 1926, with a Ph.D. in physical chemistry (!) from the University of Washington, received a joint appointment in 1952 as Assistant Professor at the Department of Chemistry and as Associate Chemist at the Ames Laboratory of the Atomic Energy Commission (now the U.S. Department of Energy), founded and guided by the late Frank H. Spedding. It is no wonder that, after some 50 years, John D. Corbett was the recipient of the 11th Frank H. Spedding Award. This honour came late, after many others, of which his election to the National Academy of Sciences (1992) and the ACS Awards in Inorganic Chemistry (1986) and for Distinguished Service in the Advancement of Inorganic Chemistry (2000), are three other prominent examples.

But first the "synthesis" had to come! John was interested in reduced metal halides, particularly for the post-transition metals cadmium, gallium, and bismuth (his Ph.D. dissertation was on anhydrous aluminum halides and mixed halide intermediates, a good start for what was to come!). However, he was not yet actively interested in rare-earth metals and their remarkable solubility in their halides. But these elements lured him one floor below where Adrian Daane headed the metallurgy section of Spedding's empire. He knew how to produce rare-earth metals with high purity and in sufficient quantity and also how to handle tantalum containers. What if one "gave it a try" and reduced some rare-earth metal halides (John insists that this term is used correctly) from their respective metals at high temperatures under appropriate conditions,

Inorganic Chemistry in Focus III.
Edited by G. Meyer, D. Naumann, L. Wesemann
Copyright © 2006 WILEY-VCH Verlag GmbH & Co. KGaA, Weinheim
ISBN: 3-527-31510-1

in tantalum or niobium containers? This soon gave rise to a whole number of rare-earth metal subhalides, both metallic and salt-like.

It is surely always difficult to imagine the unimaginable. The "eighth wonder of the rare-earth world" had yet to be discovered. It was, in fact, discovered in a tantalum ampoule as the fantastic gadolinium sesquichloride. Clusters became a passion, they were found everywhere throughout the periodic table, of course with reduced rare-earth metal halides and their "garbage chemistry", as Bob McCarley coined it, in a friendly manner. Zirconium chemistry was an obsession for while, post-transition polyanions and cations, Zintl phases, tellurides, non-carbon fullerenes and, recently, approximants and quasicrystals. John beams boyishly when he talks about his and his co-workers latest achievements. "But there is no time to rest. There is so much which is unimaginable out there". "Explore!" "Lord, grant me patience, but hurry."

John's contributions to the development of solid state chemistry are particularly noteworthy. Together with industrial and academic chemists interested in this subject, Corbett and others have encouraged and fostered this important area of chemistry. Likewise, in the role of teacher and advisor, John Corbett motivates and encourages young people towards a career in science. His influence on an entire generation of inorganic and solid state chemists *uniquely* endears John to his many friends around the world.

As a writer and author John is also well known for many "Corbett Quotables", of which we have already mentioned a few above. Here are a few other enduring examples:

"Exploratory solid state synthesis seems to be the only workable route to new phases because of a general inability to predict relative phase stabilities and thence structures or compositions", published in "$K_4La_6I_{14}Os$: A new Structure Type for Rare-Earth-Metal Cluster Compounds that Contain Discrete Tetrahedral K_4I^{3+} Units." S. Uma, J. D. Corbett, *Inorg. Chem.* **1999**, *38*, 3831–3835.

"The diverse instances in which efficient, space-filling, bonding arrangements repeat is both surprising and pleasing", published in "Synthesis, Structure, and Bonding of $BaTl_3$: An Unusual Competition between Cluster and Classical Bonding in the Thallium Layers." D.-K. Seo, J. D. Corbett, *J. Am. Chem. Soc.* **2002**, *124*, 415–420.

"The best discoveries in an unprincipled area are often those that one stumbles upon during experiments designed with plausible but incorrect or naïve ideas regarding possible compounds or structural targets", published in "Diverse Solid-State Clusters with Strong Metal–Metal Bonding. In Praise of Synthesis." J. D. Corbett, *J. Chem. Soc., Dalton Trans.* **1996**, 575–587.

"There is *much* to be discovered that cannot be imagined. It is the wonder and excitement of finding the unprecedented and unimaginable that makes research enjoyable, even exhilarating, and worthwhile." Published in "Exploratory Synthesis in the Solid State. Endless Wonders." J. D. Corbett, *Inorg. Chem.* **2000**, *39*, 5178–591.

Needless to say, John's energy and enthusiasm for chemistry have not diminished over 50-plus years of active research but, on the contrary, appear to be on the increase. His recent discoveries of large Buckyball networks of indium, his election to the National Academy of Sciences, and his Department of Energy's Division of Basic Energy Sciences Award for Sustained Outstanding Research, all attest to his achievements. We wish John a very happy birthday and look forward to many more "Corbett Quotables" in the years to come.

Evanston, IL, and Cologne   Kenneth R. Poeppelmeier & Gerd Meyer
Summer 2006

Contents

Inorganic Chemistry in Focus III.
Edited by G. Meyer, D. Naumann, L. Wesemann
Copyright © 2006 WILEY-VCH Verlag GmbH & Co. KGaA, Weinheim
ISBN: 3-527-31510-1

Preface

This third volume of *Inorganic Chemistry in Focus* is special in many ways. First, it is dedicated to Professor John D. Corbett of Iowa State University, on the occasion of his 80th birthday on March 23rd, 2006. Second, with its 21 articles, it focuses almost entirely on inorganic solid state chemistry, although it covers a wide area stretching from theoretical considerations via new syntheses, structures and physical properties to applications. Third, these articles are written exclusively by John Corbett's former graduate students and by postdoctoral associates from throughout the world, who have all entered academia. The readers of this book will certainly notice John D. Corbett's influence on the research contained in it, which in this way will be passed on to the next generation, influencing the (solid state) chemists of the future. This book, therefore, is also a documentation of how science progresses and develops over time and how the knowledge of chemistry is disseminated. It is encouraging that all of the authors have found the time to write articles for this special book.

It is to be hoped that the celebrant will enjoy perusing through the chemistry presented in his birthday present, and also that there will be a wide-ranging and appreciative readership. Finally, we all wish John D. Corbett a very happy birthday, *ad multos annos*.

Cologne and Tübingen, Germany
Summer 2006

Gerd Meyer, Dieter Naumann,
Lars Wesemann

Inorganic Chemistry in Focus III.
Edited by G. Meyer, D. Naumann, L. Wesemann
Copyright © 2006 WILEY-VCH Verlag GmbH & Co. KGaA, Weinheim
ISBN: 3-527-31510-1

List of Contributors

Ekaterina V. Anokhina
Department of Chemistry
Wake Forest University
Winston-Salem, NC 27106
USA

Claude H. Belin
Laboratoire des Agrégats Moléculaires
et Matériaux Inorganiques
Université de Montpellier II
Sciences et Techniques du Languedoc
CC15
2 Place Eugène Bataillon
34095 Montpellier Cedex 5
France

Ling Chen
State Key Laboratory of Structural
Chemistry
Fujian Institute of Research on the
Structure of Matter
Chinese Academy of Sciences
Yangqiao Xi Road 155, PO Box 143
350002 Fuzhou, Fujian
China

Peter K. Dorhout
Department of Chemistry
Delivery 1005 – 204 Student Services
Building
Colorado State University
Fort Collins, CO 80523-1005
USA

Cheng-Jun Duan
State Key Laboratory of
High Performance Ceramics
and Superfine Microstructure
Shanghai Institute of Ceramics
Chinese Academy of Sciences
Dingxi Road 1295
Shanghai 200050
China

Ashok K. Ganguli
Department of Chemistry
Indian Institute of Technology
Delhi
110016, New Delhi
India

Gunjan Garg
Department of Chemistry
Indian Institute of Technology
Delhi
110016, New Delhi
India

Franck Gascoin
Department of Chemistry
and Biochemistry
University of Notre Dame
Notre Dame, IN 46556
USA

Inorganic Chemistry in Focus III.
Edited by G. Meyer, D. Naumann, L. Wesemann
Copyright © 2006 WILEY-VCH Verlag GmbH & Co. KGaA, Weinheim
ISBN: 3-527-31510-1

Arnold M. Guloy
Department of Chemistry
University of Houston
136 Fleming Building
Houston, TX 77204-5003
USA

Shalabh Gupta
Department of Chemistry
Indian Institute of Technology
Delhi
110016, New Delhi
India

Shiou-Jyh Hwu
Department of Chemistry
Clemson University
477 Hunter Hall
Clemson, SC 29634-0973
USA

Hideo Imoto
Department of Applied Chemistry
Utsunomiya University
7-1-2 Yoto Utsonomiya
Togichiken 321-8585
Japan

Jiong Jiang
Department of Chemistry
University of California
One Shields Avenue
Davis, CA 95616
USA

David A. Johnson
Department of Chemistry
The Open University
Milton Keynes MK7 6AA
United Kingdom

Stefan Kaskel
Institut für Anorganische Chemie
Technische Universität Dresden
Mommsenstraße 6
01069 Dresden
Germany

Susan M. Kauzlarich
Department of Chemistry
University of California
One Shields Avenue
Davis, CA 95616
USA

Sang-Hwan Kim
Department of Chemistry
and Biochemistry
Arizona State University
Tempe, AZ 85287-1604
USA

Martin Köckerling
Abt. Anorganische Chemie/
Festkörperchemie
Institut für Chemie
Universität Rostock
Albert-Einstein-Str. 3 a
18059 Rostock
Germany

Abdessadek Lachgar
Department of Chemistry
Wake Forest University
Winston-Salem, NC 27109
USA

Rosa Llusar
Departament de Ciències Experimentals
Universidad Jaume I
Campus de Riu Sec
Av. Sos Baynat s/n
12071 Castelló
Spain

Paul A. Maggard
Department of Chemistry
North Carolina State University
2620 Yarbrough Drive, 422 Dabney Hall
Raleigh, NC 27695-8204
USA

James D. Martin
Department of Chemistry
North Carolina State University
2620 Yarbrough Drive, 422 Dabney Hall
Raleigh, NC 27695-8204
USA

Gerd Meyer
Institut für Anorganische Chemie
Universität zu Köln
Greinstraße 6
50939 Köln
Germany

H.-Jürgen Meyer
Institut für Anorganische Chemie
Universität Tübingen
Auf der Morgenstelle 18
72076 Tübingen
Germany

Anja-Verena Mudring
Institut für Anorganische Chemie
Universität zu Köln
Greinstraße 6
50939 Köln
Germany

Andriy Palasyuk
Institut für Anorganische Chemie
Universität zu Köln
Greinstraße 6
50939 Köln
Germany

Dong-Kyun Seo
Department of Chemistry
and Biochemistry
Arizona State University
Tempe, AZ 85287-1604
USA

Slavi C. Sevov
Department of Chemistry
and Biochemistry
University of Notre Dame
Notre Dame, IN 46556
USA

Monique Tillard
Laboratoire des Agrégats Moléculaires
et Matériaux Inorganiques
Université de Montpellier II
Sciences et Techniques du Languedoc
CC15
2 Place Eugène Bataillon
34095 Montpellier Cedex 5
France

Cristian Vicent
Serveis Centrals d'Instrumentació
Científica
Universidad Jaume I
Av. Sos Baynat s/n
12071 Castelló
Spain

Li-Ming Wu
State Key Laboratory
of Structural Chemistry
Fujian Institute of Research
on the Structure of Matter
Chinese Academy of Sciences
Yangqiao Xi Road 155, PO Box 143
350002 Fuzhou, Fujian
China

Bangbo Yan
North Carolina State University
Department of Chemistry
2620 Yarbrough Drive, 422 Dabney Hall
Raleigh, NC 27695-8204
USA

Jing-Tai Zhao
State Key Laboratory of High
Performance Ceramics and Superfine
Microstructure
Shanghai Institute of Ceramics
Chinese Academy of Sciences
Dingxi Road 1295
Shanghai 200050
China

Biographical Sketches

Claude H. Belin was born in 1946 in Bourg Saint Andeol, a small city by the Rhône river in the south of France and grew up in Alès, an industrial and historic center in the Cévennes. The Cévennes is well known for the heroic resistance of its people to King Louis XIV's authority when, in 1685, he launched a terrible repression against the Huguenot (Protestant) religion. After graduating from high school, Claude Belin studied Physics and Chemistry at Montpellier University and obtained a Doctorate in Inorganic Chemistry in 1970, followed by a Doctorate in Physical Sciences in 1973. At this time he was working in the field of acid and superacid systems and their involvement in the protonation of, and hydrogen bonding to, weak organic bases. After one year's military service and the birth of his first son Renaud, he joined Professor John D. Corbett in 1976 at Ames Laboratory as a Post Doctoral Fellow to work in the field of naked anionic clusters of post transition elements. At this time he could not have imagined that his two-year old son Renaud would also come back to Ames twentyfour years later as a Post Doctoral Fellow in the laboratory of Professor Steve Martin. Claude Belin is Director of Research at the French Scientific Research Council (CNRS) in Montpellier and also teaches crystallography to Masters students. His research activity in the domain of intermetallic compounds and Zintl phases of main group elements is internationally recognized and in 2003 he received the "prix d'Aumale" of the French Academy of Sciences.

Inorganic Chemistry in Focus III.
Edited by G. Meyer, D. Naumann, L. Wesemann
Copyright © 2006 WILEY-VCH Verlag GmbH & Co. KGaA, Weinheim
ISBN: 3-527-31510-1

Ling Chen (left) was born on April 28th, 1971, in Guiyang, Guizhou, China. She studied chemistry at Southwest Teacher's University from 1989 to 1993, and received her Master's degree at Beijing Normal University in 1996 as well as a Ph.D. at Fujian Institute of Research on the Structure of Matter (FIRSM), Chinese Academy of Sciences in 1999. From 2000 to 2003, she carried out postdoctorate studies at the Ames Laboratory with Professor John D. Corbett. From 2003 to the present, she has been working as a full Professor of Inorganic Chemistry at FIRSM.

Li-Ming Wu (right) was born on July 4th, 1973, in Jianyang, Fujian, China. He studied chemistry at Beijing Normal University from 1989 to 1996, and received his Master's degree there. In 1999, he received his Ph.D. at Fuzhou University, and then he worked as a postdoctorate with Prof. Xin-Tao Wu at the Fujian Institute of Research on the Structure of Matter (FIRSM), Chinese Academy of Sciences from 1999 until 2001 and with Dr. Dong-Kyun Seo in Arizona State University from 2001 to 2004. He worked as an associate professor of Inorganic & Physical Chemistry at FIRSM in 2004 and as a full professor of Inorganic & Physical Chemistry at FIRSM from 2005.

Peter K. Dorhout was born on 13th February 1962 in Princeton, New Jersey. He graduated from the University of Illinois in 1985 with a Bachelor of Science degree in chemistry and also worked at the DuPont Company in Wilmington, Delaware as an undergraduate research assistant. He graduated from the University of Wisconsin in 1989 with a Ph.D. in inorganic chemistry and worked at Los Alamos National Laboratory on nuclear materials chemistry during his doctoral studies. He joined the group of Dr. John Corbett at Iowa State University in the Ames Laboratory as a postdoctoral fellow before starting his independent academic career at Colorado State University, Fort Collins, Colorado, in 1991.

He is a recognized expert in solid state and materials chemistry and environmental chemistry. He has active programs in solid state f-element chemistry and nanomaterials science. His current research interests include heavy metal detection and remediation in aqueous environments, ferroelectric nanomaterials, actinide and rare-earth metal solid state chemistry, and nuclear non-proliferation. He currently maintains a collaboration in nuclear materials with Los Alamos National Laboratory and a collaboration in peaceful materials science development with the Russian Federal Nuclear Center – VNIIEF, Sarov, Russia, U.S. State Department projects. He has published over 100 peer-reviewed journal articles, book chapters, and reviews, while presenting over 130 international and national invited lectures on his area of chemistry. Dr. Dorhout currently serves as Vice Provost for Graduate Studies and Assistant Vice President for research. He has also served as the Interim Executive Director for the Office of International Programs and as Associate Dean for Research and Graduate Education for the College of Natural Sciences at Colorado State University.

Ashok K. Ganguli was born on January 25th, 1961, in New Delhi, India. After high school (Raisina Bengali Hr. Sec. School), he studied chemistry at HansRaj College, Delhi and the University of Delhi from 1979 to 1984. He joined the research team at the Solid State & Structural Chemistry Unit in the Indian Institute of Science, Bangalore, studying superconducting oxides to earn the Ph.D. degree under the supervision of Professor C. N. R. Rao in 1990. He continued working on superconducting oxides until 1991 as a visiting scientist at Dupont Company, CR&D, Wilmington, Delaware, USA in the group of Dr. M. A. Subramanian. In March 1991, he moved to Iowa as a Post-Doctoral Associate at the Ames Laboratory with Professor John D. Corbett and worked on intermetallic compounds of Sn and Sb until March 1993. He then moved to India and worked as a scientific officer in the CSIR-Centre for Excellence in Chemistry at the Indian Institute of Science, Bangalore until 1995, after which he joined the Indian Institute of Technology, Delhi, India as Assistant Professor of Inorganic Chemistry, where he currently holds the position of Associate Professor. He was a visiting scientist at Ames Laboratory in the summers of 1996 and 1997 and also spent a sabbatical there in 2004. Ashok Ganguli has published almost 100 research papers and is a co-author of chapters in three books. His current interests are mainly in the area of synthesis and structural chemistry of a variety of materials: dielectric oxides, quaternary thiospinels, intermetallic compounds and nanocrystalline materials. He was the co-convenor of the symposia on "Nanostructured Materials" and "Modern Trends in Inorganic Chemistry" held at IIT Delhi in 2002 and 2005, respectively. He has received the Sudborough Medal for best thesis (1990) and the MRSI Medal (2006) for his research in materials chemistry.

Arnold M. Guloy was born in Manila, in the Philippines. After graduation from Manila Science High School, he studied chemistry and received his BS degree at the University of the Philippines, Quezon City. He received his Ph.D. in Inorganic Chemistry from Iowa State University under the supervision of Professor John D. Corbett in 1992. After an IBM Postdoctoral fellowship at the IBM Research Center at Yorktown Heights, he joined the faculty at the University of Houston in 1994, where he is currently an Associate Professor. He has over 70 publications with emphasis on exploratory research in solid state chemistry. He has received the NSF CAREER Award (1998) and the University of Houston Award for Excellence in Research and Scholarship (1997). He was a Visiting Professor at the University of the Philippines (1998–99) and at the Max Planck Institute for Chemical Physics of Solids in Dresden (2004).

Shiou-Jyh Hwu was born on December 10th, 1954 in Taipei, Taiwan, in the Republic of China. He became interested in Science at high school and has been fascinated by chemistry and chemical research since college. He carried out his undergraduate research in organic chemistry concerning strained heterocycle compounds, under the supervision of the late Professor She-Chong Jiang. He then earned his B.S. (1978) in Chemistry from Fu-Jen Catholic University in Taipei. After graduation, he went to the US to pursue an advanced degree and acquired a Ph.D. (1985) in Inorganic Chemistry from Iowa State University. His dissertation research focused on solid-state, condensed cluster, chemistry under the guidance of Professor John D. Corbett. He conducted postdoctoral research (1985–1988) in materials chemistry and superconductivity at Northwestern University with Professor Kenneth R. Poeppelmeier. Professor Hwu served on the faculty at Rice University before resuming his current academic profession at Clemson University. He is now a full professor and has served in the capacity of Associate Chair since 2003. During his tenure, he was a visiting professor at the Institut des Matériaux de Nantes in France (2001–2002). He was also the director of NSF Summer Research Program in Solid State and Materials Chemistry (2000–2005). His current research involves the exploratory synthesis of mixed-metal oxides and chalcogenides of catalytic, optical, electronic and magnetic importance. Professor Hwu has published almost 100 research papers and co-authored a couple of book chapters. He has also been granted one patent based on the inventions in salt-inclusion chemistry that his research group made at Clemson. He received the College of Engineering and Science Award of Excellence for Teaching in 2002, the ACS Charles H. Stone Award for Outstanding Contributions to Chemistry in the South Eastern United States in 2003, and the Researcher of the Year Award from the Sigma Xi International Honorary Research Society, in 2004.

Hideo Imoto was born on October 11th, 1949, in Kyoto, Japan, and grew up in a suburb of Osaka. After graduating from Mikunigaoka High School, Osaka, he studied chemistry at the University of Tokyo from 1968 to 1972 and continued research on hydrido complexes and transition metal hydrides to earn the Dr. sc. degree under the supervision of Professor Yukiyoshi Sasaki in 1977. As a postdoctoral fellow, he worked with Professor John D. Corbett at the Ames Laboratory from 1977 to 1979, and with Professor Arndt Simon at Max-Planck-Institut für Festkörperforschung, Stuttgart, Germany from 1979 to 1981. In 1984, he became Assistant Professor of Osaka University and moved to the University of Tokyo as Lecturer in 1990. In 2000, he moved as Professor to Utsunomiya University (Department of Applied Chemistry). His main interest is the preparation and crystal structure of metal oxides including tetrahedral units. He has published about eighty research papers.

David A. Johnson was born in 1940 and studied chemistry at Jesus College, Cambridge. He graduated in 1962 and was awarded his Ph.D. in 1966 for research on the thermodynamics of transition metal ions under the supervision of Alan Sharpe. From 1965 to 1969 he was a Research Fellow of Trinity Hall, Cambridge and during that time he spent a year at Iowa State University where he worked with John Corbett on the stabilities of lanthanide oxidation states. In 1970, he joined the Science Faculty at Britain's new Open University which subsequently pioneered large-scale distance learning in Britain and elsewhere. As part of the Open University's program, he has published some 40 books and made over 30 films and videos. In 1981, his television programme, "Elements Organized; the Periodic Table" won a Blue Ribbon award in the United States for excellence in the field of educational films. David Johnson's research interests are concerned with the application of thermodynamics to the understanding of inorganic chemistry, especially in those areas that offer new insights in teaching.

Stefan Kaskel was born on February 24th, 1969, in Bonn, Germany. He studied chemistry at the Eberhard-Karls-University in Tübingen, Germany and received his PhD degree at the same University in 1997 in the group of J. Strähle. After a post-doctorate at Ames Laboratory in the group of J. D. Corbett (1998–2000), where he worked on the synthesis and structural analysis of Zintl-type compounds, he moved in 2000 to the Max-Planck-Institute for Coal Research (Mülheim a.d. Ruhr, Germany) in the group of F. Schüth and worked on the synthesis of porous and nanostructured materials.

In 2002 he became group leader at the MPI Mülheim and established a young investigator group supported by the German Ministry of Science and Education. In 2003 he received his Habilitation degree at the Ruhr University, Bochum, Germany. In 2004 he became full Professor (Chair) of Inorganic Chemistry at the Technical University of Dresden, Germany. He has received the Feodor Lynen Award of the Alexander von Humboldt Foundation, the Reimar Lüst Award of the Max-Planck-Society, and the Young Scientist Nanotechnology Award of the German Ministry of Science and Education.

Susan M. Kauzlarich was born on September 24th, 1958 in Worcester, MA (USA). She obtained her B.S. in Chemistry in 1980 at the College of William and Mary, and her Ph.D. at Michigan State University in 1985, working with Professor Bruce Averill and James L. Dye. From 1985 to 1987, she worked with Professor John D. Corbett as a postdoctoral Fellow at Ames Laboratory and Iowa State University. In 1987, she became an Assistant Professor at the University of California at Davis and later became a full Professor. She held an Invited Guest Faculty Research Participation Appointment at Argonne National Laboratory during the fall of 1992, working with Dr. Lynn Soderholm on Skutterudites. She held a visiting scholar appointment at Argonne National Laboratory from 1997 to 1998, working on rare earth compounds with Dr. Lynn Soderholm. She received the 2001 Outstanding Mentor Award given by the Consortium for Women and Research at UC Davis, along with the 2005 Outstanding Mentor Award given by the UC Davis Academic Senate. She has overseen the graduation of 24 Ph.D. and 2 M.S. students and has mentored 11 postdoctoral fellows. Susan Kauzlarich has published 140 research papers, is co-editor of a number of books, and has authored a patent. She received the Maria Goeppert Mayer Distinguished Scholar Award from Argonne National Laboratory in 1997.

Martin Köckerling was born on the 6th of July 1965 in Olsberg, North Rhine Westphalia, Germany. After graduating from the Mauritiusgymnasium, Büren he studied chemistry and physics at the Westfälische Wilhelms-Universität Münster from 1984 until 1989, where he earned his chemistry diploma. Supported by a Kekulé fellowship of the Fonds der Chemischen Industrie he moved to the University of Duisburg (Germany) where he worked in the field of bioinorganic chemistry of nickel in the group of Professor Gerald Henkel. There he earned his Dr. rer. nat. degree in 1992. Directly afterwards, he joined Professor John D. Corbett's research group at the Ames Laboratory for a two-year postdoctoral period as a Feodor-Lynen fellow (Alexander von Humboldt foundation). Back in Germany he worked on his Habilitation at the inorganic chemistry department of the University of Duisburg, which was granted in 2001. In 2003 he accepted the offer of a full professorship for Inorganic Solid State Chemistry at the Institute of Chemistry at Rostock. The research interests of Martin Köckerling are inorganic solid-state chemistry generally and metal-rich compounds of early transition elements, in particular. He received the Bennigsen-Foerder award of the State of North Rhine Westphalia, in 2001.

Abdessadele Lachgar was born on September 5th, 1958, in Rabat, Morocco. After high school (Lycee les Orangers, Rabat), he studied Chemistry and Physics at the University of Nantes, France from 1977 to 1982, graduating with a Master's degree in Physical Sciences. He joined the Institute of Materials Jean Rouxel in Nantes to pursue research in the field of phosphato-antimonates and earned a Ph.D. degree in 1987 under the supervision of Professors Michel Tournoux and Jean Rouxel, and Dr. Yves Piffard. He joined John Corbett's group in 1988 and worked on zirconium and scandium iodide systems. In 1990, he joined the University of Washington, Seattle as a lecturer and conducted research in Professor James Mayer's group where he concentrated on the use of electrochemical synthesis to prepare extended solids from molecular precursors. In 1991, he joined the Department of Chemistry at Wake Forest University as an Assistant Professor, and was promoted to Associate Professor with tenure in 1996 and then to Full Professor in 2002. Abdou has received the Reynolds Faculty Research Fellowship twice: in 1996 and in 2003, was awarded visiting faculty fellowships of the Japanese Government in 1997 and in 1998, and was a visiting Professor at the Institute of Materials Jean Rouxel in 2003.

Rosa Llusar was born on September 20th, 1960 in Almenara (Spain). She studied chemistry at the University of Valencia (Spain) where she graduated in 1983 with the highest honors. Her doctorate work was devoted to the chemistry of cubane-type molybdenum and tungsten sulfides and she received Ph.D. degrees from Valencia University in 1987 and from Texas A&M University (USA) in 1988, under the guidance of Prof. F. Albert Cotton. After working in the Research and Development Department of a Caprolactam Production Plant in Castelló (Spain) for three years, she spent one year (1992) with Prof. John D. Corbett at the Ames Laboratory (Iowa State University, USA) investigating new solid state phases based on reduced rare earth halides. Since 1993, she has held a position at the University Jaume I of Castelló (Spain) and became Associate Professor of Physical Chemistry in 1995. During the second semester of 2005, she held a visiting professor position at the Laboratory of Chemistry, Molecular Engineering and Materials of the CNRS-Universtity of Angers (France). Her research has been focussed on the chemistry of transition metal clusters with special interest in multifunctional molecular materials and the relationship between the molecular and electronic structures of these systems with their properties. She is currently coauthor of around 80 research papers on this and related topics.

Paul A. Maggard was born on September 2nd, 1973, in Independence, Missouri (USA). He studied chemistry at William Jewell College (B.A.) from 1991 to 1995 and continued on to graduate school at Iowa State University, Ames, Iowa. There he performed research in the group of Professor John D. Corbett, obtaining his Ph.D. in 2000 on the synthesis and study of scandium/yttrium-tellurides. Subsequently, he took a postdoctoral fellowship at Northwestern University, Evanston, Illinois, working with Professor Kenneth R. Poeppelmeier on the hydrothermal synthesis of metal-oxide/fluorides from 2000 to 2002. He is currently an Assistant Professor of Inorganic Chemistry at North Carolina State University, Raleigh, North Carolina. His research is on the synthesis of heterometallic solids at the convergence of oxides, organics, and reduced compounds. He is the recipient of an Arnold and Mabel Beckman Foundation Young Investigator Award and a Ralph E. Powe Award.

James D. Martin was born June 12th, 1964 in Iowa City, Iowa, USA. He grew up in North Dakota, California and Pennsylvania, and went on to attend Goshen College, Goshen, Indiana, 1982–1986, where he majored in Chemistry and Biology and conducted undergraduate research in Physics. From 1986 to 1990 he pursued doctoral studies in molecular inorganic chemistry at Indiana University, Bloomington, Indiana under the direction of Professor Malcolm H. Chisholm. In 1991 he obtained an NSF fellowship to pursue postdoctoral studies exploring the electronic structure of organic and inorganic solids with Dr. Enric Canadell at the Laboratoire de Chemie Théorique, Orsay, France. In 1992 he returned to Indiana University as a visiting scientist to work with Dr. John C. Huffman in the X-ray Molecular Structure Center. From 1992 to 1994 he joined the research group of Professor John D. Corbett at Iowa State University where he pursued synthetic solid-state chemistry research exploring structure/property relationships. In 1994 he joined the faculty of the Department of Chemistry at North Carolina State University where he is a full professor of Inorganic Chemistry pursuing synthetic, structural and mechanistic investigations in inorganic condensed matter. He has published more than 70 research papers and has several patented discoveries. He received an NSF CAREER award in 1995, was named a Cottrell Scholar of the Research Corporation in 1997, and received a Sigma Xi Research Award in 1999.

Gerd Meyer was born on June 1st, 1949, in Schadeck, Hessen, Germany. After graduation from the Gymnasium Philippinum at Weilburg, he studied chemistry at the Justus-Liebig-Universität Gießen from 1967 to 1972 and continued research on ternary oxides to earn the Dr. rer. nat. degree under the supervision of Professor Rudolf Hoppe in 1976. As an Akademischer Rat he taught freshmen chemistry and prepared for his habilitation which was granted in 1982 at Gießen. Meanwhile, in 1980, he worked as a visiting scientist at the Ames Laboratory with Professor John D. Corbett, and was a visiting scientist and guest professor there later on several occasions. In 1988, he moved as full Professor of Inorganic Chemistry to the Universität Hannover, served as a Dean of the Chemistry Department for five years, and moved on to the Universität zu Köln in 1996. Gerd Meyer has published almost 500 research papers and is co-author and co-editor of a number of books. He has received the Preis der Justus-Liebig-Universität (1987), the Carl-Duisberg-Gedächtnispreis (1987), the Gold Medal of the Uniwersytet Wroclaw (2000), and the Terrae Rarae 2005 award for his achievements in rare-earth element chemistry.

H.-Jürgen Meyer was born in 1958 in Braunschweig (Germany), and studied chemistry from 1978 to 1983 at the Technical University of Berlin, where he received his Dr. rer. nat. degree in 1987. As a post-doctoral fellowship holder he spent two years of research at the Ames Laboratory in Ames (Iowa) and at the Baker Laboratory of the Cornell University in Ithaca (New York). In 1990 he moved to the University of Hannover (Germany) as a DFG scholarship holder, where he completed his habilitation in 1993. As a Heisenberg fellow he was offered professorships at the Universities of Bonn and Tübingen. In 1996 he accepted the position at the Eberhard-Karls-Universität at Tübingen, where he is now leading the Abteilung für Festkörperchemie und Theoretische Anorganische Chemie. He is author of the solid state section of the textbook Moderne Anorganische Chemie. His current research interests include syntheses routes, reactivities, as well as structures and electronic properties of solids. Present research topics include metal-rich halides, oxocuprates, garnets, and metal-B-C-N compounds.

Anja-Verena Mudring was born on February 14th, 1971, in Bonn, Germany. After graduating from High School at Rheinbach, she studied chemistry at the Rheinische Friedrich-Wilhelms Universität at Bonn, received her Dipl.-Chem. degree in 1995 and the Dr. rer. nat. degree in 2001, both *summa cum laude* from the Universität Bonn, under the supervision of Professor Martin Jansen. Her dissertation was concerned with the chemistry of gold: aurides, aurates and auride-aurates. She received the Feodor Lynen stipend of the Alexander von Humboldt foundation and worked as a postdoctoral associate with Professor John D. Corbett at the Ames Laboratory and the Department of Chemistry of Iowa State University. She returned to Germany as a Liebig fellow to the Universität zu Köln to establish her own research program which is concerned with "chemistry with relativity" (both experimental and theoretical work on 6th row elements) and with rare-earth element chemistry in ionic liquids. She has received a number of awards, among them the H.C. Starck award 2002 and the GEFFRUB award 2003, and she is a member of a number of learned societies, among them the German and American Chemical Societies, Sigma Xi, and the German Society of Crystallographers.

Kenneth R. Poeppelmeier was born on October 6th, 1949 in St. Louis, Missouri, USA. He studied chemistry at the University of Missouri, Columbia from 1967 to 1971 (B.S. Chemistry). From 1971 to 1974 he was an Instructor in Chemistry at Samoa College in Western Samoa as a United States Peace Corps volunteer. He joined the research group of John Corbett at Iowa State University after leaving the Peace Corps, and received his Ph.D. in 1978. He then joined the research staff of Exxon Research and Engineering Company, Corporate Research Science Laboratory, where he worked with John Longo and Allan Jacobson on the synthesis and characterization of mixed metal oxides and their application in heterogeneous catalysis. He joined the chemistry faculty of Northwestern University in 1984 where he is now Professor of Chemistry and an active member of the Center for Catalysis and Surface Science and the Materials Research Science and Engineering Center. Kenneth Poeppelmeier has published over 250 research papers and supervised approximately 40 Ph.D. students in the area of inorganic and solid state chemistry. He is a Fellow of the American Association for the Advancement of Science (AAAS) and the Japan Society for the Promotion of Science (JSPS) and has been a Lecturer for the National Science Council of Taiwan (1991), Natural Science Foundation of China (1999) and Chemistry Week in China (2004), and more recently an Institut Universitaire de France Professor (2003).

Dong-Kyun Seo was born on July 4th, 1968 in Incheon, South Korea. He studied chemistry at Seoul National University and finished his Master's degree in 1992 working on the synthesis of double perovskites. From 1994 to 1997, he studied for his PhD under the supervision of Prof. M.-H. Whangbo at North Carolina State University, USA, on theoretical studies of structural and electronic factors affecting metal-versus-insulating properties of low-dimensional materials. His theoretical work continued in his postdoctoral research at Cornell University with Prof. Roald Hoffmann. In 1998, he started to work as a postdoctoral researcher with Prof. John D. Corbett in the Ames Laboratory at Iowa State University on exploratory synthesis of alkaline-earth trielides. He joined the Department of Chemistry and Biochemistry at Arizona State University in 2001 as an assistant professor and received the CAREER Award from the National Science Foundation (NSF), USA. His research interests include half metals and bulk giant magnetoresistance materials, the chemical understanding of itinerant electron magnetism, Zintl chemistry for new semiconducting materials and the development of new synthetic methodologies for inorganic solids in various length scales.

Slavi C. Sevov was born on April 14th, 1960, in Kurdjali, Bulgaria. After attending the National Mathematics High-School in Sofia and a subsequent two-year mandatory military service, he studied chemistry at the Sofia University "St. Clement of Ohrid" from 1980 to 1985 and received an M.S. degree. He worked as a Research Associate for two years in the Bulgarian Academy of Sciences. In the Fall of 1987 he entered the graduate program at Iowa State University and joined Prof. John Corbett's group. In 1993, he obtained his Ph.D. degree for research on synthesis and characterization of indium clusters in the solid state. After two years as a postdoctoral fellow at the University of Chicago, he joined the faculty at the University of Notre Dame in 1995 and was promoted to Associate and Full Professor in 1999 and 2002, respectively. Slavi Sevov is a recipient of the Camille and Henry Dreyfus New Faculty Award for 1995, and the ExxonMobil Faculty Fellowship in Solid State Chemistry for 1999. He has authored more than 100 publications.

Jing-Tai Zhao was born on April 24th, 1962, in Hebei, China. He enrolled into Fudan University in 1978 after having passed the national examination. After obtaining his BSc., majoring in Nuclear Physics in 1982, he passed the examination for graduate study and was admitted to the Institute of Physics (Beijing) Chinese Academy of Sciences, and obtained his MSc in 1985 majoring in Solid State Physics. After joining the Laboratoire de Cristallographie aux Rayons X at the University of Geneva in 1996, he broadened his interests to crystallography and obtained his Ph.D. in 1991 in Crystallography. From 1991 to 1994 he was a postdoctoral fellow at the Ames Laboratory with Professor John D. Corbett. In 1994 he moved as an Associate Professor to the Department of Chemistry at Xiamen University, China and was then promoted to full Professor in 1997 in the college of Chemistry and Chemical Engineering at Xiamen University. After spending two years as a visiting scientist at the Max-Planck-Institut für Chemische Physik fester Stoffe at Dresden, Germany in 1999 and 2001, he took his current position as a Research Professor in 2002 at the Shanghai Institute of Ceramics, the Chinese Academy of Sciences, China. Jing-Tai Zhao is the author or co-author of more than 120 published research papers. He obtained the Fund for the Distinguished Young Researchers from the NSF of China (2001), the Fund for the "Hundred Talented Researchers" from the Chinese Academy of Sciences (2004) and the Fund for the Excellent Young Leading Researchers from the Shanghai Municipal government (2005). His research interests lie in the area of the Chemistry and Physics of Solid State Inorganic Materials.

1

Inter-electron Repulsion and Irregularities in the Chemistry of Transition Series

David A. Johnson

1.1
Introduction: Irregularities in Lanthanide Chemistry

Both ligand field effects and inter-electronic repulsion produce irregularities in the chemistry of transition series. Irregularities due to inter-electronic repulsion are most obvious in the lanthanide series where ligand field effects are very small. For the first century of lanthanide chemistry, talk of irregularities would have seemed ridiculous. The laborious discovery and separation of the elements by the classical techniques of fractional crystallization and precipitation naturally led to the view that the lanthanides were all very much alike. But by 1933, Klemm had exposed inadequacies in this similarity paradigm when he made dihalides of samarium, europium and ytterbium by hydrogen reduction and thermal decomposition of trihalides [1, 2]. The compounds had crystal structures that were also to be found among the alkaline earth dihalides. On an ionic formulation, they contain Ln^{2+} ions with the configurations $[Xe]4f^6$, $[Xe]4f^7$ and $[Xe]4f^{14}$.

Klemm's work revealed important differences among some of the rare earth elements. But their full extent was made apparent by Corbett and his coworkers [2, 3 a]. Corbett devised techniques for the determination of Ln/LnX_3 phase diagrams in tantalum and molybdenum containers in the temperature range 500–1200 °C. This use of more powerful reducing agents led to the preparation of alkaline earth-like dihalides of neodymium, dysprosium and thulium. Moreover, the conditions that generated new dihalides for some lanthanide elements failed to do so for others. For example, Corbett's work suggested that, whereas dihalides of dysprosium [4, 5] and thulium [6] were stable to disproportionation, those of erbium [7] were not. This was especially interesting because Klemm had emphasized that in both halves of the lanthanide series, this stability of the +2 oxidation state increased: in the first half up to the half-filled shell configuration at europium and in the second up to the filled shell at ytterbium [1]. The instability of erbium dihalides showed that in the second half of the series this increase was broken. Indeed, by combining a survey of the success or failure of preparative attempts with metal solubilities in molten trichlorides, it was possi-

Inorganic Chemistry in Focus III.
Edited by G. Meyer, D. Naumann, L. Wesemann
Copyright © 2006 WILEY-VCH Verlag GmbH & Co. KGaA, Weinheim
ISBN: 3-527-31510-1

ble to compile a stability sequence for the dipositive state across the entire series: $La < Ce < Pr < Nd < (Pm) < Sm < Eu \gg Gd < Tb < Dy > Ho > Er < Tm < Yb$ [2]. Stoichiometric alkaline earth-like dihalides are known only for Nd, Sm, Eu, Dy, Tm and Yb, although those of Pm could almost certainly be obtained if desired (I exclude metallic diiodides, e.g., LaI_2). The neodymium, thulium and dysprosium dihalides are exceptionally powerful reducing agents and may therefore have synthetic applications. For instance, in the presence of amide or aryloxide ligands, the di-iodides dissolve in THF and reduce nitrogen gas. The reduced di-nitrogen bridges two lanthanide(III) sites in a $\mu\text{-}\eta^2$: $\eta^2\text{-}N_2^{2-}$ arrangement [8].

The stability sequence given above applies to both of the following situations:

$$MCl_2(s) + 1/2\,Cl_2(g) = MCl_3(s) \tag{1}$$

$$MCl_2(s) = 1/3\,M(s) + 2/3\,MCl_3(s) \tag{2}$$

How can it be explained? It is useful to begin with reaction (1). By constructing a thermodynamic cycle around this reaction, we obtain the equation,

$$\Delta G^0(1) = I_3 + L(MCl_3, s) - L(MCl_2, s) + C \tag{3}$$

Here I_3 is the third ionization enthalpy of the lanthanide element and $L(MCl_n,s)$ is ΔH^0 for the following reaction:

$$M^{n+}(g) + nCl^-(g) = MCl_n(s) \tag{4}$$

The term C, which includes the enthalpy of formation of the gaseous chloride ion and $-T\Delta S^0(1)$, varies very little across the series. Thus the variations in $\Delta G^0(1)$ are determined by those in the first three terms on the right of Eq. (3). The combination of smooth lanthanide contractions with negligible ligand field effects suggests that $[L(MCl_3,s) - L(MCl_2,s)]$ should change smoothly and slightly across the lanthanide series. Consequently the variations in $\Delta G^0(1)$ should be almost entirely determined by those in I_3.

When this analysis was first attempted [9–11] very few values of I_3 had been obtained from series limits in the third spectra of the lanthanides, and the first comprehensive sets were calculated from Born-Haber cycles [9]. Subsequent spectroscopic values [12] confirmed the early work and are plotted in Fig. 1.1. In all cases they refer to the ionization process

$$M^{2+}([Xe]4f^{n+1}, g) = M^{3+}([Xe]4f^n, g) + e^-(g) \tag{5}$$

The specified configurations are ground-state configurations except at $La^{2+}(g)$ and $Gd^{2+}(g)$ where the ground states are $[Xe]5d^1$ and $[Xe]4f^7 5d^1$ respectively. It can be seen that the variations in I_3 do indeed correspond to the stability sequence for the dipositive oxidation state. The correspondence can also be tested quantitatively by using estimated and experimental values of $\Delta G^0(1)$. These are also plotted in Fig. 1.1. The parallelism between the two is very close.

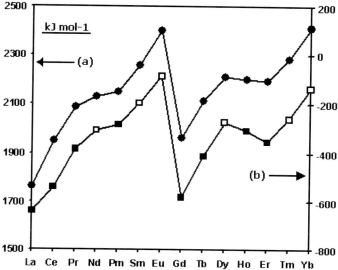

Fig. 1.1 (a) The ionization enthalpies of dipositive lanthanide ions with configurations of the type [Xe]4f^{n+1} (upper plot; left-hand axis). (b) The standard Gibbs energy change of reaction 1 (lower plot; right-hand axis; ■ estimated value; □ experimental value). Data are from Refs. [11–14].

Figure 1.1 shows that the stability sequence revealed by chemical reactions and chemical synthesis *corresponds* to thermodynamic stabilities. An *explanation* requires a theory that will explain both. To get it we apply the theory of atomic spectra [9]. The energy of the 4f electrons in an ion with the configuration [Xe]4f^n, $E(4f^n)$, can be written [$nU+E_{rep}(4f^n)$] where U, a negative quantity, is the energy of each 4f electron in the field of the positively charged xenon core, and $E_{rep}(4f^n)$ represents the repulsion between the n 4f electrons. In Table 1.1, $E_{rep}(4f^n)$ is expressed as a function of the Racah parameters E^0, E^1 and E^3. The subsequent column gives the ionization energy of each configuration, [$E(f^{n-1})-E(f^n)$].

Despite our neglect of spin-orbit coupling, the theoretical ionization energies of columns 3 and 6 account for the I_3 variation in Fig. 1.1. The term [$-U-(n-1)E^0$] leads to an overall increase across the series brought about by the increasing nuclear charge. But this increase is set back after the half-filled shell by the appearance of the quantity $-9E^1$. Finally the terms in E^3 produce irregularities in the 1/4- and 3/4-shell regions. Because the Racah parameters increase steadily across the series as the f-orbitals contract and inter-electronic repulsion rises, E^3 is greater in the second half of the series than in the first. Indeed, the terms in E^3 are then large enough to eliminate the overall increase in ionization energy between f^{10} and f^{12}, so dysprosium(II) compounds are more stable than those of erbium(II).

Table 1.1 The inter-electronic repulsion energies, E_{rep} (f^n), and the ionization energies, I (f^n), of f^n configurations according to the theory of atomic spectra.

n	E_{rep} (f^n)	I (f^n)	n	E_{rep} (f^n)	I (f^n)
0	0				
1	0	$-U$	8	$28E^0+9E^1$	$-U-7E^0-9E^1$
2	E^0-9E^3	$-U-E^0+9E^3$	9	$36E^0+18E^1-9E^3$	$-U-8E^0-9E^1+9E^3$
3	$3E^0-21E^3$	$-U-2E^0+12E^3$	10	$45E^0+27E^1-21E^3$	$-U-9E^0-9E^1+12E^3$
4	$6E^0-21E^3$	$-U-3E^0$	11	$55E^0+36E^1-21E^3$	$-U-10E^0-9E^1$
5	$10E^0-9E^3$	$-U-4E^0-12E^3$	12	$66E^0+45E^1-9E^3$	$-U-11E^0-9E^1-12E^3$
6	$15E^0$	$-U-5E^0-9E^3$	13	$78E^0+54E^1$	$-U-12E^0-9E^1-9E^3$
7	$21E^0$	$-U-6E^0$	14	$91E^0+63E^1$	$-U-13E^0-9E^1$

The *mathematics* of the theory of Table 1.1 therefore accounts for the variations in both I_3 and in the stability of alkaline earth-like dihalides. There remains the question of a *physical* explanation. The most important irregularity is the very large downward break after the half-filled shell, and the main contribution to it comes from the exchange energy [15]. This arises from the fact that electrons with parallel spins experience a smaller repulsion than do those with opposed spins. Blake [16] showed that whether one chooses the familiar real orbitals, or imaginary ones with defined m_1 values, the exchange energy contributes about 70% of the half-filled break for p^n configurations and 75% for d^n configurations. In the case of f^n configurations, Newman's coulomb and exchange integrals [17] suggest that the contribution is over 80%. From [Xe]4f^1 to [Xe]4f^7, ionization destroys 0, 1, 2, 3, 4, 5 and 6 parallel spin interactions, progressively raising I_3. At europium therefore, the +2 oxidation state reaches a stability maximum because afterwards, at [Xe]4f^8, the new electron goes in with opposed spin. Its loss then destroys no parallel spin interactions, and the 0–6 pattern is repeated from [Xe]4f^8 to [Xe]4f^{14} where a second stability maximum occurs at ytterbium(II). A more formal treatment includes an explanation of the 1/4- and 3/4-shell effects related to Hund's second rule [15].

1.2
A General Principle of Lanthanide Chemistry

Our analysis of thermodynamic stabilities has been developed through Eq. (3), but is of more general importance [18]. This is because it leads to a general principle composed of two parts. Each part deals with a particular class of reaction. The first class is typified by reaction 1. Because ligand field effects are very small, L(MCl_2,s) and L(MCl_3,s) change smoothly and slightly across the series, and the variations in ΔG^0(1) are completely dominated by those in I_3. This is apparent from the close parallelism between the two quantities. Under such circumstances, a change in the ligands leaves the irregularities largely unaffected, and the I_3-type variation in Fig. 1.1 is characteristic of *any* process in which the number of 4f electrons decreases by one.

The second class of reaction is that of processes in which the 4f electrons are conserved. The obvious examples are the complexing reactions of tripositive lanthanide ions. Here the irregularities due to changes in inter-electronic repulsion almost entirely disappear. We then get the slight smooth energy change whose consequences were so familiar to 19th century chemists, who struggled with the separation problem.

In many cases, lanthanide reactions can either be assigned exclusively to one of these two classes, or they show deviations that the classification makes understandable. In Fig. 1.2, we plot the values of ΔH^0 for the complexing of the tripositive aqueous ions by EDTA^{4-}(aq), a reaction in which the 4f electrons are conserved. The irregularities are negligible at the chosen scale. Also shown are the values of ΔH_f^0 (MCl$_3$,s) which refer to:

$$M(s) + 3/2Cl_2(g) = MCl_3(s) \tag{6}$$

Here, for nearly all of the elements, the number of 4f electrons in the metallic state and in the trichloride is the same, so we expect a largely smooth energy

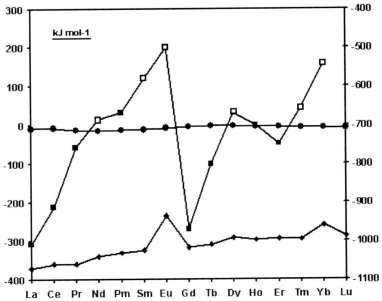

Fig. 1.2 Standard enthalpy changes of (a) the complexing of lanthanide ions in aqueous solution by EDTA^{4-} (● left-hand axis); (b) the standard enthalpy change of reaction 2, the dichloride being a di-f compound (left-hand axis; ■ estimated value; □ experimental value); (c) the standard enthalpy change of reaction 6 (◆ right-hand axis). Data are from Refs. [11, 13, 14, 18 and 19].

variation. This is what we get. The exceptions are at europium and ytterbium, which form two-electron metals. In these two cases, reaction 6 is one in which the number of 4f electrons decreases by one. We can assume that the energy variation would be smooth if europium and ytterbium were three-electron metals like the other lanthanides. The observed deviations of about 85 and 40 kJ mol^{-1} then tell us the stabilizations that europium and ytterbium metals achieve by adopting a two- rather than a three-electron metallic state [9, 19]. Finally Fig. 1.2 contains the values of $\Delta H^0(2)$, the standard enthalpy of disproportionation of an alkaline earth-like dichloride. In nearly all cases, this is a process in which the number of 4f electrons decreases by one, and we see the extreme, but characteristic, I_3-type variation. Again the exceptions are at europium and ytterbium where the occurrence of two-electron metals on the right-hand side of the equation lowers the values by about 28 and 13 kJ mol^{-1}, respectively.

The principle introduced above is best exploited by classifying lanthanide compounds not by oxidation state, but by the number of 4f electrons at the metal site. For example, the reaction

$$MS(\text{semi-conductor}) = MS(\text{metallic}) \tag{7}$$

is one in which there is no change in *formal* oxidation state. In this sense, it resembles the EDTA complexing reaction of Fig. 1.2. But the energy variation is quite different. The reaction is one in which the 4f electron population decreases by one and its energy variation parallels I_3. Thus we observe semiconductors at Sm, Eu and Yb, and metallic sulfides at the other lanthanide elements. By using these ideas, quantitative values of $\Delta G^0(7)$ have been estimated [20]. If we label species which contain the same number of 4f electrons as the [Xe]4f^{n+1} configuration of the free M^{2+} ion, di-f, and those with the same number of 4f electrons as the [Xe]4f^n configuration of the free M^{3+} ion, tri-f, then reaction 7 is one in which a di-f to tri-f transformation occurs and the number of 4f electrons decreases by one. This classification simplifies discussion of "lower oxidation states" of the lanthanide elements [2].

1.3
Extensions of the First Part of the Principle

The principle just outlined has two parts. The first part deals with redox processes and was developed here by examining the relative stabilities of the +2 and +3 oxidation states of the lanthanides. It can be extended in a variety of ways. Thus if the I_3 variation is shifted one element to the right, it tells us the nature of the I_4 variations, and accounts for the distribution of the +4 oxidation states of the lanthanides [2, 10, 15]. Their stability shows maxima at cerium(IV) and terbium(IV), decreasing rapidly as one moves from these elements across the series.

Similar principles apply to the actinides. The +2 oxidation state is present in dipositive aqueous ions and alkaline earth-like dihalides. In the first half of the series only americium, where the Am^{2+} ion has the half-shell configuration $[Rn]5f^7$, forms such a dihalide. The drop in I_3 suppresses further dihalide formation at curium and berkelium, but such compounds reappear at californium and einsteinium. By mendeleevium, the dipositive aqueous ion is more stable than $Eu^{2+}(aq)$, and at nobelium, $No^{3+}(aq)$ is a stronger oxidizing agent than dichromate.

The +4 oxidation state is most stable at thorium, which lies beneath cerium. Its stability then decreases progressively until we reach curium where aqueous solutions containing the tetra-positive state must be complexed by ligands such as fluoride or phosphotungstate. Even then, they oxidize water and revert to curium(III). The expected drop in I_4 between curium and berkelium provides $Bk^{4+}(aq)$ with a stability similar to that of $Ce^{4+}(aq)$, but the decrease in stability is then renewed, and beyond californium, the +4 oxidation state has not yet been prepared [2, 10, 15].

In the lanthanide and actinide series, arguments like these are greatly eased by the very small ligand field effects. Consider the reaction

$$M^{2+}(aq) + H^+(aq) = M^{3+}(aq) + 1/2\, H_2(g) \tag{8}$$

The *variations* in ΔG^0 across the series are given by:

$$\Delta G^0(8) = I_3 - \Delta H_h^0(M^{2+}, g) + \Delta H_h^0(M^{3+}, g) + C \tag{9}$$

Here ΔH_h^0 (M^{n+},g) is the enthalpy of hydration of the gaseous M^{n+} ion, and the entropy change is assumed to be constant. Because ligand field effects are very small, the hydration enthalpies vary smoothly and slightly across the series, and the variations in $\Delta G^0(8)$ are dominated by those in I_3. If we move to the first transition series, the values of I_3 follow the expected pattern. They increase from Sc^{2+} to Mn^{2+} where we reach the half-shell configuration $[Ar]3d^5$, and then drop steeply at Fe^{2+}. The increase is then renewed up to the full shell at zinc. But the hydration enthalpies no longer vary smoothly. They show double-bowl shaped variations explained by octahedral ligand field stabilization energies. Because H_2O is a weak field ligand, the bowls are not too deep. The $\Delta G^0(8)$ variations are therefore still dominated by those in I_3, albeit in an attenuated form. Thus at the beginning of the series, Sc^{2+} (aq) and Ti^{2+} (aq) are unknown. V^{2+} (aq) and Cr^{2+} (aq) exist but are readily oxidized by air, and Mn^{2+} (aq) is stable with respect to this reaction. The decrease in stability of the dipositive oxidation state between manganese and iron is neatly shown by the ready occurrence of the reaction

$$Mn^{3+}(aq) + Fe^{2+}(aq) = Mn^{2+}(aq) + Fe^{3+}(aq) \tag{10}$$

After iron, the tripositive ions are unstable: $Co^{3+}(aq)$ slowly oxidizes water at room temperature, and $Ni^{3+}(aq)$, Cu^{3+} (aq) and $Zn^{3+}(aq)$ do not exist [15, 33].

1.4
Extensions of the Second Part of the Principle

As Fig. 1.2 shows, thermodynamics distinguishes lanthanide reactions in which the 4f population changes from those in which the 4f population is conserved. In the latter type of reaction, the second part of our principle states that the energy variation is nearly smooth. Why do we need the qualification "nearly"? First, there are many important chemical changes to which thermodynamics is rather insensitive. Structure is often a good example. The smooth energy variation in Fig. 1.2 refers to a complexing reaction in aqueous solution. It is generally accepted that, as we move across the lanthanide series, the coordination number of the aqueous ion changes. Yet on the scale of Fig. 1.2, this produces no obvious irregularities. A more relevant structural case is that of the lower halides of La, Ce, Pr and Gd [21, 22]. These are elements which do not form di-f alkaline earth-like dihalides, and their lower halides contain significant metal–metal bonding. Again, the work was both pioneered and continued by Corbett [3]. In such compounds, the 4f populations at the metal sites seem to be identical with those in both the metallic element and the trihalide. Thus stability is determined by reactions such as

$$Gd_2Cl_3(s) = Gd(s) + GdCl_3(s) \qquad (11)$$

in which all substances are tri-f and there is no change in the 4f electron populations. With three bonding electrons per metal atom, there are six such electrons on each side of Eq. (11). Three can be allocated to the formation of bonds with chlorine, and three to the formation of multi-centred Gd–Gd bonding. So the bonding on each side of the equation is similar: on the left it is distributed over one substance; on the right over two. If correct, this suggests that $\Delta H^0(11)$ should be close to zero and, in fact, the value is only 30 ± 15 kJ mol^{-1} [23]. An important contribution to the small positive value seems to be the splitting of the 5d bands generating the metal–metal interaction in Gd_2Cl_3. This stabilizes the compound and makes it a semiconductor [24].

Unlike the di-f dihalides, such compounds differ little in energy from both the equivalent quantity of metal and trihalide, and from other combinations with a similar distribution of metal–metal and metal–halide bonding. So the reduced halide chemistry of the five elements shows considerable variety, and thermodynamics is ill-equipped to account for it. All four elements form di-iodides with strong metal–metal interaction, PrI_2 occurring in five different crystalline forms. Lanthanum yields LaI, and for La, Ce and Pr there are halides M_2X_5 where X=Br or I. The rich variety of the chemistry of these tri-f compounds is greatly increased by the incorporation of other elements that occupy interstitial positions in the lanthanide metal clusters [3b, 21, 22].

These difficulties show that the description "nearly smooth" for the energies of inter-conversion of tri-f compounds is a confession of inadequacy. But other kinds of reaction in which the 4f electrons are conserved suggest that it may be possible to refine "nearly smooth" into something more precise. To this we now turn.

1.5
The Tetrad Effect

What became known as the tetrad effect was first observed in the late 1960s during lanthanide separation experiments [25]. Fig. 1.3 shows a plot of log K_d, where K_d is the distribution ratio between the aqueous and organic phases in a liquid–liquid extraction system. There are four humps separated by three minima, first at the f^3/f^4 pair, secondly at the f^7 point, and thirdly at the f^{10}/f^{11} pair. Calls for an explanation were answered by Jorgensen and elaborated by Nugent [26]. When a lanthanide ion moves from the aqueous to the organic phase, the nephelauxetic effect leads to a small decrease in inter-electronic repulsion within the 4f shell. This decrease varies irregularly with atomic number and is responsible for the irregularities in Fig. 1.3.

This initial explanation and subsequent developments use Jorgensen's refined spin-pairing energy theory. This theory refers the repulsion energy changes to a baseline drawn through points at the f^0, f^1, f^{13} and f^{14} configurations. But, for reasons that will become apparent, I shall use a baseline through the f^0, f^7 and f^{14} points. Column 2 of Table 1.2 repeats the formulae for $E_{rep}(f^n)$ taken from Table 1.1. The baseline function g(n) in column 3 passes smoothly through the f^0, f^7 and f^{14} values and takes the form

$$g(n) = 1/2\,n(n-1)E^0 + (9\,n/14)(n-7)E^1 \qquad (12)$$

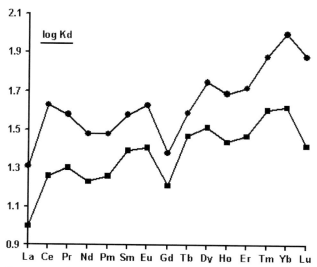

Fig. 1.3 (a) Observed values of log K_d where K_d is the distribution constant for lanthanide ions between aqueous 11.4 M LiBr in 0.5 M HBr and 0.6 M $(ClCH_2)PO(OC_8H_{17})_2$ in benzene (Ref. [26a]; upper plot). (b) A similar variation constructed by using the theory of Table 1.2 (lower plot; see text).

Column 4 is the difference between columns 2 and 3. It tells us the deviations of the total 4f inter-electronic repulsion energy from the f^0, f^7 and f^{14} baseline. In between f^0 and f^7, and again between f^7 and f^{14}, the relative sizes of E^1 and E^3 are such that the repulsion is raised. In discussing the effect upon *reactions*, the quantities E^1 and E^3 should be replaced by ΔE^1 and ΔE^3. If ΔE^1 and ΔE^3 are both negative, then the formulae of column 4 can reproduce the form of the log K_d variation. The change in ΔE^1 increases the f^1– f^6 and f^8 – f^{13} values relative to those at f^0, f^7 and f^{14}, but that in ΔE^3 moderates those increases, most notably at f^3, f^4, f^{10} and f^{11}. This can reproduce the four-hump variation of Fig. 1.3. The lower plot has been constructed by superimposing the formulae in column 4 of Table 1.2, with the values $\Delta E^1 = -28$ cm^{-1} and $\Delta E^3 = -7$ cm^{-1}, upon a linear increase of 0.03 per element. The parallelism is obvious. Effects of this sort are of interest to geochemists. Tetrad patterns in the concentrations of lanthanide elements have been used to explore the evolutionary history of igneous rocks such as granites [27]. The effect has also been invoked to explain the distribution of rare earth elements in sea water [28].

Very often, the tetrad effect is *not* clearly discernible in the energies of processes in which 4f electrons are conserved. It may, for example, be obscured by irregularities caused by structural variations in either reactants or products. This is especially likely given the willingness of lanthanide ions to adopt a variety of coordination geometries. There is, however, no doubt that tetrad-like patterns are often observed. But does Table 1.2 provide a convincing explanation of what is seen?

Imagine a thoroughly convincing test of the explanation. We begin with a reaction in which the 4f electrons are conserved. In the sequence La → Lu, each

Table 1.2 The excess inter-electronic repulsion for f^n configurations (column 4), relative to a smoothly varying baseline function, g(n), drawn through the formulae for f^0, f^7 and f^{14}.

n	E_{rep} (f^n)	g (n)	[E_{rep} (f^n) − g (n)]
0	0	0	0
1	0	$-(54/14)E^1$	$(54/14)E^1$
2	E^0-9E^3	$E^0-(90/14)E^1$	$(90/14)E^1-9E^3$
3	$3E^0-21E^3$	$3E^0-(108/14)E^1$	$(108/14)E^1-21E^3$
4	$6E^0-21E^3$	$6E^0-(108/14)E^1$	$(108/14)E^1-21E^3$
5	$10E^0-9E^3$	$10E^0-(90/14)E^1$	$(90/14)E^1-9E^3$
6	$15E^0$	$15E^0-(54/14)E^1$	$(54/14)E^1$
7	$21E^0$	$21E^0$	0
8	$28E^0+9E^1$	$28E^0+(72/14)E^1$	$(54/14)E^1$
9	$36E^0+18E^1-9E^3$	$36E^0+(162/14)E^1$	$(90/14)E^1-9E^3$
10	$45E^0+27E^1-21E^3$	$45E^0+(270/14)E^1$	$(108/14)E^1-21E^3$
11	$55E^0+36E^1-21E^3$	$55E^0+(396/14)E^1$	$(108/14)E^1-21E^3$
12	$66E^0+45E^1-9E^3$	$66E^0+(540/14)E^1$	$(90/14)E^1-9E^3$
13	$78E^0+54E^1$	$78E^0+(702/14)E^1$	$(54/14)E^1$
14	$91E^0+63E^1$	$91E^0+(882/14)E^1$	0

reactant must be isostructural, as must each product. Uncertainties in the energy variation for the reaction must be smaller than the irregularities attributed to the tetrad effect. Finally, the spectra of each reactant and product must be analyzed to provide values of ΔE^1 and ΔE^3, and the auxiliary changes in ligand field stabilization and spin-orbit coupling energies. The size of the humps can then be evaluated, auxiliary contributions subtracted and the residues compared with the values predicted using the values of ΔE^1 and ΔE^3. Published tests are impressive, but fall short of this standard [29]. The difficulty is the small size of lanthanide nephelauxetic effects compared with uncertainties in the input data.

1.6
The Diad Effect

Quantitative tests of the effect of inter-electronic repulsion on the energies of reactions in which 4f electrons are conserved are therefore very difficult. But they are possible for reactions in which 3d electrons are conserved, and the standards of proof set out in the previous section can then be applied. In the first transition series, variations in lattice and hydration enthalpies take the form of double-bowl shapes. Standard texts have long attributed these irregularities to what George and McClure called inner-orbital splitting [30]. This splitting is induced by the symmetry of the ligand field. George and McClure noted that in some cases, especially the hydration enthalpies of the M^{3+} ions, the size of the bowls was too large to be consistent with spectroscopic values of the orbital splitting parameter Δ. They thought that the discrepancy might be explained by changes in the spin-orbit coupling energy, or by a relaxation energy that allows for the effect of changes in bond length induced by the ligand field. These additional terms, however, only increased the disagreement. In the 1990s, the discrepancy was attributed to the nephelauxetic effect, using an explanation of the kind embodied in Table 1.2 [31].

In d^n series, we use a d^0, d^5 and d^{10} baseline, and the values analogous to those given in column 3 of Table 1.2 are expressed in terms of the Racah parameters B and C. At d^1, d^4, d^6 and d^9, the values are $(7B+2.8C)$; at d^2, d^3, d^7 and d^8, they are $(6B+4.2C)$. When a gaseous ion becomes coordinated, the nephelauxetic effect ensures that ΔB and ΔC are both negative. The relative sizes of ΔB and ΔC are such as to give rise to a bowl-shaped contribution to the binding energy in each half of the series. So, whereas in the lanthanide series there was a tetrad effect, here we have a *diad* effect that supplements the orbital stabilization energies of the ligand field. In the first transition series, the method recommended in the previous section has been applied to the lattice enthalpies of K_3MF_6 [31], MF_2, MCl_2 and MI_2 [32], and to the hydration enthalpies of M^{2+} and M^{3+} [33, 34]. The four contributions to the irregularities were calculated. These are the orbital stabilization energies, ΔE_{os}, the irregularities due to the nephelauxetic effect, ΔE_{rep}(irreg), spin-orbit coupling, ΔE_{so}, and the relaxation energy, ΔE_{rlx}. Along the chosen baseline, these four quantities are all zero and this is its main advantage.

Table 1.3 Estimated values of the four components of the contribution made by ligand field stabilization energy to the lattice enthalpy of K_3CuF_6, to the hydration enthalpy of Ni^{2+}(aq), $\Delta H_h^0(Ni^{2+},g)$, and to the standard enthalpy change of reaction 13.

	ΔE_{os}	ΔE_{rep}(irreg)	ΔE_{so}	ΔE_{rlx}	Total
			kJ mol^{-1}		
$L(K_3CuF_6)$	−202	−156	+16	+14	−328
$\Delta H_h^0(Ni^{2+}, g)$	−123	−18	+12	+10	−119
$\Delta H^0(13)$	+0.3	−3.3	+0.8	0	−2.2

Table 1.3 contains values for two $3d^8$ cases. At K_3CuF_6, ΔE_{rep}(irreg) contributes about 40% of the total stabilization, but at Ni^{2+}(aq) only 15%. This is because in the first transition series, the nephelauxetic effect increases substantially when the oxidation state increases from +2 to +3. The relatively small contribution for the M^{2+}(aq) ion explains why text books use this example to explain the double bowl shapes: ΔE_{rep}(irreg) is almost exactly cancelled by the sum of ΔE_{so} and ΔE_{rlx}, so the total stabilization is nearly equal to the orbital stabilization energy. In most other cases, ΔE_{rep}(irreg) is much more important and may play an important role in sustaining the Irving-Williams rule in complexing reactions [32, 33].

In the lanthanide series, the equivalent values are much reduced by the retreat of the 4f electrons into the xenon core. This is so whether we consider processes that involve the condensation of gaseous ions, or conventional reactions. Table 1.3 includes data for the change

$$NdF_3(s) + 3/4\,O_2(g) = 1/2\,Nd_2O_3(s) + 3/2\,F_2(g) \tag{13}$$

These have been calculated from Caro's spectroscopic analyses [35]. The ligands come from opposite ends of the nephelauxetic series, so for a lanthanide reaction, ΔE_{rep}(irreg) should be relatively large. Even so, although it proves to be the largest contributor to the overall change, ΔE_{os} and ΔE_{so} are significant. Quantitative analyses of claimed examples of the tetrad effect must take such terms into account.

It is striking that, despite its small size, the tetrad effect was discovered before the diad effect. This is because the diad effect occurs in d-electron systems and is therefore masked by the orbital stabilization energies produced by the stronger ligand field.

References

1 Jantsch, G.; Klemm, W. *Z. Anorg. Allg. Chem.*, **1933**, *216*, 80.
2 Johnson, D. A. *Advan. Inorg. Chem. Radiochem.* **1977**, *20*, 1.
3 Corbett, J. D. (a) *Rev. Chim. Miner.* **1973**, *10*, 239; (b) *J. Chem. Soc. Dalton Trans.* **1996**, 575.
4 Corbett, J. D.; McCollum, B. C.; *Inorg. Chem.* **1966**, *5*, 938.
5 Johnson, D. A.; Corbett, J. D. *Colloq. Int. CNRS* **1970**, *180*, 429.
6 Caro, P. E.; Corbett, J. D. *J. Less-Common Met.* **1969**, *18*, 1.
7 Corbett, J. D.; Pollard, D. L.; Mee, J. E. *Inorg. Chem.* **1966**, *5*, 761.
8 Evans, W. J.; Zucchi, G.; Ziller, J. W. *J. Amer. Chem. Soc.* **2003**, *125*, 10.
9 Johnson, D. A. *J. Chem. Soc. A* **1969**, 1525.
10 Johnson, D. A. *J. Chem. Soc. A* **1969**, 1529.
11 Johnson, D. A. *J. Chem. Soc. A* **1969**, 2578.
12 *http://physics.nist.gov/PhysRefData/ASD/ levels*; see also Spector, N.; Sugar, J.; Wyart, J. F. *J. Opt. Soc. Am. B* **1997**, *14*, 511.
13 Morss, L. R. *Standard Potentials in Aqueous Solution* (Bard, A. J.; Parsons, R.; Jordan, J. eds.); Marcel Dekker, New York, 1985; p. 587.
14 Cordfunke, E. H. P.; Konings, R. J. M. *Thermochim. Acta* **2001**, *375*, 17.
15 Johnson, D. A. *Some Thermodynamic Aspects of Inorganic Chemistry*, 2nd ed; Cambridge University Press, 1982; Chapter 6 and problems 6.6–6.9.
16 Blake, A. B. *J. Chem. Educ.* **1981**, *58*, 393.
17 Newman, J. B. *J. Chem. Phys.* **1967**, *47*, 85.
18 Johnson, D. A. *J. Chem. Educ.* **1980**, *57*, 475.
19 Johnson, D. A. *J. Chem. Soc. Dalton Trans.* **1974**, 1671.

20 Johnson, D. A. *J. Chem. Soc. Dalton Trans.* **1982**, 2269.
21 Meyer, G. *Chem. Rev.* **1988**, *88*, 93.
22 Meyer, G.; Wickleder, M. S. *Handbook on the Physics and Chemistry of the Rare Earths* (Gschneidner, K. A.; Eyring, L. eds.), North Holland, Amsterdam, **2000**, *28*, 53.
23 Morss, L. R.; Mattausch, H.; Kremer, R.; Simon, A.; Corbett, J. D. *Inorg. Chim. Acta* **1987**, *140*, 107.
24 Bullett, D. W.; *Inorg. Chem.* **1985**, *24*, 3319.
25 Peppard, D. F.; Mason, G. W.; Lewey, S. *J. Inorg. Nucl. Chem.* **1969**, *31*, 2271.
26 (a) Peppard, D. F.; Bloomquist, C. A.; Horowitz, E. P.; Lewey, S.; Mason, G. W. *J. Inorg. Nucl. Chem.* **1970**, *32*, 339. (b) Jorgensen, C. K. *ibid.*, 3127. (c) Nugent, L. J. *ibid.*, 3485.
27 Veksler, I. V.; Dorfman, A. M.; Kamenetsky, M.; Dulski, P.; Dingwell, D. B. *Geochem. Cosmochim. Acta* **2005**, *69*, 2847.
28 Kawabe, I.; Toriumi, T.; Ohta, A.; Miura, N. *Geochem. J.* **1998**, *32*, 213.
29 Kawabe, I. *Geochem. J.* **1992**, *26*, 309.
30 George, P.; McClure, D. S. *Prog. Inorg. Chem.* **1959**, *1*, 381.
31 Johnson, D. A.; Nelson, P. G. *Inorg. Chem.* **1995**, *34*, 3253.
32 Johnson, D. A.; Nelson, P. G. *J. Chem. Soc. Dalton Trans.* **1995**, 3483.
33 Johnson, D. A.; Nelson, P. G. *Inorg. Chem.* **1995**, *34*, 5666.
34 Johnson, D. A.; Nelson, P. G. *Inorg. Chem.* **1999**, *38*, 4949.
35 (a) Caro, P.; Derouet, J.; Beaury, L.; Soulie, E. *J. Chem. Phys.* **1979**, *70*, 2542. (b) Caro, P.; Derouet, J.; Beaury, L.; Teste de Sagey, G.; Chaminade, J. P.; Aride, J.; Pouchard, M. *J. Chem. Phys.* **1981**, *74*, 2698.

2

Stereochemical Activity of Lone Pairs in Heavier Main-group Element Compounds

Anja-Verena Mudring

2.1
Introduction

Thallium, lead and bismuth, when compared with their lighter homologues, show a pronounced preference for an oxidation state of two less than the respective group number. This has been attributed to the effect of the "inert electron pair" which was initially introduced by Sidgwick [1]. Since then it is generally used in introductory textbooks to explain the particular behavior of these elements [2]. The lone-pair effect itself originates from a combination of shell-structure and relativistic effects [3]. Because of to this effect, the element thallium might be regarded as a relativistic alkali metal as not only does it strongly prefer the oxidation state +I in its compounds, but the chemistry of thallous compounds also bears a close resemblance to the chemistry of alkali metals. For example, the basicity of the oxides and hydroxides is similar as well as their tendency to absorb carbon dioxide. Thallium carbonate is readily soluble in water unlike the other heavy metal carbonates. Thallium is able to replace potassium ions in biological systems [4], where it has an affinity which is about ten times superior to that of potassium [5]. This is also the main reason for the large biological toxicity of thallium. In analogy, lead can be viewed as a relativistic alkaline-earth metal and bismuth as a relativistic rare-earth metal.

The classical view of the lone pair is that, after mixing of the s and p orbitals on the heavy metal cation, the lone pair occupies an inert orbital in the ligand sphere [6]. This pair of electrons is considered chemically inert but stereochemically active [7]. However, this implies that the lone pair would always and in any (chemical) environment be stereochemically active, which is not the case. For example, TlF [8] adopts a structure, which can be considered as a NaCl type of structure which is distorted by a stereochemically active lone pair on thallium. In contrast TlCl [9] and TlBr [10] adopt the undistorted CsCl type of structure at ambient temperature, and at lower temperatures the (again undistorted) NaCl type of structure. The structure of PbO [11] is clearly characterized by the stereochemically active lone pair. In all the other 1:1 compounds of lead with

Inorganic Chemistry in Focus III.
Edited by G. Meyer, D. Naumann, L. Wesemann
Copyright © 2006 WILEY-VCH Verlag GmbH & Co. KGaA, Weinheim
ISBN: 3-527-31510-1

higher chalcogenides, the stereochemical influence of the lone pair on the crystal structure is absent (under ambient conditions) [12].

Understanding the factors that control the stereochemical activity of lone pairs in thallium(I), lead(II) and bismuth(III) compounds is important for many reasons. For example, in the design of complexing agents that can remove the toxic metal selectively from biological systems [13–15] or in the development of materials with ferroelectric [16], piezoelectric [17] properties or nonlinear optical materials [18], which affords the synthesis of compounds crystallizing in non-centrosymmetric space groups. One strategy to introduce the desired non-centrosymmetry is to incorporate main group metals that have an s^2 free electron pair, which tends to destroy inversion symmetry even in simple coordination polyhedra, such as the octahedron, by off-center distortions. Such building units usually exhibit very high dipole moments as the counter-anions are unevenly distributed in the coordination sphere of the heavy metal. In summary, there is a real need to understand how and, even more importantly why, a lone pair of electrons becomes stereochemically active in a certain compound or material.

2.2
When Does a Lone Pair of Electrons Become Stereochemically Active? – Observations

So far, many rules of thumb have appeared in the literature and have also found their way into general and inorganic chemistry textbooks [19]. Unfortunately, all these rules of thumb lack deeper explanation or quantification, and their predicting power is generally low. The most common rules (together with some examples) of when a lone pair is expected to become stereochemically active are:

The stereochemical activity of the lone pair of a central atom:

A *decreases, the heavier the ligand atom*

Tl(I), Pb(II) and Bi(III) are mainly found in natural minerals predominantly in the form of sulphide ores and rarely as silicates [20, 21].

B *is enhanced by the basicity of the counter-ion* [22]

Heavier main-group elements like Tl(I), in complex compounds, generally exhibit a stereochemically active lone pair when the donor atoms are nitrogen, oxygen or fluorine [23].

C *is enhanced by ligands that form more covalent bonds to the central atom*

The tendency for the lone pair, for example, on Pb(II) to become stereochemically active grows as the tendency of the donor atoms on the ligand increases to form more covalent metal–ligand bonds [24].

D *is favoured by a low coordination number of the central atom*

Because of the comparatively large space requirements of a lone pair, low coordination numbers favor the expression of a stereochemically active lone pair [25]. Specific ligand design can trigger the stereochemical activity of a lone pair in complex compounds.

E *is favoured in the presence of small ligands*

A stereochemically active lone pair is found more in the presence of F^- than Br^-. Recall, for example, the above mentioned example of TlF vs. TlBr. Interligand repulsion of the methyl groups in the di-methyl-substituted tri(pyrazolyl)borate complex in the ligand hemisphere of Pb^{2+} suppresses the stereochemical activity of the Pb-$6s^2$ electron in $\{Pb[HC(3,5$-$Me_2pz)_3]_2\}\{B[3,5$-$(CF_3)_2C_6H_3]_4\}_2$ and $\{Pb[HC(3,5$-$Me_2pz)_3]_2\}(BF_4)$ when compared with $\{Pb[HC(pz)_3]_2\}\{B[3,5$-$(CF_3)_2C_6H_3]_4\}_2$ and $\{Pb[H(pz)_3]_2\}(BF_4)$ which carry unsubstituted pyrazolates as ligands [26].

F *can be suppressed by the high site symmetry of the central atom*

In many perovskite-like structures of the ABO_3 type the lone pair of the B-cation leads not to a structural distortion. In $CsPbF_3$ under ambient conditions no lone-pair activity observed [27], but upon cooling a phase transition is observed that leads to less symmetrical surrounding of Pb^{2+} by fluoride [28].

G *grows, the lighter the central atom*

Lighter elements show a stronger tendency to develop a stereochemically active lone pair than their heavier homologues. For instance, for antimony(III) more distorted structures are known than for bismuth(III) [29].

2.3
Theoretical Concepts

Traditional considerations state that a stereochemically active electron pair is generated by mixing or hybridization – to use the terms of valence bond theory – of the 6s and 6p orbitals on the central ion. The thus-generated orbital, which has gained its angular direction from a certain degree of admixing p-orbitals, then behaves space-wise as a ligand in the coordination sphere of the central atom [30]. In consequence it is widely believed that s–p mixing is a prerequisite for a lone pair to become stereochemically active [31] and the outlined explanation for the stereochemical activity of a lone pair is commonly found in chemistry textbooks [32].

This view somehow seems dubious in the case of heavier elements like 6th row metals. The high energy separation, as well as the very different spatial distribution of the 6s/6p wavefunctions, which are found for these elements because of the strong influence of relativity, stand against an efficient s–p hybridization. The first excited state of Tl^+ (in the gas phase), s^1p^1 lies 7.4 eV above the s^2 ground state [33]. An even larger energy separation between s and p states is found for lead and bismuth, which is also well reflected by the ionization potentials [34] and by commonly applied Hückel parameters [35].

In order to establish the true nature of lone-pair distortions for 6th row elements, a number of theoretical calculations were undertaken recently.

2.3.1
Molecular/Complex Compounds

Comparative studies of the host-guest chemistry of alkali-metal and pseudo alkali-metal cations such as of potassium (K^+) and thallium (Tl^+) show that thallium and analogous alkali-metal salts seldom crystallize isotypically. Rather, Tl^+ is included by the host in such a way that a stereochemically active electron pair may be assumed. Figure 2.1 shows the relevant local surroundings of Tl^+ in the [Tl@kryptofix5^{TM}] [TlI$_4$], [Tl@18-crown-6] [TlI$_4$] and [Tl@crypt2.2.2] [TlI$_4$]. Figure 2.2 likewise shows the surroundings for [Tl@18-crown-6] [ClO$_4$] in comparison with [K@18-crown-6] [ClO$_4$] [36]. The observation that thallium is found in a rather unsymmetrical coordination, when compared with analogous alkali-metal compounds, is made for a large number of salts with various other macrocyclic polyethers such as podands, crown ethers or cryptands, as our own experimental results and an evaluation of structural data in the CCDC database [37], clearly show.

An asymmetric surrounding of thallium(I) in a crown ether (or cryptand) is especially surprising as these polyethers generally provide a highly symmetric surrounding for the coordinated cation. In fact, alkali-metal cations like sodium,

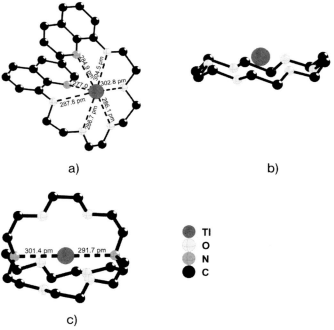

Fig. 2.1 Coordination spheres of Tl^+ in (a) [Tl@kryptofix5^{TM}] [TlI$_4$], (b) [Tl@18-crown-6] [TlI$_4$] and (c) [Tl@crypt2.2.2] [TlI$_4$]; hydrogen atoms omitted for clarity.

potassium and even rubidium (the latter cation being even larger than thallium (I) [38]) show a pronounced preference for a symmetric complexation by 18-crown-6, although strong interactions with better, and hence stronger, coordinating ligands might pull them out of the crown ether plane. This influence can be neglected if the respective counter-anion is a weakly coordinating anion, such as ClO_4^- or TlI_4^-.

In [Tl@18-crown-6] [ClO$_4$] (Fig. 2.1), the thallium(I) cation resides 75 pm above the plane defined by the oxygen atoms of the crown ether and 66 pm in [Tl@18-crown-6] [TlI$_4$] [39] whereas, for example, in the analogous potassium compound [K@18-crown-6] [ClO$_4$] [40] the cation is found right in the middle of the crown ether (Fig. 2.2).

Geometry optimization of the structure of [Tl$^+$@18-crown-6] at the MP2 as well as at the DFT/B3LYP levels of theory clearly shows that Tl$^+$ favors a position above the oxygen plane of the crown ether. The MP2 level of theory gives a distance of 66 pm as found experimentally for [Tl@18-crown-6] [TlI$_4$], and at the DFT/B3LYP level a distance of 77 pm is calculated. In contrast, similar calculations show that K$^+$ (and even Rb$^+$ which is larger than Tl$^+$ [42]) prefers a symmetrical coordination within the cavity of the crown ether – as it is commonly observed in potassium crown ether complexes. A closer inspection of the molecular orbitals involved reveals a deeper insight into the chemical bonding both in the thallium and in the potassium complex. A simplified interaction diagram (Fig. 2.3) of the relevant orbitals shows a bonding and an anti-bonding interaction of the ligand oxygen 2p lone pairs with the complexed monovalent cation. In the case of the alkali-metal cation the bonding orbital combinations are occupied and the anti-bonding combinations are empty. Thus, in addition to the Coulombic interactions of the metal cation with the crown ether, energy is gained from covalent bonding. In contrast to alkali metals, thallium(I) has two more valence electrons and an occupied 6s^2 valence level. In consequence, not

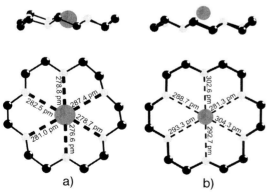

Fig. 2.2 Comparison of the coordination sphere of the monovalent cations in (a) [K@18-crown-6] [ClO$_4$] [40] and (b) [Tl@18-crown-6] [ClO$_4$] [41].

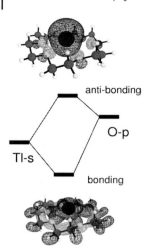

Fig. 2.3 Simplified interaction scheme of Tl^+ with 18-crown-6.

only are the bonding metal-cation orbital combinations occupied but the anti-bonding combination is also, leading to a net destabilization. This destabilization of the compound by covalent forces can be minimized if the thallium(I) cation moves out of the crown ether. Of course, Coulombic forces (in the case of the solid compound lattice forces) work in the opposite direction and hold the cation back. In consequence, the structural distortion will occur to such an extent that both forces are optimized.

A comparison of the total energies for a hypothetical complex with Tl^+ in the middle of the crown ether and with the total energies for the optimized geometries with Tl^+ above the crown indicates, as expected, that the latter geometry is more stable [40]. A study of Pb(II) complexes also shows higher total energies for undistorted, compared to distorted, structures [43].

The repulsion of the thallium and oxygen lone pairs lead to a distortion of the Tl-6s lone pair, which is no longer totally spherical as one would intuitively expect for an s orbital. The thallium contributions involved in the anti-bonding combination are indeed of 97.7% s-, 1.8% p- and 0.5% d-character. Thus, a much less than expected p-character is found for the thallium orbital, which shows that extensive s–p mixing or hybridization is not essential for the lone pair to become stereochemically active.

2.3.2
Solid Materials

Basically, when analysing the band structures, the equivalent observations apply to typical solid state compounds like thallium halides and lead chalcogenides. In studies on the origin of distortion in a-PbO, it was found that "the classical theory of hybridization of the lead 6s and 6p orbitals is incorrect and that the "lone pair" is the result of the lead-oxygen interaction" [44]. It was also noted that computed crystal orbital Hamiltonian populations (COHP) between cation s and ligand p states can be used as an indicator of trends in lone-pair stereo-chemical activity [45].

Thallous halides offer a unique possibility of studying the stereochemistry of the (chemically) inert electron pair, since their structures and their pressure and temperature-dependent phase transitions have been well established. Thallium (I) fluoride under ambient conditions, adopts an orthorhombic structure in the space group *Pbcm* which can be regarded as a distorted rocksalt structure (Fig. 2.4). In contrast to TlF, the thallium halides with heavier halogens, TlCl, TlBr and TlI, adopt the highly symmetric cubic CsCl structure type under ambient conditions [46]. Both TlCl and TlBr, at lower temperatures, undergo phase transitions to the NaCl type of structure [47].

While the structure of TlF is strongly influenced by the lone pair under ambient pressure, the thallous compounds of higher halogens are not. This can be quite well explained by theoretical calculations at the DFT-LMTO-ASA level [48].

The density of states (DOS) for TlF in its own structure type as well as in an idealized NaCl type of structure is displayed in Fig. 2.5 from –6 eV to +6 eV (with respect to the highest occupied state). The DOS of TlCl in the CsCl and NaCl types of structure are displayed in Fig. 2.6. All densities of states show principally the same features. They can be partitioned into four significant sections, a region above the Fermi level, which is mainly of Tl-6p character (IV) and a more complex area below the Fermi level, which shows three different segments (I–III). A region at lower energies around –4/–5 eV which is of Tl-6s and X-np (X=F, Cl) character (I) is followed by a region of high density of states (DOS) mainly made up by X-np (X=F, Cl) with small Tl-6p contributions (II). Then up to the highest occupied states a region is found which again shows Tl–6s as well as X-np contributions (III). The COHP curves (Figs. 2.7 and 2.8)

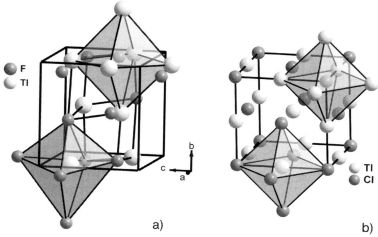

Fig. 2.4 Crystal structures of TlF under ambient conditions (a) and TlCl (b) at low temperature.

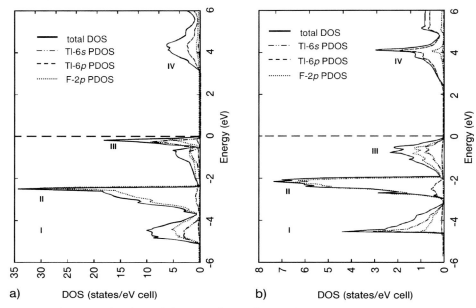

Fig. 2.5 Total and partial densities of states for TlF: *Pbcm*, (a); and *Fm-3m* (b).

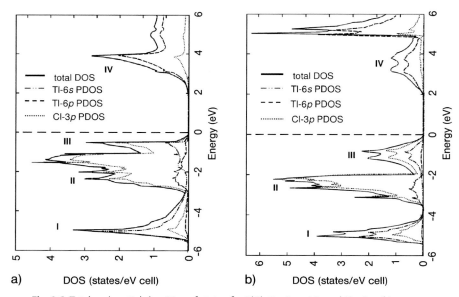

Fig. 2.6 Total and partial densities of states for TlCl: *Pm-3m*, (a); and *Fm-3m* (b).

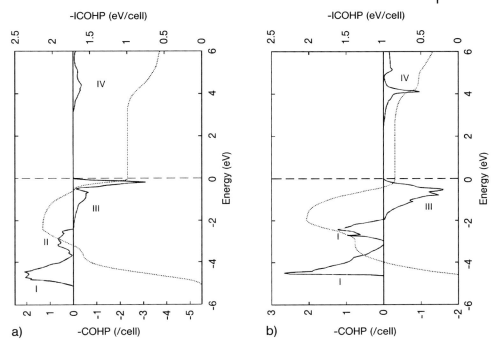

Fig. 2.7 COHP and ICOHP of the Tl–F interaction in TlF: *Pbcm*, (a); and *Fm-3m* (b).

reveal the regions I and II to be Tl–X bonding and region III right below the
Fermi level Tl–X to be *anti*-bonding.

This makes the ambivalent bonding situation in thallous halides clear: On the
one hand there is a strong Coulombic attraction between cations and anions in
the solid but, on the other hand, right at the Fermi level an unfavorable *anti*-
bonding Tl-6s/X-np interaction takes place. In the case of TlF, the Tl-6s and the
F-2p states are located in the same energy region, which leads to a strong interaction producing strongly *anti*-bonding Tl-F states immediately below the Fermi
level, which destabilize the solid. If these were reduced in the same way as was
previously noted for Tl(I) macrocyclic compounds, the whole solid would be en-
ergetically stabilized. This can be, and is, accomplished by a structural distortion. Upon distortion in TlF, the low-lying region I loses some of its Tl-6s character to region III right below the Fermi level. In exchange, region I gains F-2p
character and region III loses F-2p. Accordingly, the high-lying *anti*-bonding Tl-
6s/F-2p states, in particular, come closer in energy to the Tl-6p states, which
now can mix into the *anti*-bonding region III. This, on the one hand, introduces
some additional Tl-6p/F-2p bonding as noted earlier [49] and, more importantly,
moves *anti*-bonding states partially above the Fermi level so that a net gain in
bonding is achieved. This can be illustrated by a simplified interaction scheme
(Fig. 2.9a and b). Indeed, the ICOHP value for the Tl–F bond is higher for the

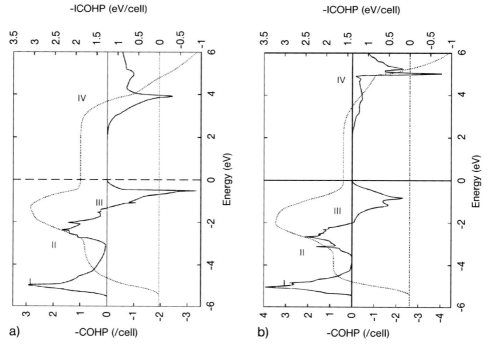

Fig. 2.8 COHP and ICOHP of the Tl–Cl interaction in TlCl: *Pbcm* (a); and *Fm-3m* (b).

distorted structure than for the symmetric structure. This mixing cannot occur in highly symmetric (cubic) structures such as the rocksalt type of structure, as it would be symmetrically forbidden.

In contrast to TlF, for TlCl and TlBr no lone-pair distortions have been observed so far. This becomes clear by comparing the DOS and COHP curves for the thallium–halogen bond for hypothetically cubic TlF (Figs. 2.5 and 2.7) with that for TlCl (Figs. 2.6 and 2.8). As for TlF in the rocksalt structure, the DOS and COHP curves can be separated for TlCl into four distinct regions and the principal features are still present. In the case of the CsCl type of structure, regions II and III have grown together. But compared to TlF, the low-lying states have gained substantially more Tl-6s character and the states right below the Fermi level more Cl-3p character. This can be illustrated by a simple interaction scheme (Fig. 2.9). As the Cl-3p states are located at higher energies compared to the F-2p states, they are farther apart from the Tl-6s states and, thus, the covalent Tl-6s/Cl-3p interaction is less and, in consequence, the energy separations between the distorted and undistorted structures diminish. Thus, in compounds with heavier halogens, the necessity for a structural distortion from a highly symmetric structure, with a large lattice energy (Madelung factor), to minimize unfavorable covalent interactions, becomes less important. Consequently, typical ionic structures like the rocksalt or cesium chloride types of

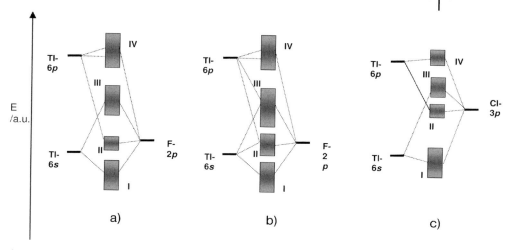

Fig. 2.9 Simplified interaction scheme of Tl⁺ with X⁻ in the solid. (a) Hypothetical TlF in the NaCl type of structure, (b) TlF in its own structure type, (c) TlCl in the CsCl type of structure.

structure are adopted. Similar observations are made in the case of group 14 chalcogenides [50].

2.4
Conclusions

In summary, one can state that s–p-hybridization on the heavier main group metals is not responsible for the stereochemical activity of a lone pair. Instead, the general conclusion can be drawn that *anti*-bonding metal ns–ligand np interactions lead to structural distortions in order to minimize these unfavorable interactions.

Altogether, recent theoretical considerations reveal not only the background for the origin of lone-pair distortions in heavier main group compounds, but also help to understand many of the rules of thumb (see above).

The heavier the ligand atom, the larger the energy and spatial mismatch between the central atom and the ligand orbitals and, thus, the lesser the unfavourable covalent interaction between the filled valence levels and in consequence the less the energy gain by a structural distortion (**A**). In contrast, more basic (**B**), lighter (**G**), and smaller (**E**) ligand atoms like nitrogen, oxygen and fluorine have energy levels that can interact much better with the occupied $6s^2$ valence states of Tl(I), Pb(II) and Bi(III) leading to a destabilization by covalent bonding (**C**). Structural distortions will now occur in order to minimize these destabilizing interactions. It can be expected that the lower the coordination number the lower the counterforces that prevent such a distortion (**D**). On the other hand, a large Coulombic counterforce will be present in the case of a highly symmetrical solid as these generally have high Madelung energies, which would be reduced dramatically by symmetry reducing distortions (**F**).

In conclusion, it can be stated that the observation of structural distortions of compounds with a $6s^2$ lone pair is strongly dependent on the electronic states of the bonding partners and the overall structural surroundings in the solid material.

Acknowledgments

This work was supported by the Federal Ministry of Education and Research (BMBF) and the Fonds der Chemischen Industrie, through a Liebig stipend. AVM gratefully acknowledges the support of Prof. Dr. Gerd Meyer, Universität zu Köln, Germany.

References

1 (a) N.V. Sidgwick, *Some Physical Properties of the Covalent Link in Chemistry*, Cornell University Press, Ithaca, New York, **1933**; (b) N.V. Sidgwick, *Ann. Rep.* **1933**, *30*, 120; (c) N.V. Sidgwick, H.M. Powell, *Proc. Roy. Soc. London Ser. A* **1940**, *176*, 153; (d) N.V. Sidgwick, *The Chemical Elements and their Compounds*, Vol. 1, Clarendon Press, Oxford, **1950**.

2 (a) N.N. Greenwood, A. Earnshaw, *Chemistry of the Elements*, Pergamon, Oxford, **1997**; (b) J.G. Huheey, E.A. Keiter, R.L. Keiter, *Inorganic Chemistry: Principles of Structure and Reactivity*, Pearson Education, New York, **1997**; (c) F.A. Cotton, G. Wilkinson, C.A. Murillo, M. Bochmann, *Advanced Inorganic Chemistry*, 6th ed, Wiley, New York, **1999**.

3 (a) P. Pyykkö, J.P. Desclaux, *Acc. Chem. Res.* **1979**, *12*, 276; (b) P. Pyykkö, *Chem. Rev.* **1988**, *88*, 563; (c) P. Schwerdtfeger, G.A. Heath, M. Dolg, M.A. Bennett, *J. Am. Chem. Soc.* **1992**, *114*, 7518; (d) M. Seth, M. Dolg, P. Fulde, P. Schwerdtfeger, *J. Am. Chem. Soc.* **1995**, *117*, 6597.

4 (a) J.S. Britten, M. Blank, *Biochem. Biophys. Acta*, **1968**, *159*, 160; (b) J.F. Kayne, *Arch. Biochem. Biophys.* **1971**, *143*, 232; (c) F.A. Cotton, G. Wilkinson, *Advanced Inorganic Chemistry*, 5th edn, John Wiley and Sons, New York, **1980**.

5 P.J. Gehring, P.B. Hammond, *J. Pharmacol. Exp. Ther.* **1967**, *155*, 187; (b) C.E. Inturrisi, *Biochem. and Biophys. Acta* **1969**, *178*, 630.

6 L.E. Orgel, *J. Chem. Soc.* **1959**, 3815.

7 See for example: (a) I. Lefebvre, M. Lannoo, G. Allan, A. Ibanaz, J. Fourcade, J.C. Jumas, E. Beaurepaire, *Phys. Rev. Lett.* **1987**, *59*, 2471; (b) I. Lefebvre, M. Lannoo, J. Olivier-Fourcade, J.C. Jumas, *Phys. Rev. B* **1991**, *44*, 1004.

8 P. Berastegui, S. Hull, *J. Solid State Chem.* **2000**, *150*, 266.

9 W.P. Davey, F.G. Wick, *Z. Phys. Chem.* **1925**, *117*, 51; (b) M. Blackman, I.H. Khan, *Proc. Phys. Soc.* **1961**, *77*, 471.

10 A.E. van Arkel, *Physica* **1924**, *4*, 33; M. Blackman, I.H. Khan, *Proc. Phys. Soc.* **1961**, *77*, 471.

11 R.G. Dickinson, J.B. Friauf, *J. Am. Chem. Soc.* **1924**, *46*, 2457.

12 PbS: L.S. Ramsdell, *Am. Min.* **1925**, *10*, 281; PbSe: L.S. Ramsdell, *Am. Min.* **1925**, *10*, 281; T.K. Chattopadhyay, H.G. von Schnering; W.A. Grosshans, W.B. Holzapfel, *Physica B and C* **1986**, *139*, 356; PbTe: (a) Y. Noda, S. Ohba, S. Sato,

Y. Saito, *Acta Cryst.* **1983**, *39*, 312; (b) Y. Fujii, K. Kitamura, A. Onodera, Y. Yamada, *Solid State Comm.* **1984**, *49*, 135.

13 L. Shimoni-Livny, J. P. Glusker, C. W. Bock, *Inorg. Chem.* **1998**, *37*, 1853.

14 R. D. Hancock, *Pure & Appl. Chem.* **1993**, *65*, 941.

15 R. D. Hancock, M. S. Shaikjee, S. M. Dobson, J. C. A. Boeyens, *Inorg. Chim. Acta* **1988**, *154*, 229.

16 (a) H. Chiba, T. Atou, Y. Syono, *J. Solid State Chem.* **1997**, *132*, 139; (b) N.A. Hill, K. M. Rabe, *Chem. Mat.* **2001**, *13*, 2892; (c) R. Seshadri, N. A. Hill, *Chem. Mat.* **2001**, *13*, 2899; (d) A.M. dos Santos, S. Parashar, A. R. Raju, Y. S. Zhao, A. K. Cheetham, C.N.R. Rao, *Solid State Commun.* **2002**, *122*, 49.

17 P. Baettig, Ch. F. Schelle, R. LeSar, Z. V. Waghmare, N. A. Spalding, *Chem. Mat.* **2005**, *17*, 1376.

18 J. Goodey, J. Broussard, P. S. Halasyamani, *Chem. Mat.* **2002**, *13*, 3174; J. Goodey, K. M. Ok, C. Hofmann, J. Brossard, F. V. Esobedo, P. S. Halasyamani, *J. Solid State Chem.* **2003**, *175*, 2; H.-S. Ra, K. M. Ok, P. S. Halasyamani, *J. Am. Chem. Soc.* **2003**, *125*, 7764.

19 See e.g. Holleman, Wiberg, Lehrbuch der Anorganischen Chemie, 101. Aufl., de Gruyter, Berlin, **1995**.

20 A. Kyono, M. Kimata, M. Shimizu, *Am. Min.* **2000**, *85*, 1287.

21 M. Mehring, D. Mansfeld, M. Schürmann, *Z. Anorg. Allg. Chem.* **2004**, *630*, 452.

22 I. D. Brown, R. Faggiani, *Acta Cryst.* **1980**, *B36*, 1802.

23 Sh.-Y. Wan, J. Fan, T. Okamura, H. F. Zhu, X.-M. Ouyang, W.-Y. Sun, N. Ueyama, *Chem. Commun.* **2002**, 2520.

24 R. D. Hancock, *Pure & Appl. Chem.* **1993**, *65*, 941.

25 R. J. Gillespie, *Angew. Chem.* **1996**, *108*, 539.

26 D. L. Reger, T. D. Wright, C. A. Little, J. J. S. Lama, M. D. Smith, *Inorg. Chem.* **2001**, *40*, 3810.

27 O. Schmitz-DuMont, G. Bergerhoff, E. Hartert, *Z. Anorg. Allg. Chem.* **1956**, *283*, 314.

28 P. Berastegui, S. Hull, S. G. Eriksson, *J. Phys.: Cond. Mat.* **2001**, *13*, 5077.

29 ICSD Database, National Institute of Standards and Technology (NIST) and Fachinformationszentrum Karlsruhe (FIZ); Cambridge Crystallographic Data Centre (CCDC, 12 Union Road, Cambridge CB2 1EZ, fax: (+44)1223-336-033; e-mail: *deposit@ccdc.cam.ac.uk*)

30 L. E. Orgel, *J. Chem. Soc.* **1959**, 3815; J. D. Dunitz, L. E. Orgel, *Adv. Inorg. Chem. Radiochem.* **1960**, *2*, 42.

31 (a) J. D. Dunitz, L. E. Orgel, *Adv. Inorg. Chem. Radiochem.* **1960**, *2*, 45; (b) L. E. Orgel, *J. Chem. Soc.* **1959**, 3815; (c) E. A. Boudreaux, *Coord. Chem. Rev.* **1967**, *2*, 117; (d) E. A. Boudreaux, L. D. Dureau, H. B. Jonassen, *Mol. Phys.* **1963**, *6*, 377; (e) G. A. Samara, *Phys. Rev.* **1968**, *165*, 959.

32 See for example: A. G. Massey, Chemistry of Main Group Elements, Wiley, Chichester, **2000**.

33 C. E. Moore, *Atomic Energy Levels*, Nat. Bur. Stand. (US) Cr. No 467; GPO, Washington, D.C., **1958**.

34 J. Emsley, Die Elemente, de Gruyter, Berlin, New York, **1994**.

35 A. Herman, *Modelling Simul. Mater. Sci. Eng.* **2004**, *12*, 21.

36 F. Rieger, A.-V. Mudring, unpublished research, **2005**.

37 Cambridge Crystallographic Data Centre (CCDC, 12 Union Road, Cambridge CB2 1EZ, fax: (+44)1223-336-033; e-mail: *deposit@ccdc.cam.ac.uk*).

38 R. D. Shannon, *Acta Cryst.* **1976**, *A32*, 751.

39 P. Luger, C. Andre, R. Rudert, D. Zobel, A. Knochel, A. Krause, *Acta Cryst.* **1992**, *B48*, 33.

40 F. Rieger, A.-V. Mudring, *Inorg. Chem.* **2005**, *44*, 6240.

41 P. Luger, C. Andre, R. Rudert, D. Zobel, A. Knochel, A. Krause, *Acta Cryst.* **1992**, *B48*, 33.

42 (a) R. D. Shannon, *Acta Cryst.* **1976**, *A32*, 751; (b) Landolt-Börnstein, Neue Serie, Ed. K.-H. Hellwege, Gruppe III, Bd. 7, Teil a, Springer, Berlin, **1973**, p. 480.

43 L. Shimoni-Livny, J. P. Glusker, C.W. Bock, *Inorg. Chem.* **1998**, *37*, 1853.

44 (a) J. M. Raulot, G. Baldinozzi, R. Seshadri, P. Cortona, *Solid State Sci.* **2002**, *4*, 467; (c) U. V. Waghmare, N. A. Spaldin,

H. C. Kandpal, R. Seshadri, *Phys. Rev. B* **2003**, *67*, 125111.

45 R. Seshadri, N. A. Hill, *Chem. Mat.* **2001**, *13*, 2899.

46 W. P. Davey, F. G. Wick, *Phys. Rev.* **1921**, *17*, 403; *Z. Phys. Chem.* **1925**, *117*, 51; A. E. van Arkel, *Physica* **1924**, *4*, 33; *Norsk Geologisk Tidsskrift* **1925**, *8*, 217.

47 M. Blackman, I. H. Khan, *Proceedings of the Physical Society*, London **1961**, *77*, 471.

48 (a) R. W. Tank, O. Jepsen, A. Burckhardt, O. K. Andersen, *TB-LMTO-ASA Program*, Vers. 4.7, Max-Planck-Institut für Festkörperforschung, Stuttgart, Germany, **1998**; (b) H. L. Shriver, *The LMTO Method*, Springer, Berlin, Germany, **1984**; (c)

O. Jepsen, M. Snob, O. K. Andersen, *Linearized Band-structure Methods in Electronic Band-structure and its Applications*, Springer Lecture Note, Springer Verlag, Berlin, Germany, **1987**; (d) O. K. Anderson, O. Jepsen, *Phys. Rev. Lett.* **1984**, *53*, 2571.

49 U. Häussermann, P. Berastegui, St. Carlson, J. Haines, J.-M. Léger, *Angew. Chem. Int. Ed.* **2001**, *40*, 4624.

50 (a) G. W. Watson, *J. Chem. Phys.* **2001**, *114*, 758; (b) A. Walsh, G. W. Watson, *Phys. Rev. B* **2004**, *70*, 2535114; (c) A. Walsh, W. Graeme, *J. Phys. Chem. B* **2005**, *109*, 18868; (d) A. Walsh, G. W. Watson, *J. Solid State Chem.* **2005**, *178*, 1422.

3
How Close to Close Packing?

Hideo Imoto

3.1
Introduction

"Close packing" is one of the fundamental concepts in crystal chemistry. More than 70% of metallic elements ($Z \leq 83$) as well as all of the noble gas elements, crystallize in one of the close packing structures like cubic close packing (ccp) and hexagonal close packing (hcp) at least in some temperature ranges, under ambient pressure. The importance of close packing lies also in the fact that many inorganic structures can be interpreted as derivatives of the close-packing structures [1]. In some structures like sodium chloride and zinc blende, each of the components makes the perfect close-packing arrangement. On the other hand, in the fluorite or cuprite structures, only cations are arranged in close packing. All of these structures are usually described as close packing of one of the components with the other component in the tetrahedral or octahedral interstices of the close packing. Unfortunately, except for some simple structures, neither of the components forms perfect close packing and the arrangement of one component is interpreted as a distorted close-packing arrangement. However, simple distortions of close packing can lead to many other definite structures including the body-centered cubic (bcc) and the simple cubic structures as will be discussed below. There is no definite boundary between deformed close-packing structures and others. Even if an arrangement can be interpreted as a distortion of close packing, it can be closer to another arrangement.

To evaluate the degree of distortion from close packing, we need some quantitative parameter. The elongation or contraction along the c axis of hcp metals is the simplest and best-known example of the distortion of close packing and this distortion can be quantitatively measured by the deviation of the axial ratio ($= c/a$) from the ideal value ($= \sqrt{8/3} \approx 1.6330$). In zinc metal, due to the elongation (13.7%) of the c axis, the longest Zn–Zn bond distance is 9.3% longer than the shortest. On the other hand, the arrangement of the oxygen atoms in corundum Al_2O_3 can also be considered as a deformed hcp arrangement, although the deformation is much more complex. The deviation of the z coordinates from the

Inorganic Chemistry in Focus III.
Edited by G. Meyer, D. Naumann, L. Wesemann
Copyright © 2006 WILEY-VCH Verlag GmbH & Co. KGaA, Weinheim
ISBN: 3-527-31510-1

ideal value indicates the degree of distortion from hcp but we cannot imagine the degree of distortion directly from this figure. Furthermore, to compare the degree of distortions in different structures, like Zn metal and corundum, we need common parameters.

In this article, two parameters that measure the degree of the deviation from close packing are introduced. The comparison of these parameters in many inorganic structures, including some which are very different from close packing, will give some insight into the concept of close packing.

3.2
Essential Features of Close Packing

To find the parameters that indicate proximity to close packing, we have to consider the essential features of close packing. First, it is the densest possible structure. This feature indicates the global nature of close packing. It is related to the so-called "Kepler's conjecture", which can be formulated as "Close packing arrangements like ccp and hcp are the densest possible sphere packing" [2]. A very complex proof, which depended on the results of the computer calculations, was published very recently. Since close-packing structures are the densest, the parameter indicating how densely the atoms are packed in the structure is a good parameter for our purpose.

The second feature of close packing is local: each atom has twelve neighbors. This feature is related to another mathematical problem, known as the "kissing number problem": The maximum number of spheres that can touch the central sphere is twelve if all spheres have the same radius. This problem was solved by three mathematicians, independently, in the mid 1950s and the answer is twelve. In the close-packed structures, each atom has twelve neighbors at an equal distance. Twelve neighboring equal spheres around an atom can be realized not only as a local structure of the close packing but also in many other geometries, one of the examples being an icosahedron centered on an atom. Icosahedra with five-fold symmetry, however, cannot fill the space and cannot constitute any periodic structure. Except for close packing, no atomic arrangement has been found in which all atoms have twelve neighbors at an equal distance. Therefore, the feature that all atoms have twelve neighbors is unique for close-packing structures and we will adopt the number of neighboring atoms as the second parameter.

3.3
Parameter Definitions

The first parameter, which indicates how densely the atoms are packed, will be denoted by D here since it may be called the "Density parameter". It should be proportional to the number density (= number of atoms in a unit volume) but must be independent of the size of the atoms, depending only on the relative arrangement. This means that the density parameter can be obtained by dividing the number density by the cube of some parameter measuring the size of the atoms. Then, the simplest definition that can be applied to any arrangements can be given by using the shortest interatomic distance as the size parameter and can be formulated as follows.

$$D = \frac{n_D r_s^3}{\sqrt{2}V}$$

Here n_D is the number of atoms in a unit cell, the volume of which is V, and r_s is the shortest interatomic distance in the arrangement. The definition contains a division by $\sqrt{2}$ so that the parameter D becomes unity for close-packing structures. Kepler's conjecture ensures that the parameter D is always less than or equal to unity. The fraction of space occupied (f) in the rigid-sphere model, which is often used in the discussion of metallic structures, is proportional to the parameter D and the relation is as follows.

$$f = (\sqrt{2}\pi/6)D \ (\approx 0.74048 \ D)$$

The second parameter, which indicates the number of neighbors, is denoted by N here. It cannot be defined as easily as D. If the atoms surrounding one central atom are at the same distance from the center and other atoms are far from the central atom, the number of neighboring atoms is clear. However, in many structures the interatomic distances have a somewhat continuous distribution and it is not possible to define the border between the neighboring atoms and the distant atoms. Then, a practical definition of the parameter N is to count the neighboring atoms with weights. The nearest atom has a weight of unity, an atom a little distant from the central atom has a weight of a little less than unity, and those with larger distances have smaller weights. The weight function is supposed to decrease with the interatomic distance very rapidly around the vague boundary between neighboring and distant atoms. On the other hand, it must change slowly around the shortest interatomic distance because the number of neighbors is supposed not to change sensitively with small deviations of the distances of the nearest neighbors. Amongst the functions that satisfy these conditions, the Gaussian function may be the simplest choice and then the definition of the parameter N can be formulated as follows.

$$N = \sum_{j(d_j < 2d_s)} m_j \exp\left(-a\left(\frac{d_j}{d_s} - 1\right)^2\right)$$

Here, n_j is the number of neighboring atoms at the distance d_j. Because the atoms separated by more than twice the shortest interatomic distance d_s cannot be counted as neighboring, the summation is limited only to the interatomic distances less than twice d_s. Without this limitation, calculation of the parameter could be troublesome. If the atomic arrangement contains non-equivalent atoms, the neighboring parameter for each type of atom is calculated where the shortest interatomic distance d_s is common in the arrangement. Then the values calculated are averaged with the weight proportional to the compositional ratio. The parameter a, which is related to the vague boundary between neighboring atoms and distant atoms, cannot be determined by any a priori argument. In Table 3.1, the values of N calculated with various a values are given for some typical arrangements. If the parameter a is less than 15, the parameter N for the close-packing structures or the simple-cubic structure becomes much larger than the generally accepted number of neighbors, 12 and 6, respectively. However, if it is more than 30, the parameter N for zinc, which is usually considered as an elongated hcp structure, becomes as low as 10.6. If the a parameter is set around 20, the resulting N parameters fit reasonably well with the usual discussion of the number of neighbors, although for highly symmetric arrangements the values of N are slightly larger than the number of neighbors in the strict sense, due to the contribution of longer interatomic distances. As shown in Table 3.1, the neighboring parameter N does not depend much on a for some structures like diamond but it varies greatly with a for structures $Mn(a)$ and Te. For these structures, it may be impossible to define the number of neighbors in any significant way. In the following discussion, the parameter a is always set to 20. With this setting, the distance where the weights of the atom is equal to 0.5 is $1.186\,d_s$ while the distance where the weight is 0.1 is $1.339\,d_s$ as shown in Table 3.1. Therefore, the range $1.2\,d_s \sim 0.3\,d_s$ is approxi-

Table 3.1 N parameter and boundary radius calculated for several a values.

Structure	Neighboring parameter N					Density parameter D
	$a=10$	$a=15$	$a=20$	$a=30$	$a=40$	
ccp	13.1919	12.4653	12.1946	12.0349	12.0063	1.0000
hcp-ideal	13.2031	12.4683	12.1951	12.0349	12.0063	1.0000
hcp-Mg	13.1725	12.4511	12.1845	12.0298	12.0021	0.9942
hcp-Ti	13.0469	12.3636	12.1129	11.9606	11.9184	0.9725
hcp-Zn	12.1437	11.4542	11.1005	10.6282	10.2381	0.8794
bcc	12.9468	12.2198	11.7217	10.9265	10.3036	0.9185
s.c. [1]	8.1956	6.9177	6.3883	6.0698	6.0126	0.7071
diamond	4.2211	4.0295	4.0040	4.0001	4.0000	0.4593
$Mn(a)$	9.7578	8.5164	7.5339	6.0593	5.0228	0.6720
Te	6.8891	5.7509	5.0624	4.1977	3.6996	0.6324
Boundary radius [2] at the level p						
$p = 0.5$	1.263	1.215	1.186	1.152	1.132	
$p = 0.1$	1.480	1.392	1.339	1.277	1.240	

1) simple cubic structure
2) The d_j/d_s value where the weight of the atom in the calculation of the N parameter is equal to p (here 0.5 or 0.1).

mately the boundary between neighboring and distant atoms with the setting adopted here.

3.4
Correlation Between *D* and *N*

We now have two parameters which measure how close to close packing an atomic arrangement may be. Figure 3.1 shows the plot of *D* vs. *N* for selected structures of elements and binary compounds. The structures used are listed in the Appendix with their ICSD (Inorganic Crystals Structure Database) collection codes. For each binary compound, two points are plotted, one for cations and the other for anions, and they are distinguished by filled circles (cation) and open circles (anion). The two parameters have good correlation except for some scattered points. In particular, the correlation is excellent in the region with larger *D* and *N*. However, the values of *N* for the atomic arrangements having the *D* parameter ranging from 0.9 to 1 are all approximately twelve. This means that the *N* parameter is not sensitive to a slight distortion of close packing. On the other hand, for the arrangements that are very different from close packing, the *N* parameter seems useful to describe the differences of atomic arrangements as we can expect from its definitions.

The plot shown in Fig. 3.1 reminds us that the left half of a Gaussian curve and the observed correlation curve can be approximated by the Gaussian function having its maximum at the point for close packing ($D=1$, $N=12.195$). Then the only parameter to be adjusted by curve fitting is the parameter defining the width of the Gaussian curve. In Fig. 3.1, the points designated by the diamond marks ($\blacklozenge$ and $\lozenge$) are for the atoms with homoatomic bonds, like cationic components of cluster halides and anionic components of polyanionic compounds. Many of these points, especially those for the cluster compounds, deviate largely from other points and were excluded from the data for curve fitting. The fitting function obtained was $N \approx 12.195 \exp(-\{(D-1)/0.476\}^2)$. The slope of this function is zero at $D=1$, corresponding to the insensitive nature of the parameter N for the atomic arrangements very close to close packing. This good fitting clearly indicates that the *D* parameter is better for measuring small distortions of atomic arrangements from close packing.

In Fig. 3.1, several ideal structures are also plotted with the '+' mark. All of these structures have no adjustable parameter and most of them lose some of the symmetry elements when they are distorted. As shown in the figure, most of the ideal structures have some deviation from the fitting curve. It may be related to the fact that some of these ideal structures are deformed in real binary compounds.

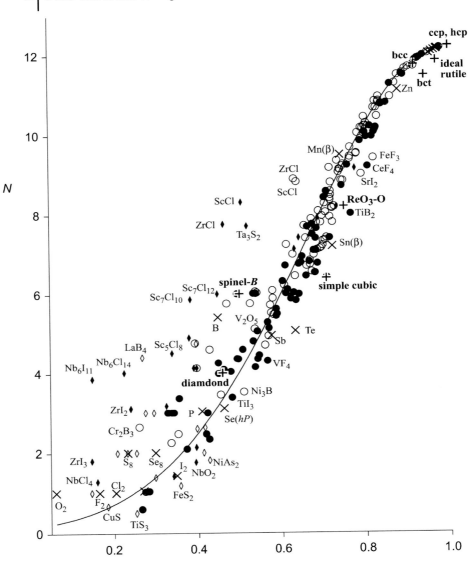

Fig. 3.1 The plot of parameter D vs. parameter N. The plot contains elements (×), cations without homoatomic bonding (•), anions without homoatomic bonding (O), cations with homoatomic bonding (◆), anions with homoatomic bonding (◇), and ideal structures (+). The definitions of some of the ideal structures are as follows. "ideal rutile": the oxygen positions of the ideal rutile structure where all MO_6 octahedra are regular. "bct": the structure obtained by compressing the bcc structure along the tetragonal axis so that all atoms have ten neighbors at the same distances. "ReO_3-O": the oxygen positions in the ReO_3 structure. "Spinel-B": the B sites (octahedral sites) in the spinel structure.

3.5
Transformation of Close-packing Arrangements

If close-packing structures are deformed in a special way, they can be transformed into other definite structures. In this section, the variations of the D and N parameters during the transformation will be examined.

First, we discuss the trigonal compression of the ccp arrangement. In this transformation, we regard the ccp structure as the stacking of close-packing layers in the ABC sequence and adopt the hexagonal unit cell belonging to the space group $R\bar{3}m$. Then, the axial ratio (c/a) of the ccp structure is $\sqrt{6}$ (≈ 2.4495), which is 1.5 times larger than hcp. When the axial ratio is reduced to the half ($\sqrt{3/2} \approx 1.2247$), the structure becomes simple cubic as shown in Fig. 3.2. Further reduction of the axial ratio to the half ($\sqrt{3/8} \approx 0.6124$) gives the bcc structure. The variation of the D and N parameters against the axial ratio is illustrated in Fig. 3.3. The bcc structure is a sharp local maximum in either parameter, while the simple cubic structure corresponds to the local minimum. Therefore, when the simple cubic structure is either compressed or expanded along the c axis, keeping the shortest interatomic distance constant, the structure always has higher density.

The tetragonal distortion of the ccp structure also leads to the bcc structure. In this transformation, the a and b axes of the unit cell are defined as the half diagonals of the square face of the ordinary cubic cell of the ccp structure. The c axis is the same as for the ccp structure. The structure during the transformation belongs to the space group $P4/mmm$. Starting from the ccp structure $(c/a = \sqrt{2})$, the structure is compressed along the c axis and leads to the bcc structure when the axial ratio is unity. The bcc structure gives the local minimum of the D parameter as illustrated in Fig. 3.4, although it is the local maximum for the trigonal transformation. Further compression gives the local maximum of the D parameter when the axial ratio is equal to $\sqrt{2/3}$ (≈ 0.8165). This structure corresponds to the "body centered tetragonal" (bct) structure, where each atom has

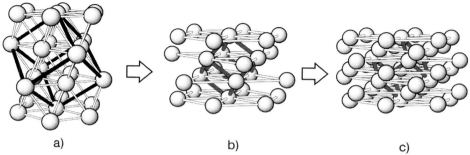

a)   b)   c)

Fig. 3.2 Trigonal transformation of ccp. (a) ccp $(a/c \approx 2.4495)$. (b) The simple cubic structure $(a/c \approx 1.2247)$. (c) bcc $(a/c \approx 0.6124)$.

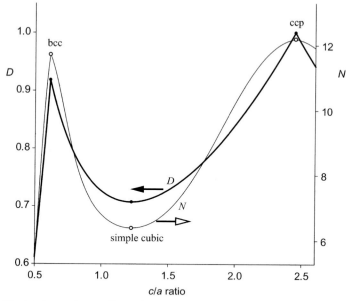

Fig. 3.3 Dependences of *D* and *N* parameters on the axial ratio *a/c* in the trigonal transformation of ccp.

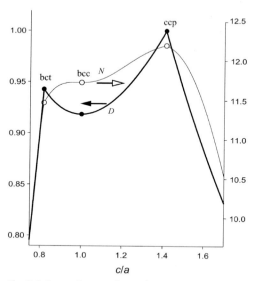

Fig. 3.4 Dependences of *D* and *N* parameters on the axial ratio *a/c* in the tetragonal transformation of ccp.

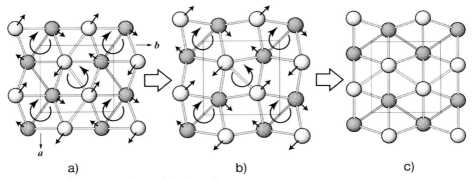

Fig. 3.5 Transformation of hcp to the ideal rutile structure.
(a) hcp (rotation angle=0). (b) Ideal rutile structure (rotation
angle=9.74°). (c) hcp (rotation angle=19.47°).

10 neighbors at equal distances [3]. The D parameter of this structure is 0.9428, larger than the D parameter for bcc (0.9186). On the other hand, the N parameter has a very shallow local minimum for the bcc structure but no extremum for the bct structure.

In the tetragonal transformation, the bcc structure is not the local maximum of the D parameter. Although the bcc structure has a higher density if it is compressed along the tetragonal axis, many elements take this structure. On the other hand, the bct structure is the local maximum of the D parameter in the tetragonal distortion but no metal takes this structure. These results indicate that density is not an important factor for the choice of crystal structure.

The third transformation discussed here is not a simple elongation or compression but an assembly of concerted rotations keeping the local octahedral arrangements as shown in Fig. 3.5, which is often discussed for the rutile structure. By the concerted rotations, the hcp structure is converted to the arrangement corresponding to the positions of oxygen atoms in the ideal rutile structure, where all cations are surrounded by perfect oxygen octahedra. The symmetry during the transformation is *Pnnm*. By further rotations, the structure recovers the hcp structure placed in a different orientation as illustrated in Fig. 3.5c. The variations of the D and N parameters in Fig. 3.6 show simple local minimum at the ideal rutile structure between two hcp structures. In real rutile structures, the rotation angle is the same as for the ideal structure, which means it has the density of locally minimum. Therefore, it is again suggested that higher density does not contribute to the stability of the structure.

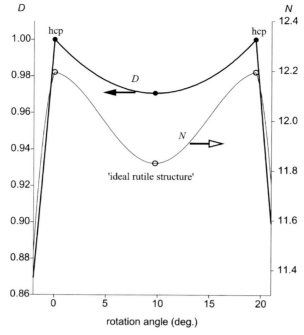

Fig. 3.6 Dependences of the D and N parameters on the rotation angle in the transformation illustrated in Fig. 3.5.

3.6
Close-packing of Cations or of Anions?

In the interpretation of binary structures, a common approach is to regard the arrangement of one component as close packing with the other component in its interstitial vacancies. In these descriptions, anionic components are generally considered to have the close-packing arrangements except for the structures with very large cations. For example, in describing the NaCl structure, the explanation that the chloride anions are arranged in close packing is preferred to the close packing of sodium cations. This preference comes from the assumption that larger ions tend to make close-packing arrangements and anions are generally considered to be the larger components.

In the NaCl or ZnS(cF) structures, both components are arranged in the ideal close-packing structure and it cannot be judged which component has the stronger tendency for close packing. However, for other structures where at least one of the components does not have the ideal close-packing structures, we can compare the D or N parameters of the two components. Here we will use the D parameter since it is more sensitive for measuring the proximity to close packing as discussed above. In Fig. 3.7, the D parameters of the cations (D_c) are

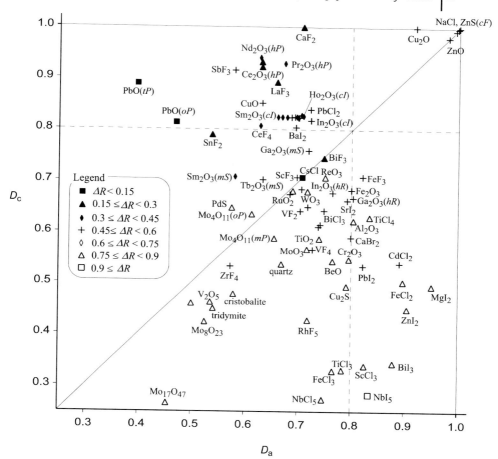

Fig. 3.7 Plot of the D parameter of cations (D_c) against the D parameter of anions (D_a) for selected binary structures. Compounds with homoatomic bonds are excluded.

Compounds are classified by ΔR ($= R_a - R_c$) as shown in the legend, where R_a and R_c are the crystal radius of a cation and an anion, respectively.

plotted against those of the anions (D_a) for many binary compounds, where compounds with homoatomic bonds are excluded. In this plot, the structures with anions in approximate close-packing arrangements come close to the right end (generally $0.8 < D_a$) while those with close-packed cations are near the top of the plot (generally $0.8 < D_c$). The traditional view suggests that many structures would appear in the lower right region due to the stronger preference of the anions for close packing. However, the plot shows approximately symmetrical distributions around the diagonal line ($D_c = D_a$). This distribution indicates that, on the whole, cations and anions prefer close packing to a similar degree.

In Fig. 3.7 the difference in the sizes between cations and anions for each compound is distinguished by markers and the plot shows the general trend that larger ions have a stronger tendency for close packing. The structures plotted in the right lower region have a structure conforming with the traditional view: anions are arranged in close packing with cations in the interstices. Those in the left upper region have the reverse arrangements. Then, the distance from the diagonal line, which is equal to the difference of the D parameters ($\Delta D = D_a - D_c$) will indicate the difference in the tendency towards close packing between cations and anions. As illustrated in Fig. 3.8, the difference ΔD has a positive correlation with the difference of the radii ($\Delta R = R_a - R_c$), where R_a and R_c are crystal radii of the cation and the anion, respectively, by Shannon. The correlation coefficient between ΔD and ΔR is 0.82. If anions are much larger than the other ions ($\Delta R \gg 0$), the anions have a stronger tendency to have a close-packing arrangement ($\Delta D \gg 0$) than the cations. This is what we can expect. However, if the ionic sizes of cations and anions are approximately equal ($\Delta R \approx 0$), Fig. 3.8 shows that the cations have a much stronger tendency toward close packing. In fact, the least-squares line of the ΔD vs. ΔR plot indicates that the tendencies towards close packing become equal if the crystal radius of anions is 0.40 Å larger than that of cations. In other words, the radii to be used for comparing the tendency towards close packing are 0.20 Å larger for cations

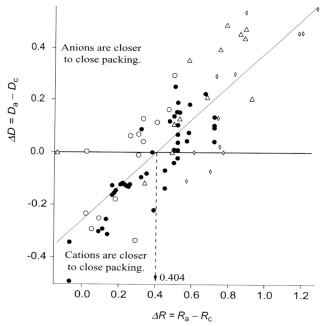

Fig. 3.8 Plot of ΔD ($= D_a - D_c$) vs. ΔR ($= R_a - R_c$) for selected binary compounds. D_a and D_c are the D parameters of anions and cations, respectively, while R_a and R_c are Shannon's crystal radii of anions and cations, respectively. Compounds with homoatomic bonds are excluded. Compounds are classified by the anions: oxides ($\bullet$), fluoride (O), chlorides and bromides ($\triangle$), iodides and sulfides ($\diamond$).

Table 3.2 Close-packing radii [a].

Ion	Radius	Ion	Radius	Ion	Radius	Ion	Radius
Si^{4+}	0.74	O^{2-}	1.06	Lu^{3+}	1.201	Pb^{2+}	1.53
Be^{2+}	0.79	Mg^{2+}	1.06	Y^{3+}	1.24	Br^-	1.62
Al^{3+}	0.875	Zr^{4+}	1.06	Ca^{2+}	1.34	Se^{2-}	1.64
Mo^{6+}	0.93	Zn^{2+}	1.08	Na^+	1.36	Ba^{2+}	1.69
Sb^{5+}	0.94	Sc^{3+}	1.085	Bi^{3+}	1.37	K^+	1.72
Ti^{4+}	0.945	Li^+	1.10	La^{3+}	1.372	Rb^+	1.86
Nb^{5+}	0.98	Sb^{3+}	1.10	Cl^-	1.47	I^-	1.86
Fe^{3+}	0.985	Fe^{2+}	1.12	S^{2-}	1.50	Cs^+	2.01
F^-	0.99	In^{3+}	1.14	Sr^{2+}	1.52		

a) Calculated from Crystal Radii by Shannon [4].

and 0.20 Å smaller for anions than the crystal radii by Shannon. Here we will call such radii "close-packing radii", and the value for selected ions are shown in Table 3.2. This is an unexpected result. For example, the close-packing radius of the potassium ion (1.72 Å) is much larger than that of the isoelectronic chloride anion (1.47 Å). Apparently these results force us to reconsider the relation between inorganic structures and close packing.

3.7
What Determines the Structure?

An ordinary interpretation of inorganic structure using the concept of the close packing is that larger ions make close-packing arrangements with the other ions in the interstices. This is often a convenient method to view binary structures. However, in the previous section, cations have a stronger tendency for close packing than anions of a similar size if the size is measured by Shannon's crystal radius. This result indicates that the principal factor that determines binary structures is not the close-packing principle.

The energy difference between possible structures of a binary compound is mainly determined by the local cation–anion interactions. Generally, smaller ions have a stronger request for its local structure. For example, an oxide anion in a tetrahedral environment is much more stable than one in a square planner environment, while cesium ions are observed in various coordination environments. The fulfillment of the local structure is seldom compatible with the close-packing arrangement. In many cases, one kind of ion selects the stability of the local structure and the other makes the close-packing arrangement. Then the close-packing radius indicates the degree of the request for its local environment in the reverse direction: ions with the smaller radii have a stronger request. From this viewpoint, the sequence given in Table 3.2 seems reasonable. The potassium ion has very low preference in its local structure while the chloride ion has stronger preference for the appropriate projection of its four electron pairs.

The analyses of the inorganic structures with the D parameter have led us to the following conclusion. The ions that have a stronger preference in their environment determine the local structure. The other ions will be arranged in close packing if the arrangement is consistent with the local structure. In other words, close packing does not determine the structure but is a consequence of the cation–anion interaction.

In this article, we have focused on the arrangements of one component, neglecting interactions between cations and anions, and examined how close they are to close packing. However, the analyses have suggested that "close packing" is not the principle that determines the structure. In compounds, close packing is a simple motif that appears as a consequence of complex local interactions.

Appendix. ICSD Codes, D and N Parameters of the Structures Used

Table A1 Ideal structures.

Compound	D	N	Compound	D	N
bcc	0.919	11.72	simpleCubic	0.707	6.39
ccp	1.000	12.19	spinel-B	0.500	6.00
diamond	0.459	4.00	ideal rutile	0.971	11.83
hcp-ideal	1.000	12.20	bct	0.943	11.47
ReO_3-O	0.750	8.19			

Table A2 Structures of elements.

Compound	Code	D	N	Compound	Code	D	N
B(hR)	56992	0.449	5.40	P(oS,black)	23836	0.409	3.04
Ba(hP)	52680	0.967	12.08	Ru(hP)	40354	0.955	12.01
Be(hP)	1425	0.961	12.05	S_8(cF)	63082	0.233	2.00
Br_2(oS)	201692	0.270	1.05	Sb(hR)	64695	0.576	4.94
Cl_2(oS)	201696	0.204	1.00	Se(hP,metallic)	22251	0.462	3.11
F_2(mS)	16262	0.165	1.00	Se_8(mP,a)	2718	0.299	2.01
I_2(oS)	67611	0.349	1.42	Sn(tI,β)	43415	0.722	7.19
Mg(hP)	44923	0.994	12.18	Tb(hR)	52497	0.974	12.13
Mn(cI,a)	42774	0.672	7.53	Te(hP)	65692	0.632	5.06
Mn(cP,β)	41775	0.741	9.49	Ti(hP)	43416	0.973	12.11
O_2(mS)	31163	0.062	1.00	Zn(hP)	52543	0.879	11.10

Table A3 Structures of binary compounds.

Compound	Code	D(c)	N(c)	D(a)	N(a)	Compound	Code	D(c)	N(c)	D(a)	N(a)
$AgI(hP)$	62789	0.998	12.19	0.998	12.19	$CrB(oC)$	40792	0.776	9.15	0.232	2.00
$Al_2O_3(hR)$	92629	0.623	5.88	0.804	10.28	$Cs_3As(hP)$	409668	0.610	6.71	0.872	11.27
$BaI_2(oP)$	15707	0.804	9.93	0.694	7.79	$CsBi(mP)$	55067	0.507	4.60	0.208	2.00
$BeO(tP)$	18147	0.543	4.36	0.765	9.44	$CsBi_2(cF)$	55070	0.459	4.00	0.500	6.00
$Bi_{12}Cl_{14}(oP)$	9225	0.325	3.17	0.428	4.60	$CsCl(cP)$	22173	0.707	6.39	0.707	6.39
$Bi_2Cs_3(mS)$	240017	0.542	5.06	0.146	1.00	$Cu_2O(cP)$	63281	1.000	12.19	0.919	11.72
$BiCl_3(oP)$	41179	0.642	5.99	0.748	9.25	$Cu_2S(mP)$	100333	0.494	4.37	0.791	10.15
$BiF_3(oP)$	1269	0.745	8.71	0.748	8.84	$CuO(mS)$	16025	0.851	10.78	0.631	5.97
$BiI_3(hR)$	78791	0.343	3.00	0.879	11.47	$CuS(hP)$	67581	0.682	7.39	0.185	0.67
$CaBr_2(oP)$	56763	0.590	5.49	0.799	10.43	$Dy_2O_3(cI)$	82421	0.828	10.16	0.705	7.35
$CaCl_2(tP)$	56421	0.611	6.01	0.738	9.05	$Er_2O_3(cI)$	39521	0.822	10.05	0.699	7.22
$CaF_2(cF)$	60559	1.000	12.19	0.707	6.39	$Eu_2O_3(cI)$	40472	0.824	10.08	0.669	6.84
$CdCl_2(hP)$	86440	0.539	6.00	0.890	11.52	$EuI_2(oP)$	56816	0.684	7.63	0.706	8.04
$CdS(hP)$	60629	0.994	12.18	0.994	12.18	$Fe_2O_3(hR)$	22505	0.683	7.09	0.800	10.21
$Ce_2O_3(hP)$	61379	0.922	11.84	0.630	5.98	$Fe_2P(hP)$	70115	0.732	8.18	0.800	10.00
$CeF_4(mS)$	89621	0.807	9.19	0.628	6.99	$FeCl_2(hR)$	4059	0.503	6.00	0.897	11.62
$CoF_2(tP)$	73460	0.650	6.93	0.718	8.57	$FeCl_3(hR)$	39764	0.328	3.00	0.766	9.67
$Cr_2B_3(mC)$	44195	0.687	7.90	0.261	2.64	$FeF_3(hR)$	41120	0.708	6.41	0.821	9.41
$Cr_2O_3(hR)$	75577	0.546	4.46	0.796	10.15	$FeP(oP)$	43248	0.570	5.13	0.560	4.71
$FeP_2(oP)$	15027	0.373	2.10	0.413	2.00	$ReO_3(cP)$	77679	0.707	6.39	0.750	8.19
$FeP_4(oP)$	2413	0.419	2.47	0.397	2.60	$RhF_5(mP)$	10173	0.427	2.34	0.719	8.48
$FeP_4(oS)$	2442	0.659	6.44	0.417	2.62	$RuO_2(tP)$	84575	0.674	7.49	0.684	7.78
$FeS_2(cP)$	316	1.000	12.19	0.357	1.17	$SbF_3(oS)$	16142	0.915	11.69	0.579	5.89
$Ga_2O_3(hR)$	27431	0.667	6.81	0.803	10.21	$Sc_5Cl_8(mS)$	36403	0.339	4.51	0.828	10.66
$Ga_2O_3(mS)$	83645	0.759	9.22	0.718	8.79	$Sc_7Cl_{10}(mS)$	1018	0.383	5.86	0.716	9.16
$Gd_2O_3(cI)$	40473	0.824	10.08	0.678	6.98	$Sc_7Cl_{12}(hR)$	36021	0.379	4.89	0.833	10.79
$Ho_2O_3(cI)$	82422	0.826	10.13	0.708	7.41	$ScCl(hR)$	1004	0.505	8.31	0.637	8.81
$In_2O_3(cI)$	14388	0.818	9.94	0.722	7.79	$ScCl_3(hR)$	74517	0.338	3.00	0.825	10.82
$In_2O_3(hR)$	16086	0.677	7.09	0.764	9.38	$ScF_3(hR)$	36011	0.707	6.39	0.697	7.95
$KBi(mP)$	55065	0.497	4.36	0.254	2.00	$SiO_2(hP)$	89276	0.536	4.15	0.669	6.74
$La_2O_3(hP)$	100204	0.932	11.90	0.629	5.97	$SiO_2(mS)$	408281	0.449	3.99	0.541	6.02
$LaB_4(tP)$	2360	0.686	6.76	0.269	4.41	$SiO_2(oP)$	75300	0.478	4.04	0.579	6.08
$LaF_3(hP)$	74729	0.892	11.47	0.659	7.39	$SiO_2(cF)$	35536	0.459	4.00	0.500	6.00
$LaI_2(tI)$	72191	0.397	4.13	0.721	8.16	$Sm_2O_3(cI)$	40475	0.824	10.08	0.661	6.70
$Lu_2O_3(cI)$	40471	0.823	10.07	0.700	7.33	$Sm_2O_3(mS)$	202903	0.708	8.57	0.581	5.47
$MgB_2(hP)$	93925	0.716	7.40	0.275	3.00	$SnF_2(mS)$	10485	0.789	9.83	0.538	6.05
$MgF_2(tP)$	8120	0.615	6.11	0.741	9.11	$SrI_2(oP)$	203137	0.662	7.68	0.792	9.00
$MgI_2(hP)$	281551	0.494	6.00	0.951	12.01	$Ta_3S_2(oS)$	71143	0.518	7.70	0.393	4.76
$Mo_{17}O_{47}(oP)$	4111	0.266	0.60	0.454	3.46	$Ta_6S(aP)$	202564	0.631	7.12	0.336	2.25
$Mo_4O_{11}(mP)$	82363	0.588	5.62	0.655	7.71	$Ta_6S(mS)$	16041	0.642	7.41	0.352	2.47
$Mo_4O_{11}(oP)$	201573	0.635	5.83	0.613	7.23	$TaS_2(hP)$	68488	0.447	6.00	0.744	8.84
$Mo_8O_{23}(mP)$	68809	0.423	3.00	0.525	5.76	$TaS_3(mP)$	15251	0.355	3.37	0.253	0.50
$MoO_2(mP)$	80830	0.341	1.41	0.862	10.93	$Tb_2O_3(cI)$	40474	0.824	10.08	0.686	7.10
$MoO_3(oP)$	35076	0.566	5.28	0.717	8.40	$Tb_2O_3(mS)$	28172	0.702	8.41	0.633	6.44
$Na_3As(hP)$	81566	0.603	6.31	0.918	11.68	$TiB_2(hP)$	56723	0.766	8.01	0.295	3.00
$NaCl(cF)$	18189	1.000	12.19	1.000	12.19	$TiCl_3(hR)$	39426	0.330	3.00	0.784	10.03

Table A3 (continued)

Compound	Code	D(c)	N(c)	D(a)	N(a)	Compound	Code	D(c)	N(c)	D(a)	N(a)
$Nb_6Cl_{14}(mS)$	28535	0.225	4.01	0.734	9.34	$TiCl_4(mP)$	280981	0.630	6.03	0.834	10.83
$Nb_6I_{11}(oP)$	32736	0.149	3.86	0.672	7.88	$TiO_2(tP)$	62679	0.587	5.44	0.739	9.12
$NbCl_4(mS)$	1010	0.160	1.28	0.607	6.22	$TlI_3(oP)$	61349	0.481	3.39	0.298	1.38
$NbCl_5(mS)$	300102	0.274	1.03	0.747	9.20	$Tm_2O_3(cI)$	78582	0.824	10.09	0.696	7.16
$NbI_5(aP)$	10457	0.283	1.04	0.834	10.93	$V_2O_5(oP)$	60767	0.462	4.10	0.536	5.12
$NbO_2(tI)$	28500	0.394	1.76	0.839	10.75	$VF_2(tP)$	62768	0.642	6.74	0.704	8.27
$Nd_2O_3(hP)$	32514	0.940	11.96	0.627	5.94	$VF_4(mP)$	65785	0.566	4.31	0.727	8.19
$NdI_2(tI)$	72190	0.390	4.13	0.716	7.93	$WO_3(aP)$	80053	0.681	6.53	0.689	8.07
$Ni_3B(oP)$	75794	0.801	10.05	0.509	3.52	$WO_3(mP)$	80056	0.680	6.53	0.717	8.35
$NiAs(oS)$	66120	0.395	2.13	0.830	10.44	$Yb_2O_3(cI)$	78583	0.824	10.09	0.692	7.13
$NiAs_2(oP)$	34851	0.674	7.74	0.427	1.81	$ZnI_2(tI)$	2404	0.449	4.26	0.905	11.67
$PbCl_2(oP)$	27736	0.839	10.75	0.722	8.15	$ZnO(hP)$	65119	0.980	12.14	0.980	12.14
$PbI_2(hP)$	68819	0.533	6.01	0.822	10.66	$ZrB_{12}(cF)$	409634	1.000	12.19	0.396	4.75
$PbO(oP)$	60135	0.813	10.20	0.469	5.75	$ZrCl(hR)$	869	0.463	7.75	0.631	8.88
$PbO(tP)$	62840	0.889	11.49	0.397	4.13	$ZrF_4(tP)$	35100	0.532	4.81	0.573	5.72
$Pd_2B(oP)$	10487	0.861	11.25	0.569	4.94	$ZrI_2(mP)$	26418	0.240	3.10	0.781	9.52
$PdS(tP)$	61063	0.647	6.78	0.575	5.52	$ZrI_2(oP)$	24807	0.240	3.11	0.780	9.51
$Pr_2O_3(hP)$	75481	0.929	11.88	0.673	6.56	$ZrI_3(oP)$	74648	0.147	1.80	0.927	11.87
$RbBi_2(cF)$	55069	0.459	4.00	0.500	6.00						

References

1 B.G. Hyde and S. Andersson, *"Inorganic Crystal Structures"*, Wiley, New York, 1989. The book is a great collection of inorganic structures described mostly from the viewpoint of close packing and bcc.

2 G.G. Szpiro, *"Kepler's Conjecture: How Some of the Greatest Minds in History* *Helped Solve One of the Oldest Math Problems in the World"*, 2003, Wiley.

3 M. O'Keeffe and B.G. Hyde, *"Crystal Structures I. Patterns and Symmetry"*, chapter 6, Mineralogical Society of America, Washington DC, 1996.

4 R.D. Shannon, Acta Crystallogr., A32, 751 (1976).

4
Forty-five Years of Praseodymium Di-iodide, PrI$_2$

Gerd Meyer and Andriy Palasyuk

Foreword

Forty-five years ago, in 1961, John D. Corbett published two papers in which he established the existence of four rare-earth metal di-iodides, LaI$_2$, CeI$_2$, PrI$_2$ [1] and NdI$_2$ [2] which he thereafter coined as metallic (the first three) and salt-like di-iodides, respectively (see for example Corbett's overview of 1973 [3]). It took a while and afforded one of Hartmut Bärnighausen's students, Eberhard Warkentin [4], the opportunity to discover that PrI$_2$ is special in that it may appear in five modifications, contrary to LaI$_2$ and CeI$_2$ which crystallize with the CuTi$_2$ structure only over the whole temperature range and also contrary to NdI$_2$ which is a salt just like the heavier alkaline-earth iodides. Only under high pressures does it transform to a metallic di-iodide [5].

I (GM) have avoided this chemistry for a long time. First I thought everything had already been "said and done". Then because the five modifications of PrI$_2$ puzzled me and I thought it was too complicated a matter to consider, although with Pr$_2$Br$_5$ and Pr$_2$I$_5$ we were quite close. It needed Niels Gerlitzki to cause me to go deeper into "praseodymium di-iodide". He reduced numerous rare-earth iodides with alkaline-earth metals and all of a sudden there were single crystals of PrI$_2$ (IV, V) and Pr$_2$I$_5$ [6]. Fortunately, he discovered hitherto unseen compounds like BaLaI$_4$ [7] and Ba$_6$Pr$_3$I$_{19}$ [8] so that I could again avoid any deeper involvement with PrI$_2$. However, subconsciously I was still interested and in 2003 I found myself in the freedom and quiteness of Ames, Iowa, while Anja-Verena Mudring was a postdoctoral associate with John D. Corbett. I was visiting for a few weeks, and I talked her into using her recently increased knowledge of quantum chemistry to shed some light on the electronic structures of PrI$_2$, and we then involved John (see [9]). Indeed, PrI$_2$ is unique among the rare-earth di-iodides in that the configurational crossover from the [Xe]6s^{0}5d^{1}4f^2 to the [Xe]6s^{0}5d^{0}4f^3 electronic configuration of "Pr^{2+}" takes place gradually and is dependent upon the actual crystal structure. A higher d-orbital contribution results in stronger Pr–Pr interactions. The optimization of bonding aimed at a balance between Pr–Pr bonding and Pr–I antibonding interactions

Inorganic Chemistry in Focus III.
Edited by G. Meyer, D. Naumann, L. Wesemann
Copyright © 2006 WILEY-VCH Verlag GmbH & Co. KGaA, Weinheim
ISBN: 3-527-31510-1

close to the Fermi level appears to be responsible for the stability of each modification. But what is the thermodynamic driving force for their formation?

Fortunately, Andriy Palasyuk from Lviv, with his vast background in thermochemistry, came to Cologne as a postdoctoral associate and took on the tedious task of investigating the influence of temperature on the formation of "the praseodymium di-iodides". Section 4.3 reports his results. It is preceded, after an introduction, by a section on synthesis and structures of PrI_2. We have largely excluded a report on the electronic structures, as this can be found in [9].

4.1
Introduction

Almost all of the rare-earth metal/rare-earth metal tri-iodide systems, R/RI_3, contain binary phases with the rare-earth element in an oxidation state lower than +3 ("reduced" rare-earth metal iodides) [3, 7, 10–13]. More common is the oxidation state +2. Elements that form di-iodides RI_2 are illustrated in Fig. 4.1.

These di-iodides may be roughly divided into two groups. First, salt-like di-iodides $RI_2 = (R^{2+})(I^-)_2$ that are very reminiscent, at least structurally, of alkaline-earth iodides. Secondly, "metallic" di-iodides, $RI_2 = (R^{3+})(e^-)(I^-)_2$. In these a configuration crossover from $6s^0 5d^0 4f^n$ ("R^{2+}") to $6s^0 5d^1 4f^{n-1}$ ("$R^{3+}e^-$") has taken place. It is, therefore, the electron in the 5d state that makes these iodides interesting. This $5d^1$ electron may be delocalized into a conduction band with (semiconductor) or without a band gap (metallic). More interestingly, clusters may be formed through attractive interactions of such orbitals, or long-range magnetic effects may occur. For the latter, GdI_2 is a paramount example as it not only exhibits ferromagnetism but also a huge magnetoresistance [14].

Praseodymium di-iodide, PrI_2, appears to be a special case. Five modifications were reported almost thirty years ago [4]. Judging from the crystal structures alone (below), one is tempted to assign these to the "metallic" (PrI_2-I, $CuTi_2$-type of structure, PrI_2-II, -III, MoS_2-type structures) and to the "salt-like" (PrI_2-IV, $CdCl_2$-type) groups of rare-earth metal di-iodides. In PrI_2-V a tetrahedral cluster {Pr_4} is observed, surrounded by and connected through iodide ions according to the formulation {Pr_4}$I_4 I_{12/3}$. So far, PrI_2-V represents the only case of

Sc						
Y						
La	Ce	Pr	Nd	Pm	Sm	Eu
Gd	Tb	Dy	Ho	Er	Tm	Yb
Lu						

Fig. 4.1 Rare-earth elements that form di-iodides, RI_2.
Light grey: "Metallic" di-iodides, $RI_2 = (R^{3+})(e^-)(I^-)_2$. Dark grey:
Salt-like di-iodides, $RI_2 = (R^{2+})(I^-)_2$. Ambient conditions.
The scandium di-iodide is non-stoichiometric, $Sc_{0.9}I_2$ [13].

a binary reduced rare-earth iodide with the formation of a cluster. There are, however, numerous compounds with octahedral and trigonal-bipyramidal clusters, interstitially stabilized by carbon atoms, di-carbon units, and a large number of main-group and transition-metal atoms [12, 15].

Taking the simple view, salt-like $R^{2+}(I^-)_2$ and metallic $R^{3+}e^{-+}(I^-)_2$ iodides with the involvement of metal–metal bonding in PrI_2-V need, however, a refinement with respect to their electronic structures. Indeed, a gradual configurational crossover from $6s^0 5d^0 4f^3$ (PrI_2-IV at low temperatures) to $6s^0 5d^1 4f^2$ (metallic PrI_2-I) can be observed. Between these extremes, a higher d orbital contribution results in stronger Pr–Pr interactions. The optimization of Pr–Pr bonding is important as is the avoidance of Pr–I antibonding interactions [9].

One major drawback for the elucidation of the physics of the five (or even more) modifications of praseodymium di-iodide, PrI_2, is the complicated phase behavior and therefore the extreme difficulty in producing, not only PrI_2 free of other compounds (such as Pr_2I_5, PrI_3, PrOI, or others) but also PrI_2 as a single phase ("one structure").

4.2
Phases and Structures in the System Praseodymium-Iodine

4.2.1
Synthesis Generalities

The binary system praseodymium-iodine (Pr/I) contains three compounds, PrI_3, Pr_2I_5, and PrI_2 (see Fig. 4.2). According to early phase diagram determinations [1, 16] these melt at 738 °C (PrI_3, congruently), 676 °C (Pr_2I_5, congruently), and 758 °C (PrI_2, incongruently).

Praseodymium tri-iodide, PrI_3, as the starting material for reduction reactions, might be easily produced by the oxidation of praseodymium metal with elemental iodine [17]. With catalytic amounts of hydrogen dissolved in praseodymium metal powder, the reaction temperature can be as low as 230 °C [18]. Sublimation in high vacuum in tantalum tubes yields pure PrI_3.

Dipraseodymium penta-iodide, Pr_2I_5, is obtained either by a conproportionation reaction [4, 19],

$$5PrI_3 + Pr = 3Pr_2I_5 \tag{1}$$

or by metallothermic reductions of PrI_3 with alkali or alkaline-earth metals [6], for example,

$$Sr + 4PrI_3 = 2Pr_2I_5 + SrI_2 \tag{2}$$

Both reaction types are carried out in sealed tantalum containers at temperatures around 700 °C, above the melting point of Pr_2I_5 to speed up the reaction and to assure crystal growth upon slow cooling.

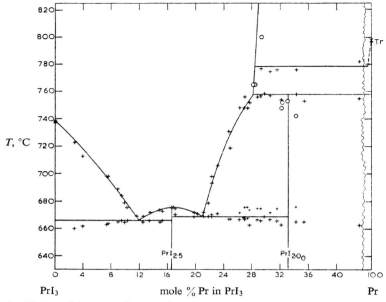

Fig. 4.2 Part of the phase diagram of the system praseodymium-iodine (Pr/I); the abscissa is drawn as molar fractions of Pr in PrI$_3$. Reproduced from Ref. [1].

Praseodymium di-iodide, PrI$_2$, can essentially be made in the same way. If sufficient care is taken to exclude air and moisture, oxidic impurities can be avoided. To avoid the formation of Pr$_2$I$_5$, praseodymium metal is used in excess as chunks to easily remove the unreacted metal after the reaction is completed. The pure compound PrI$_2$ is thus obtained, with a reaction temperature well above the peritectic temperature, around 800 °C. Reaction times seem not to matter much, a few days are usually sufficient, perhaps even less. The cooling procedure, however, is crucial as it determines the phases (I through V) that are formed and their relative quantities. Section 4.3 will deal with this issue.

Praseodymium di-iodide is moisture-sensitive under ambient conditions and reacts at higher temperatures with the constituents of air. It should be handled and stored under dry box conditions.

4.2.2
Structural Principles

In the crystal structure of Pr$_2$I$_5$ there are two crystallographically independent praseodymium sites, Pr1 and Pr2, which have coordination numbers of 7 and 8 with mean Pr-I distances of 322.0 and 335.3 pm, respectively [19]. These distances are much like the respective Pr-I distances for coordination number 7 in K$_2$PrI$_5$ (d = 321.2 pm [20]) and are as expected for coordination number 8 in

PrI_3 (PuBr$_3$-type of structure, for LaI$_3$ d=334 pm (6×), 340 pm (2×) [21]). Pr$_2$I$_5$ is a semiconductor which orders antiferromagnetically below T_N=37(1) K [19, 22]. In a localized picture, the electronic structure of Pr$_2$I$_5$ [23] and also of Pr$_2$Br$_5$ [24] can be understood as if triangular clusters {Pr$_3$} contained one electron in a tricentric orbital. These triangles share common corners to chains {Pr$_{2/2}$Pr$_{1/1}$}.

The structural principles of PrI$_2$ can be derived either from 4^4 nets (PrI$_2$-I) or 3^6 nets (all other modifications) of iodine atoms that are stacked along a prominent crystallographic direction, in most cases the [001] direction. Between these layers, half of the respective interstices are filled with praseodymium atoms (but see PrI$_2$-V below). Please note that 4^4 and 3^6 nets are closely related to each other, it only needs a shear procedure to transform one net to the other (Fig. 4.3). In the iodine layers I-I distances are even shorter in the 4^4 net (386 in PrI$_2$-I [4]) than in the 3^6 net of PrI$_2$-IV (426.5 pm [6, 9]).

In PrI$_2$-I (as in LaI$_2$ and CeI$_2$) the praseodymium atoms are also arranged in 4^4 nets such that praseodymium is eight-coordinate (Fig. 4.4). The coordination polyhedron is a cube [PrI$_8$], the cubes share four common faces such that iodine is four-coordinate, in accordance with the formulation PrI$_{8/4}$. In other words, I–Pr–I slabs are stacked in the [001] direction of the tetragonal body-centered structure, usually called the CuTi$_2$ type of structure.

Iodine 3^6 nets as two-dimensional closest packings of spheres can be stacked perpendicularly, for example, in the manner of the cubic closest packing of spheres as observed for PrI$_2$-IV (CdCl$_2$-type of structure). The simple alternative, the hexagonal closest packing (CdI$_2$-type) has not been observed for PrI$_2$, but for Sc$_{0.9}$I$_2$, with the smaller scandium atoms [13, 25]. Less dense packings are represented by primitive stackings, AA or the like, observed for PrI$_2$-II and -III (2H- and 3R-MoS$_2$ types of structure) [4]. Considerable stacking disorder can occur, and such "phases" are frequently observed (below). In any of these cases, between every second iodine layer, all interstices are filled with praseodymium atoms, which results in all octahedral (IV) or all trigonal prismatic (II, III) holes (Fig. 4.5). It has been assumed, so far, that PrI$_2$ is a line compound as the phase diagram suggests, at least at sufficiently low temperatures (see Fig. 4.2).

PrI$_2$-V is unique. Tetrahedral {Pr$_4$} clusters may reasonably be drawn (Fig. 4.6). These are capped on all four faces and have three terminal iodine

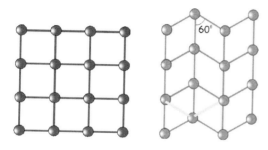

Fig. 4.3 4^4 and 3^6 nets of iodine atoms as observed in the modifications of PrI$_2$.

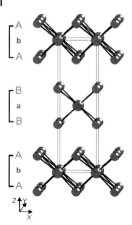

Fig. 4.4 Crystal structure of PrI₂-I.

atoms at each vertex that they share with other clusters according to $\{Pr_4\}I_4I_{12/3}$ (Fig. 4.6). It may also be understood as a structural variant of the $CdCl_2$-type PrI₂-IV in that, respectively, three-quarters and one-quarter of the octahedral holes between every iodine layer are occupied by praseodymium atoms which, properly ordered, form a tetrahedron. The close relation between PrI₂-V and the spinel type of structure has been stressed [4]. Its structure is also closely related

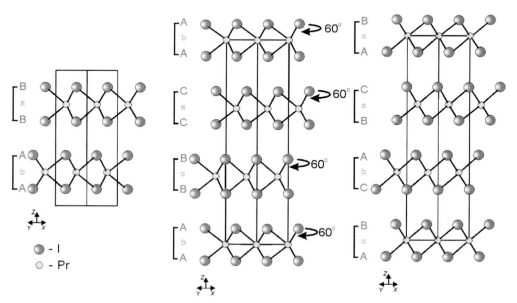

Fig. 4.5 Crystal structures of PrI₂-II, -III, and -IV.

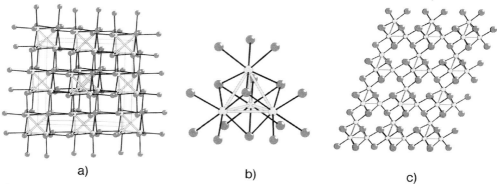

Fig. 4.6 Crystal structure of PrI$_2$-V. (a) The cubic face-centered unit cell. (b) One cluster {Pr$_4$}I$_4$I$_{12}$. (c) Hexagonal setting of the structure exhibiting the close relationship to the CdCl$_2$-type of structure.

to that of LiTiO$_2$ [26] and, more importantly, to NaPrTe$_2$ [27], according to NaPrTe$_2$ = □PrI$_2$, which also have cubic structures.

From the crystal structures of PrI$_2$-IV and -V, as they were refined recently [6, 9], Pr-I distances are 318.8(2) pm in IV and 327.8(1) and 313.6(1) pm (average: 320.7 pm) in PrI$_2$-V are calculated. These distances compare very well with the mean Pr-I distances for seven-coordinate Pr in Pr$_2$I$_5$ (above, 322 pm). Thus, by size, we essentially "see" the Pr^{3+} ion with the excess electron being available for bonding interactions. Their strength is reflected by the Pr-Pr distances: 386.4(2) in PrI$_2$-I [4], 391.3(2) pm in PrI$_2$-V and 426.5(1) pm in PrI$_2$-IV [9]. For PrI$_2$-I and -V the configuration 6s^{0}5d^{1}4f^2 best describes the situation. In PrI$_2$-I the 5d^1 electron is delocalized in a conduction band, which makes the phase metallic [25]. In PrI$_2$-V the 5d^1 electron is engaged in bonding interactions. The respective band is flat and well below the Fermi level [9]. For PrI$_2$-IV the configuration is more like 6s^{0}5d^{0}4f^3, at least at sufficiently low temperatures, for which there is some evidence from preliminary magnetic measurements. There is a non-negligible bonding interaction at higher temperatures (thus the 5d band is partially populated). With long Pr–Pr separations (426.5 pm) these interactions are rather weak but counterbalance the Pr–I antibonding interactions that also occur close to the Fermi level.

4.3
PrI$_2$: Phases and Phase Analysis

Physical properties, such as magnetism and electronic conduction, are largely unexplored for PrI$_2$. This is not because pure PrI$_2$ cannot be made (see 4.2). Rather, that it is extremely difficult, if not impossible, to produce single-phase, i.e., one-modification samples on a scale sufficiently large for these measurements. Growth of single crystals for X-ray structure determination is, however, not a problem (although structure refinement is often obscured by disorder). Therefore, we have put much effort into the elucidation of synthesis conditions and phase analysis in recent months.

To begin with, we followed the procedure as described by Warkentin [4] for the synthesis of PrI$_2$. The maximum temperatures reported were 1000 °C and 850 °C, much higher than the peritectic transformation, $L+a\text{-Pr} \leftrightarrow \text{PrI}_2$ at 758 °C [1]. X-ray phase analysis [28] of such samples revealed mixtures of three modifi-

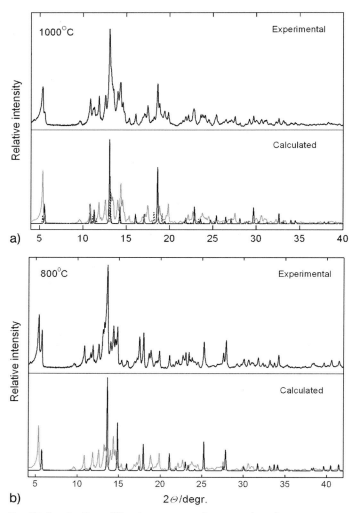

Fig. 4.7 Powder X-ray diffraction patterns of two samples of PrI$_2$. (a) Initial maximum temperature 1000 °C. (b) 800 °C. Calculated patterns: I, V: black lines, IV: dashed, Pr$_2$I$_5$: grey.

cations of PrI_2, I, IV, and V. Their relative quantities depend on the maximum temperature of 1000 °C, followed by slow cooling to 750 °C, just below the peritectic temperature, and then quenching of the samples. PrI_2-V was the major phase, where initial heating to 800 °C mostly produced PrI_2-I (Fig. 4.7).

These initial samples contained some Pr_2I_5. To avoid Pr_2I_5 impurities, an excess of praseodymium metal was applied as chunks and the maximum temperature was decreased to 750 °C, slightly below the peritectic temperature at 758 °C. As Fig. 4.8 attests, pure single-phase PrI_2-IV was obtained when the sample was quenched after prolonged heating (five days) at 750 °C. Rietveld refinement for $CdCl_2$-type PrI_2-IV resulted in a = 424.24(5), c = 2236.62(8) pm, z(I) = 0.0799(3) with R_{Bragg} = 0.095. These values match quite well with recent single-crystal X-ray data: a = 426.5(1), c = 2247.1(8) pm, z(I) = 0.0766(2), R_{all} = 0.0930 [6, 9].

It had been suspected that at least PrI_2-II and -III could be contaminated with hydrogen and thereby stabilized, a problem that frequently occurs. It has been well investigated for the LaI_2/H_2 [31] and GdI_2/H_2 [32] systems. Other than the di-iodides RI_2 that, to the best of our knowledge, exist free of hydrogen or other interstitially stabilizing atoms, it is now commonly believed that no hydrogen-free rare-earth di-chlorides and di-bromides exist. $LaBr_2$ which has been reported [33] does obviously contain hydrogen. In the systems $Pr/PrCl_3$ [34] and $Pr/PrBr_3$ [35] the only reduced phases that exist are $PrCl_{2.31} = Pr_{0.29}PrCl_3$ [36] and Pr_2Br_5 [37] and these appear to be free of hydrogen.

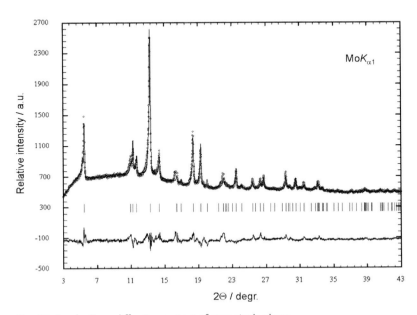

Fig. 4.8 Powder X-ray diffraction pattern of pure single-phase PrI_2-IV and Rietveld refinement.

A pure single-phase sample of PrI$_2$-IV was hydrogenated in a sealed tantalum tube at 700 °C. As Fig. 4.9 shows, hydrogen uptake causes PrI$_2$-IV to disorder and to form polytypic intergrowths of 2H-MoS$_2$ and 2H-NbS$_2$ structural fragments with stacking faults. If only the sharp reflections of PrI$_2$H$_n$ are indexed according to the 2H-MoS$_2$ type of structure, one may determine lattice constants of a = 413.2(1) and c = 1539.2(3) pm which would attest for a slight contraction of the lattice in the (001) plane and an elongation of the c axis.

Still, the question has to be addressed as to which of the many modifications of PrI$_2$ is thermodynamically stable under standard conditions. So far, it is clear that PrI$_2$-IV must be a high-temperature phase as it is produced in pure and single-phase by annealing just below the peritectic temperature (with an excess of praseodymium metal in order to avoid the formation of Pr$_2$I$_5$) and rapid cooling to ambient temperature.

Thus, phase-pure samples of PrI$_2$-IV were annealed at different temperatures below the peritectic temperature and phase analyses were undertaken by X-ray powder diffraction. Each sample was heated at a certain temperature ranging from 200 to 750 °C for five days and then quenched. Figure 4.10 gives an impression of how the phase changes proceed. Already temperatures as low as 200 °C, over such a long period of time, produce PrI$_2$-I and -V with I being the main phase. At higher temperatures, below 400 °C, PrI$_2$-V is the main product. With increasing temperature, the amount of a disordered phase increases, with a powder pattern which is very much reminiscent of the hydrogenated phase, PrI$_2$H$_x$. We conclude that this is a hydrogen-free disordered phase consisting of an intergrowth of those phases that were addressed as isostructural to the 3R- and 2H-MoS$_2$ structure types. These and others may occur dependent upon the

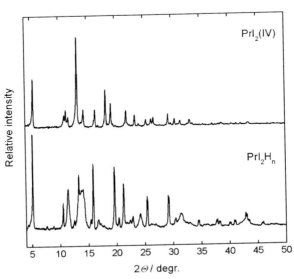

Fig. 4.9 Powder X-ray diffraction pattern of PrI$_2$-IV before and after hydrogenation.

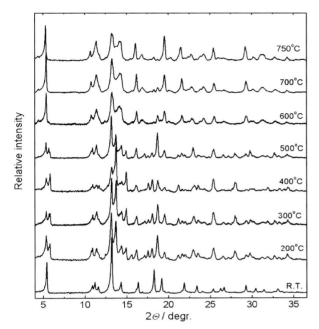

Fig. 4.10 Powder X-ray diffraction patterns for samples of phase-pure PrI₂-IV annealed at different temperatures.

actual reaction conditions. We shall call this "phase" PrI₂-D (D stands for disorder). Although PrI₂Hₓ and PrI₂-D have very similar powder patterns, they differ in a number of ways. Their colours are different, the former tends to grey-black, the latter to a brown-black colour. The lattice parameters are different: the hydrogenated samples show the same trends as recently seen for the LaI₂/H₂ system [31] whereas the lattice parameters for PrI₂-D are very similar to those reported for PrI₂-II (2H-MoS₂-type; for example: a = 418.2(6), c = 1518(2) pm).

Paralleling these powder X-ray investigations, thermoanalytical investigations (DSC [38]) with, again, phase-pure samples of PrI₂-IV were carried out. These show small thermal effects at onset temperatures of 330 °C (exothermic) and at 484 °C (endothermic) (Fig. 4.11). The exothermic effect, sluggish over a temperature range of 330–430 °C, must be interpreted as the transformation of, at that temperature, thermodynamically unstable PrI₂-IV to the thermodynamically more stable modifications PrI₂-I and -V and some -D, as X-ray diffraction patterns show. The higher temperature effect, although less sluggish, hence with better kinetics at higher temperatures, may be interpreted as the ongoing transformation of PrI₂-I in PrI₂-V and, increasingly, to PrI₂-D. On cooling, no thermal effects were detected which is certainly due to the rather high cooling rates and bad kinetics. The low-temperature effect at 330–430 °C cannot be expected on cooling as it would mean the transformation of thermodynamically more stable phases (I, V, II/III/D) to the thermodynamically unstable PrI₂-IV.

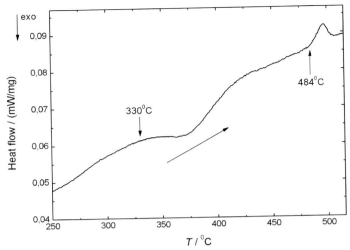

Fig. 4.11 Differential scanning thermogram (DSC) of PrI₂-IV.

One further observation, made some years ago, sheds additional light on both the true nature of PrI₂-IV and the phase behavior of praseodymium di-iodide in general. A thorough X-ray diffraction single crystal structure determination [39] revealed that PrI₂-IV does indeed crystallize with the basic $CdCl_2$-type of structure. However, additional praseodymium atoms are incorporated in octahedral voids between iodide layers, which are empty in the basic structure. There must be, in addition, an incommensurable superstructure present, attested by satellite reflections and diffuse scattering. A re-evaluation of our data that we had recently published [9] led to similar results. Therefore, PrI₂-IV is, most likely, $Pr_{1+x}I_2$ or $Pr(1)_xPr(2)_1I_2$ with $x=0.08(1)$, $0.11(2)$, and $0.13(3)$ according to refinements for three crystals (with the additional Pr(1) on the Wyckoff site 3a, not 18f as in [39]). This result, of course, reflects the composition of "PrI₂-IV" just below the peritectic temperature. One may conclude that there is a miscibility gap at higher temperatures, which opens to about 35 mol% praseodymium. Provided that the occupation factor in [39] of 0.126(9) must be divided by 3 to yield x, hence the composition of the crystal investigated was $Pr_{0.04}PrI_2$, then one would be tempted to conclude that, first, there is a miscibility gap and PrI₂ is not a line phase at high temperatures and, secondly, that our crystals would constitute the praseodymium-rich end of the miscibility gap. Note that we have applied excess praseodymium in our reactions.

If the miscibility gap of $Pr_{1+x}I_2$ is temperature dependent (which they usually are), then PrI₂ could be a line phase at, say, ambient temperature, i.e., x = 0. Hence, on quenching or when the annealing process progresses, the metastable $Pr_{1+x}I_2$ phase must release praseodymium metal. This surplus praseodymium metal can be released from both the 3a (heavily under-occupied) and 3b (fully occupied) positions combined with site changes from 3b to 3a, or not. These processes result, during the annealing process, in the formation of crystal nuclei of PrI_2-I, PrI_2-V and of PrI_2-II,III-D. As mixtures of these "three" are always observed, although in different quantities, somehow dependent upon temperature, the probabilities for the formation of nuclei of these crystals must be similar. Their formation affords the following.

• In the case of PrI_2-I, starting from the structure of $Pr_0Pr_1I_2$ ("pure" $CdCl_2$-type structure) the re-formation of 4^4 nets for both praseodymium and iodine. As the estimated quantity of PrI_2-I at lower temperatures is somewhat higher, one may conclude that this is the thermodynamically stable form under standard conditions. There are two arguments, which further support this assumption. First, LaI_2 and CeI_2 both crystallize with this structure, the $CuTi_2$ type, alone. No phase transitions have been observed. Furthermore, NdI_2 crystallizes under high pressures with the $CuTi_2$ type of structure and is a salt under ambient conditions, an $SrBr_2$ type of structure [5]. Secondly, the Pr–Pr distances in PrI_2–I are considerably shorter than in the modifications IV and II/III/D, 386 vs. 427 and 418 pm, respectively. Shorter Pr–Pr distances, however, result in stronger bonding interactions which are likely to result in higher thermodynamic stability.

• In the case of PrI_2-V, structurally speaking, further praseodymium atoms have to be transported from the octahedral holes of one layer to the neighboring layers until the ratio 0.75:0.25 is reached. This results in the formation of tetrahedral clusters {Pr_4} with Pr–Pr distances of 391 pm, only slightly longer than the Pr–Pr distances in the two-dimensional 4^4 net of PrI_2-I. Pr–Pr bonding in these {Pr_4} clusters is evident from band-structure calculations [9] and must result in an overall stabilization of this phase. It is interesting to note in this connection that the additional praseodymium atoms in "PrI_2-IV"~$Pr_{0.1}$-PrI_2 occupy around 10% (not 25% as in $Pr_{0.25}Pr_{0.75}I_2 = PrI_2$-V) of the octahe-

Fig. 4.12 Crystal structure of "PrI_2-IV" = $Pr_{0.08}PrI_2$.

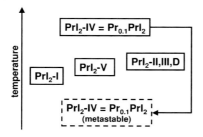

Fig. 4.13 Schematic overview of the appearance of the modifications of PrI₂ with increasing temperature.

dral voids, in such a way that tetrahedral clusters can be drawn (Fig. 4.12) (which are either statistically distributed or ordered in a way that an incommensurable superstructure is adopted).

- In the case of the PrI₂-II/III/D "phases" occur in a rotation of iodine layers against each other by 60° in the double-layer packages from, for example, AB to AA. As the AA stacking is less favorable than the more dense AB stacking, one is tempted to "understand" that these modifications are slightly less stable than the others. Disorder in the sense of formation of polytypes ("D") is more likely at higher temperatures, which is in fact observed qualitatively. Hydrogen has a stabilizing effect on these modifications, which also becomes obvious by the observation that LaI₂ (structurally identical with PrI₂-I) transforms to the same "mixture" of phases on hydrogen uptake.

Figure 4.13 summarizes, in a qualitative sketch, the dependence on temperature of the existence of the modifications of PrI₂.

4.4
Conclusions

Pure PrI₂ is obtained by a coproportionation reaction of praseodymium tri-iodide and praseodymium (2 PrI₃ + Pr) with an excess of praseodymium metal to avoid the formation of Pr₂I₅. The reaction temperature should be slightly below the peritectic temperature (at 750 °C). Upon rapid cooling, single-phase "PrI₂-IV" ~ Pr₀.₁PrI₂ is obtained as single crystals. It is metastable under ambient conditions. Thus, as the least thermodynamically stable modification it transforms on annealing to PrI₂-I, -V and -II/III/D with a high tendency to "phase D" with increasing temperature. It can be concluded that I is slightly more stable than V and this is slightly more stable than II/III/D. Apparently, the activation energies for both transition pathways (3⁶ to 4⁴ or different stackings of 3⁶) must be very close. If so, it is a matter of nucleation probabilities (maybe triggered by small but ubiquitous impurities) which of these phases nucleate and grow to (small) crystals.

Acknowledgments

This work was supported by the Deutsche Forschungsgemeinschaft (Sonderforschungsbereich 608: Komplexe Übergangsmetallverbindungen mit Spin- und Ladungsfreiheitsgraden und Unordnung). The generous support of the State of Nordrhein-Westfalen and the Universität zu Köln is also gratefully acknowledged.

References

1 J. D. Corbett, L. F. Druding, W. J. Burkhard, C. B. Lindahl, *Disc. Faraday Soc.* **1961**, *32*, 79.

2 L. F. Druding, J. D. Corbett, *J. Am. Chem. Soc.* **1961**, *83*, 2462.

3 J. D. Corbett, *Rev. Chim. Miner.* **1973**, *10*, 239.

4 E. Warkentin, *Dissertation*, Universität Karlsruhe, **1977**; E. Warkentin, H. Bärnighausen, *Z. Anorg. Allg. Chem.* **1979**, *459*, 187.

5 H. P. Beck, *Z. Naturforsch.* **1976**, *31b*, 1548; H. P. Beck, M. Schuster, *J. Solid State Chem.* **1992**, *100*, 301.

6 N. Gerlitzki, *Dissertation*, Universität zu Köln, **2002**.

7 N. Gerlitzki, G. Meyer, *Z. Anorg. Allg. Chem.* **2002**, *628*, 915; G. Meyer, N. Gerlitzki, S. Hammerich, *J. Alloys Compd.* **2004**, *380*, 71.

8 N. Gerlitzki, A.-V. Mudring, G. Meyer, *Z. Anorg. Allg. Chem.* **2005**, *631*, 381.

9 N. Gerlitzki, G. Meyer, A.-V. Mudring, J. D. Corbett, *J. Alloys Compd.* **2004**, *380*, 211.

10 Gmelin Handbuch der Anorganischen Chemie, Sc, Y, La und Lanthanide, System-Nr. 39, Teil C6, Springer, Berlin – Heidelberg – New York, **1978**.

11 G. Meyer, *Chem. Rev.* **1988**, *100*, 93.

12 A. Simon, H. J. Mattausch, G. J. Miller, W. Bauhofer, R. K. Kremer, *Handbook on the Physics and Chemistry of Rare Earths* (K. A. Gschneidner, Jr., L. Eyring, eds.), Elsevier Science Publ., **1991**, *15*, 191.

13 G. Meyer, L. Jongen, A.-V. Mudring, A. Möller, "Divalent Scandium", in: *Inorg. Chem. in Focus II*, Wiley-VCH, **2005**, *2*, 105.

14 C. Felser, K. Ahn, R. K. Kremer, R. Seshadri, A. Simon, *J. Solid State Chem.* **1999**, *147*, 19.

15 G. Meyer, M. S. Wickleder, *Handbook on the Physics and Chemistry of Rare Earths* (K. A. Gschneidner, Jr., L. Eyring, eds.), Elsevier Science Publ., **2000**, *28*, 53.

16 K. E. Mironov, R. V. Abdulin, E. N. Balyakina, *Zh. Neorg. Khim.* **1974**, *19*, 1411.

17 J. D. Corbett, *Inorg. Synth.* **1983**, *22*, 36.

18 G. Meyer, in: *Synthesis of Lanthanide and Actinide Compounds* (G. Meyer, L. R. Morss, eds.), Kluwer Acad. Publ., Dordrecht, **1991**, p. 143.

19 K. Krämer, G. Meyer, P. Fischer, A. W. Hewat, H. U. Güdel, *J. Solid State Chem.* **1991**, *95*, 1.

20 G. Meyer, J. Soose, A. Moritz, V. Vitt, Th. Holljes, *Z. Anorg. Allg. Chem.* **1985**, *521*, 161.

21 W. H. Zachariasen, *Acta Cryst.* **1948**, *1*, 265.

22 L. Keller, P. Fischer, A. Furrer, K. Krämer, G. Meyer, H. U. Güdel, A. W. Hewat, *J. Magn. Magnet. Mat.* **1992**, *104–107*, 1201.

23 H.-J. Meyer, R. Hoffmann, *J. Solid State Chem.* **1991**, *95*, 14.

24 G. Meyer, H.-J. Meyer, *Chem. Mat.* **1992**, *4*, 1157.

25 B. C. McCollum, D. S. Dudis, A. Lachgar, J. D. Corbett, *Inorg. Chem.* **1990**, *29*, 2030.

26 R. J. Cava, D. W. Murphy, S. M. Zahurak, *J. Solid State Chem.* **1984**, *53*, 64.

27 F. Lissner, Th. Schleid, *Z. Anorg. Allg. Chem.* **2003**, *629*, 1895.

28 For powder X-ray diffraction, the samples were put in glass capillaries (⌀ 0.3 mm) under an argon atmosphere

(dry box) and examined with the aid of a Huber image plate camera G 670 (MoK$_{\alpha 1}$-radiation). The STOE WinXPOW program package [29] was used for phase analysis. Rietveld refinement for PrI$_2$-IV was performed using the FullProf program [30].

29 WIN XPow 1.07, STOE & Cie, Darmstadt, **2000**.

30 J. Rodriguez-Carvajal, An introduction to the program FullProf 2000, Laboratoire Leon Brillouin (CEA-CNRS), **2001**.

31 M. Ryazanov, A. Simon, H. J. Mattausch, *Z. Anorg. Allg. Chem.* **2004**, *430*, 104.

32 C. Michaelis, H. J. Mattausch, H. Borrmann, A. Simon, J. K. Cockcroft, *Z. Anorg. Allg. Chem.* **1992**, *607*, 29; C. Michaelis, H. J. Mattausch, A. Simon, *Z. Anorg. Allg. Chem.* **1992**, *610*, 23.

33 K. Krämer, Th. Schleid, M. Schulze, W. Urland, G. Meyer, *Z. Anorg. Allg. Chem.* **1989**, *575*, 61.

34 L. F. Druding, J. D. Corbett, B. N. Ramsey, *Inorg. Chem.* **1963**, *2*, 869.

35 R. A. Sallach, J. D. Corbett, *Inorg. Chem.* **1963**, *2*, 457.

36 G. Meyer, Th. Schleid, K. Krämer, *J. Less Common Met.* **1989**, *149*, 67.

37 Th. Schleid, G. Meyer, *Z. Anorg. Allg. Chem.* **1987**, *552*, 97.

38 DSC (differential scanning calorimetry) was performed using a Netzsch Phoenix F1 apparatus. Typically, samples of about 20 mg were applied in aluminium cold-sealed crucibles with heating/cooling rates of 5 °C/min.

39 Th. Hegenscheidt, *Dissertation*, Universität Karlsruhe (TH), **1998**.

5
Centered Zirconium Clusters: Mixed-halide Systems

Martin Köckerling

Foreword

A German saying goes "You can't stand upright on one leg only". Looking at the achievements of John D. Corbett it is clear that his scientific body stands solidly on two legs: Solid-state Zintl compounds and centered cluster phases of early transition metals.

Within the family of interstitially centered cluster phases the zirconium halides comprise a special field of materials, because it has been developed solely by J. D. Corbett and coworkers. This is truly "Corbetts chemistry".

In this article we will review some recent results in mixed-halide zirconium cluster chemistry, which were developed starting from John Corbett's protocols.

5.1
The Basics of Zirconium Cluster Chemistry

Of course, the chemistry of zirconium cluster phases has been well described and reviewed in the literature [1–4]. Apart from a very few examples, mostly in the binary halides, almost all reduced zirconium halides contain octahedra of zirconium atoms centred on an interstitial atom Z. Several possible and experimentally realized Z include H, Be–N, K, Al–P, and the transition metals Mn–Ni. All these compounds have the general formula $A_x^{I,II}[(Zr_6Z)X_{12}^iX_n^a]$, with $A^{I,II}$ = alkali or alkaline earth metal cation, X = Cl; Br or I, X^i = inner; edge-bridging halide [5], X^a = outer; exo-bonded halide, and $0 \leq x,n \leq 6$ (compare Fig. 5.1).

As holds for other cluster systems, certain "magic" cluster electron counts exist, which indicates for a certain cluster-halide ratio and interstitial present the filling of all bonding molecular orbitals and therefore the thermodynamically most stable situation. For main group interstitial atoms these are 14 cluster-based electrons whereas for transition-metal interstitials the magic number is 18 [1, 10–12]. All of these phases are synthesized by high-temperature solid-state chemical methods. A remarkable variety of different structure types has been

Inorganic Chemistry in Focus III.
Edited by G. Meyer, D. Naumann, L. Wesemann
Copyright © 2006 WILEY-VCH Verlag GmbH & Co. KGaA, Weinheim
ISBN: 3-527-31510-1

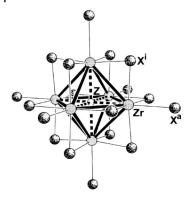

Fig. 5.1 [$(Zr_6Z)X_{18}$] cluster unit in reduced zirconium halides.

established, most of them being network structures, depending on different combinations of possible values of x and n, on the size and charge of the A cation, the size of the halide atoms, and the number of valence electrons of Z. Chlorides and bromides have been found for all $n=0–6$; whereas iodides are so far limited to $n=0$ and 2 [see 1–4 and references cited therein].

5.2
Motivation

Exploratory investigations in zirconium bromide cluster systems have revealed, in addition to cluster phases with structure types which are already known, cluster connectivities and arrangements, that have thus far not been seen in chloride or iodide systems [6–9]. Among these the $A_5^I[(Zr_6Be)Br_{15}]$ phases (A^I = Rb or Cs) exhibit a novel tunnel structure [6] whereas compounds with the composition $(A_4^I Br)_2[(Zr_6Z)Br_{18}]$ with A^I = Na–Cs, and Z = Be, B, H, or Mn contain unprecedented $[A_4^I Br]^{3+}$ counter-ions [8]. These findings suggest that the size and basicity of the anions play an important role for the stability of specific cluster phases with respect to others. Therefore, a combination of two different halide ligands was considered a possible route to either new structure types or compounds with structural properties different from those with only one type of halide. Within niobium cluster systems this concept has already been shown to be successful. The phases $[Nb_6Cl_{12-x}I_{2+x}]$ ($x<2$) and $[Nb_6I_{11-x}Br_x]$ ($x<2.7$) exhibit a new halide bridging mode for the $[M_6X_{14}]$ stoichiometry and a rarely seen magnetic spin-crossover transition, respectively [13, 14].

5.3
Mixed-Chloride-Iodide Zirconium Cluster Phases with a 6:12–Metal:Halide Ratio

Some first exploratory reactions in mixed halide chloride-iodide zirconium cluster systems with boron or beryllium as source for interstitial atoms and without any alkaline or alkaline earth metal cation present gave reaction products which are derivatives of the rhombohedral $[(Zr_6C)I_{12}]$ structure [15, 16]. The random substitution of fractional Cl at specific I sites always occurs at two-bonded X^i sites, so that single types of halogen are left in sites that interconnect clusters and generate the three-dimensional array. Reaction products from iodine-rich starting compositions gave crystals, which turned out to have compositions of $[(Zr_6Be)Cl_{1.65(4)}I_{10.35}]$, and $[(Zr_6B)Cl_{1.27(3)}I_{10.73}]$, as determined by single-crystal diffraction at room temperature [17]. Cluster connectivity in the rhombohedral $[(Zr_6C)I_{12}]$ structure is achieved through I^{a-i} and I^{i-a} bridges, respectively, only, as expressed by $[(Zr_6Z)I^{i-a}_{6/2}I^{a-i}_{6/2}(I,Cl)^i_6]$. Figure 5.2 gives a view of this structure.

Further effects seen with the substitution of Cl for I^i in the $[(Zr_6Z)I_{12}]$ type are a reduction of inter-cluster $I \cdots I$ repulsions which allows for a reduction in $Zr-I^{a-i}$ inter-cluster bond lengths. The [110] section of this rhombohedral structure is shown in Fig. 5.3 and illustrates that the clusters can be described as a cubic-close-packed array, with the $\bar{3}$ axes running vertically through Z. Phase widths found are $0 \le x \le 1.4$ for $[(Zr_6Z)Cl_xI_{12-x}]$ (Z = B, Be) [17].

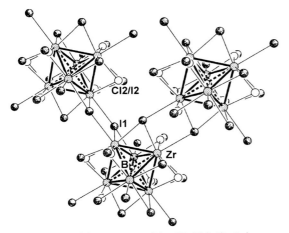

Fig. 5.2 View of the structure of the $[(Zr_6Z)(I,Cl)_{12}]$-phases showing the mixed occupied inner Cl2/I2 site, and the cluster-interconnecting $I1^{a-i}$, and $I1^{i-a}$ atoms, respectively

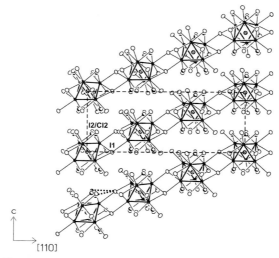

Fig. 5.3 [110] section of the rhombohedral [(Zr$_6$Z)I$_{12}$] type mixed halide structure (*c* vertical). The iodine atoms are drawn with open ellipsoids, and chlorines are shaded. Close-packed layers of halogen (and interstitials) run horizontally. One pair of the short I1···I2 inter-cluster contacts is marked at the lower left.

5.4
Mixed Chloride–Iodide Zirconium Cluster Phases with a 6:13 Metal:Halide Ratio

Similar exploratory solid-state chemical reactions to those given above, using boron as source for interstitial atoms, again are alkaline and alkaline-earth free but with a much larger chlorine/iodine ratio than those which gave the 6-12 phases, resulting in a phase which belongs to the 6-13 family of Zr-clusters. The structure which emerged from the refinement of single crystal X-ray data is a derivative of the [(Zr$_6$B)Cl$_{13}$] structure which contains unusual chains of Cl^{i-i} connected clusters, running parallel to $\vec{c}$. They gain their three-dimensional network through Cl^{a-a-a}-connections [1, 2, 18], ([(Zr$_6$B)Cl$_{2/2}^{i-i}$Cl$_{6/2}^{a-a-a}$(Cl,I)$_{10}^{i}$]), as depicted in Fig. 5.4.

The mixed halide structure, for which the composition was refined to [(Zr$_6$B)Cl$_{11.47(2)}$I$_{1.53}$], crystallizes in the tetragonal space group P4$_2$/mnm, contrary to the only-chloride parent-type, which crystallizes orthorhombically [17]. This increase of symmetry is achieved through a random iodine substitution of 19.1(3)% of the Cl4^{i}-site. Thereby the Zr$_3$Cl^{a-a-a}-units become planar, as can be seen from Fig. 5.5.

Contrary to the 6-12 mixed halide phase, the three-dimensional cluster network of this structure is based on chlorine bridges only. As detailed in [17] an interesting relationship exists between this 6-13- and the [(Zr$_6$B)Cl$_{14}$] ([Nb$_6$Cl$_{14}$]) structure [19], which is shown in Fig. 5.6. The transformation of Nb$_6$Cl$_{14}$

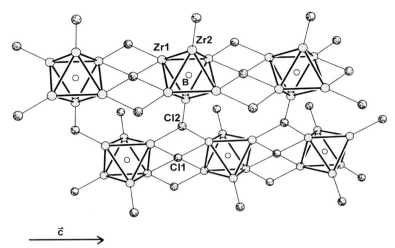

Fig. 5.4 View of the structure and inter-cluster connectivity of the $[(Zr_6B)Cl_{13}]$ structure, and mixed halide derivative, respectively. Edge-bridging chlorines are omitted for clarity. The $Cl1^{i-i}$ atoms trans-bridge clusters into chains, while $Cl2^{a-a-a}$ atoms bound exo at all cluster vertices inter-connecting the chains.

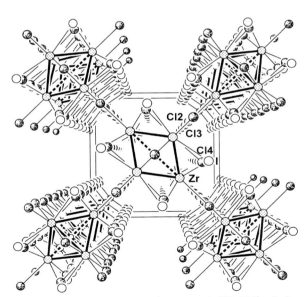

Fig. 5.5 [001] view of the network structure in $[(Zr_6B)Cl_{11.5}I_{1.5}]$. The $Cl2^{a-a-a}$ atoms, that interconnect chains alternately in the projection, with $Cl3^i$ atoms that bridge side edges of the clusters along the chains.

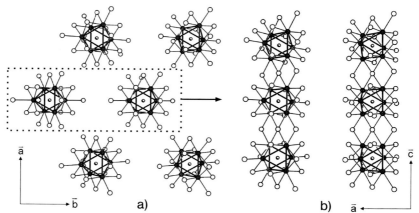

Fig. 5.6 Schematic representation of the relationship between clusters in (a) the $[(Zr_6B)Cl_{14}]$ ($[Nb_6Cl_{14}]$) structure ($[001]$ projection on $z = 1/2$) and (b) the $[(Zr_6B)Cl_{13}]$ structure ($[010]$ projection on $y = 0$).

$[Nb_6Cl_{10}^iCl_{4/2}^{i-a}Cl_{4/2}^{a-a}]$ into $[(Zr_6B)Cl_{13}] = [(Zr_6B)Cl_{10}^iCl_{2/2}^{i-i}Cl_{6/3}^{a-a-a}]$ is accomplished by a translation of every other row of clusters in the former structure along $\vec{b}$ by $b/2$ and insertion of Z.

The phase width has been found to cover approximately $8 \leq x \leq 13$ for $[(Zr_6B)Cl_xI_{13-x}]$. In between the range for $0 \leq x \leq 2$ for $[(Zr_6B)Cl_xI_{12-x}]$ and $8 \leq x \leq 13$ for $[(Zr_6B)Cl_xI_{13-x}]$ another phase exists, which has a new structure in the zirconium cluster chemistry. In this structure both atom types, chlorine and iodine, participate in the cluster connectivity. Iodine atoms take the function of I^{a-a-a} connecting atoms, whereas chlorine is found as Cl^{i-i}. Even though the connectivity, in a similar way to the chlorine-dominated structure, is expressed by $[(Zr_6B)(Cl,I)_{10}^iCl_{2/2}^{i-i}I_{6/3}^{a-a-a}]$, the structure differs since the Cl^{i-i} atoms do not bridge opposite edges but form zigzag chains of clusters, which are further three-dimensionally connected by triply bridging iodine atoms. Figure 5.7 depicts this cluster arrangement.

The non-inter-cluster bridging halogen sites (to X-rays) are again statistically occupied by a mixture of chlorine and iodine. This phase appears always to crystallize in the form of twinned crystals. The refinement of single-crystal X-ray data (orthorhombic, space group Pbcn) gave a composition of $[(Zr_6B)Cl_{6.44(7)}I_{6.56}]$ [20]. The different phases and phase ranges found in the quasi-ternary Zr-(Cl,I)-B system are put together schematically in a phase diagram in Fig. 5.8.

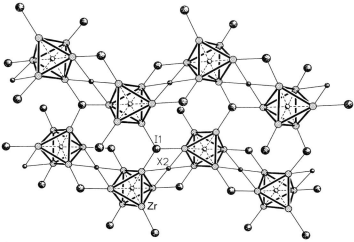

Fig. 5.7 View of the inter-cluster connectivity in the
[(Zr$_6$B)Cl$_{11-x}$I$_{2+x}$] structure (c and the zigzag chains run
horizontally). The Zr–Zr bonds are emphasized, and the *inner*
halides are omitted for clarity.

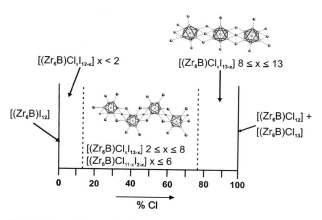

Fig. 5.8 Phase diagram for the quasi-ternary Zr-(Cl, I)-B cluster system.

5.5
Mixed Chloride–Iodide Zirconium Cluster Phases with a 6:14 Metal:Halide Ratio

With boron and an appropriate amount of some sort of alkaline metal halide
present in the starting materials for the solid-state reactions, then we obtain zirconium cluster materials belonging to the 6-14 family. Single-crystal X-ray data
of products from iodine-rich reactions were used to determine the crystal struc-
tures of Na[(Zr$_6$B)Cl$_{3.87(5)}$I$_{10.13}$], and Cs[(Zr$_6$B)Cl$_{2.16(5)}$I$_{11.84}$] [21]. Both phases

crystallize in a stuffed version of the $[Nb_6Cl_{14}]$ structure type, orthorhombic, with space group Cmca. As mentioned above, a larger number of zirconium cluster compounds as well as several others crystallize in this structure type [19]. In both mixed halide structures, chlorine and iodine atoms are randomly (to X-rays) distributed on the inner non-cluster-interconnecting ligand positions, whereas those sites which bridge metal octahedra are solely occupied by iodine. The cluster connectivity is expressed by $[(Zr_6B)(I,Cl)^i_{10}I^{a-i}_{2/2}I^{i-a}_{2/2}I^{a-a}_{4/2}]$, pictorially presented in Fig. 5.9 [see also 22–24]. The phase widths for both phases have been found to cover $0 \leq x \leq 4$ for $A^I[(Zr_6B)Cl_xI_{14-x}]$. Whereas the sodium cations in $Na[(Zr_6B)Cl_xI_{14-x}]$ occupy 25% of a site which is octahedrally surrounded by halogen atoms, the larger cations in the cesium-containing phase occupy a 12-coordinate site within the cluster network. Figure 5.10 shows the environment for the Cs cation (a) and the Na cation (b).

Similar to the mixed-halide (Cl, I) 6-13 system, where more chlorine-rich reactions produced a new structure type, materials with an unprecedented zirconium cluster structure are obtained in the Na-Zr-(Cl/I)-B system (also with other cations, see below), when larger Cl/I ratios are used than above. Compounds characterized are $Na[(Zr_6B)(Cl,I)_{14}]$ and $A^{II}_{0.5}[(Zr_6B)(Cl,I)_{14}]$ (with A^{II}=Ca, Sr, Ba) [25, 26]. Single crystals of the cubic $Na[(Zr_6B)Cl_{10.94(1)}I_{3.06}]$ and

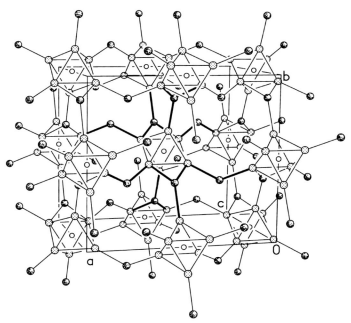

Fig. 5.9 The inter-cluster halogen (i.e., iodine) bridging between the octahedral metal clusters in the structures of both $Na[(Zr_6B)Cl_{3.9}I_{10.1}]$ and $Cs[(Zr_6B)Cl_{2.2}I_{11.8}]$ (Zr are depicted regularly dotted, I irregularly and B shaded). Cations, and those halogen atoms that only bridge octahedral edges, are omitted for clarity.

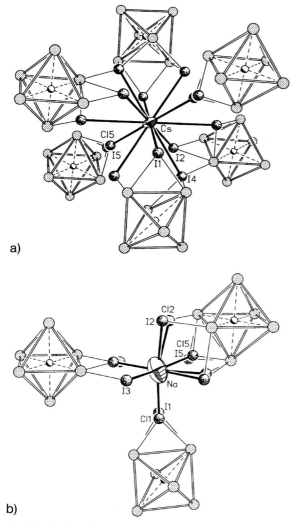

a)

b)

Fig. 5.10 (a) Coordination environment of the Cs cation in
$Cs[(Zr_6B)Cl_{2.2}I_{11.8}]$. (b) Distorted octahedral halide
environment of the Na cation in $Na[(Zr_6B)Cl_{3.9}I_{10.1}]$ (a),
thermal ellipsoids with 50% probability.

$Sr_{0.5}[(Zr_6B)Cl_{11.34(2)}I_{2.66}]$ have been characterized by X-ray diffraction. The Zr_6
octahedra in these phases are three-dimensionally connected by exo-iodide
atoms, I^{a-a-a}, which simultaneously bridge three octahedra. The resulting connectivity between clusters can be described as $[(Zr_6B)X^i_{12}I^{a-a-a}_{6/3}]$, a stuffed version
of the $[Nb_6Cl_{12-x}I_{2+x}]$ structure $(x < 2)$ [27], which is depicted in Fig. 5.11.

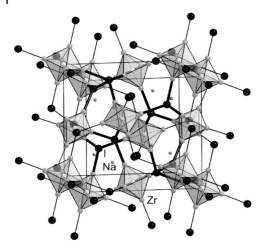

Fig. 5.11 Perspective of the unit cell of the cubic 6-14 mixed halide phases, emphasizing the I^{a-a-a} connectivity between the clusters (with the inner halides being omitted for clarity).

Six out of the twelve inner halogen atoms X (with one symmetry-independent site) which bridge the edges of each octahedron, but are not involved in inter-cluster bridging, can be completely substituted by iodine. Such a substitution is not possible for the remaining six inner halogen atoms on the second symmetry-independent site, because the larger iodine atom on this site would experience strong repulsive forces from short anionic contacts. This limits the phase width of $Na[(Zr_6B)Cl_{12-x}I_{2+x}]$ to $x \leq 6$. The size of the voids that cover the alkaline or alkaline earth cations limits this structure to members with A = Na, Ca, Sr, and Ba (at least for the boron-centered members). The environment of the alkaline (alkaline earth) cations can be described as distorted octahedral which is depicted in Fig. 5.12 for the Na cation in $Na[(Zr_6B)Cl_{10.9}I_{3.1}]$.

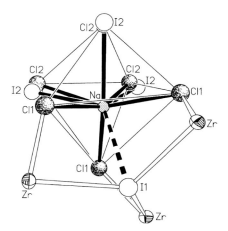

Fig. 5.12 Halide environment of the Na cations in crystals of $Na[(Zr_6B)Cl_{10.9}I_{3.1}]$.

This structure type evidently requires the simultaneous existence of two differently sized halide types, and thus exists only in mixed halide systems. Recent investigations revealed this cubic structure also for cluster phases with other interstitial atoms, i.e., the cation-free Si-centered $[(Zr_6Si)Cl_{12-x}I_{2+x}]$ [28].

5.6
Mixed Chloride–Iodide Zirconium Cluster Phases with a 6:15 Metal:Halide Ratio

Using Rb or Cs as larger cations than Na in mixed-halide solid-state chemical exploratory reactions, results in several different phases which belong to the 6-15 family of zirconium clusters. All of the examples of this family of compounds known so far are made up of cluster networks in which all the outer halides of each cluster are connected to neighboring clusters according to $[(Zr_6Z)X^i_{12}X^{a-a}_{6/2}]$. At the chlorine-rich end of the phase diagram two new members of a double salt, containing halide-supported zirconium octahedral, were obtained. Single-crystal X-ray data refinements established the isotypic rhombohedral compounds as $Cs[ZrCl_5] \cdot Cs_2[(Zr_6B)Cl_{15}]$ (pure chloride) and the mixed halide as $Cs[ZrCl_5] \cdot Cs_2[(Zr_6B)Cl_{14.57(2)}I_{0.43}]$, respectively [30]. The structure contains a network of boron-centered $[(Zr_6B)Cl^i_{12}X^{a-a}_{6/2}]$ clusters interbridged three-dimensionally by 6/2 X^{a-a} atoms at all zirconium vertices (X=Cl, or Cl+I, respectively). The only previously known parent compound contains Mn-centered clusters in $Cs[ZrCl_5] \cdot Cs_2[(Zr_6Mn)Cl_{15}]$ [29]. Within voids of the cluster network, Cs cations and $[ZrCl_5]^-$ ions are located, the latter allowing for description as a double salt. Figure 5.13 gives a view of the arrangement of the different structural units in the unit cell. The $[ZrCl_5]^-$ ion (D_{3h} symmetry) has so far only been found to exist in this structure type. The surrounding of this unusual ion is shown in Fig. 5.14.

The mixed chloride–iodide structure reveals another interesting feature besides the unusual $[ZrCl_5]^-$ ion. Contrary to many other mixed halide structures, the substitution of one halogen type by the other is not observed on the non-interconnecting halide sites X^i, but it is on the cluster interconnecting site $X3^{a-a}$. This seems to be structurally possible because the Zr-X^{a-a}-Zr bridge is bent, allowing for Cl atoms with a shorter Zr–X bond length (2.746(3) Å) and a more obtuse Zr-X-Zr angle (136.2(3)°) to be partially (and statistically) substituted by I atoms with a longer Zr–X bond length (2.905(7) Å) and a more acute Zr-X-Zr angle (122.5(5)°). This situation is depicted in Fig. 5.15. The phase width for $Cs[ZrCl_5] \cdot Cs_2[(Zr_6B)Cl_{15-x}I_x]$ has been found to cover $0 \leq x \leq 1$.

The cesium cations in crystals of $Cs[ZrCl_5] \cdot Cs_2[(Zr_6B)Cl_{14.6}I_{0.4}]$ are located on the Wyckoff site 18d. They are positionally, slightly disordered (statistically to X-rays) with 16(4) % being moved out of the 18d site. The other 84% (as in the iodine-free phase) are surrounded by a total of 12 chlorine atoms in a distance range of 3.526–3.881 Å (average 3.632 Å). The remaining 16% have an environment of 10 Cl + 2 I atoms, with a Cs1B–I3 distance of 3.77(6) Å. The halide environment of the major fraction of the Cs atoms is depicted in Fig. 5.16.

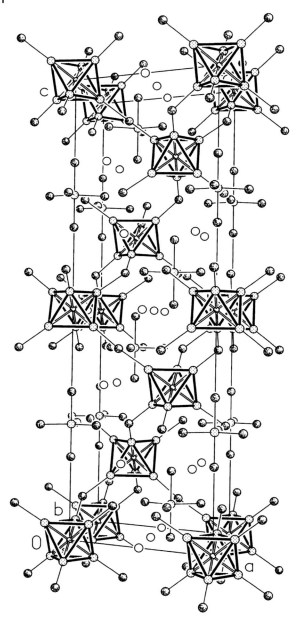

Fig. 5.13 Cs[ZrCl$_5$]·Cs$_2$[(Zr$_6$B)X$_{15}$] (X=Cl or Cl + I): Perspective of the unit cell (Zr: regularly dotted, halogen atoms: irregularly dotted, B: shaded, Cs: empty circles; inner halides are omitted for clarity).

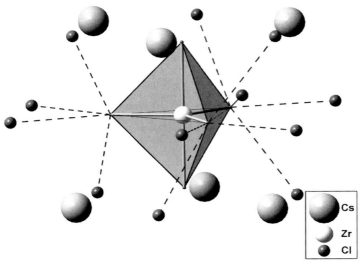

Fig. 5.14 Environment of the $[ZrCl_5]^-$ ion in crystals of $Cs[ZrCl_5] \cdot Cs_2[(Zr_6B)X_{15}]$, $X = Cl$ or $Cl + I$.

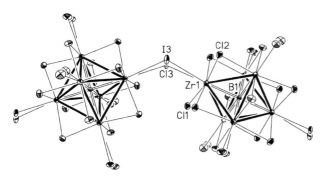

Fig. 5.15 Occupation of the inter-cluster-connecting halide site X3 with a mixture of Cl and I in crystals of $Cs[ZrCl_5] \cdot Cs_2[(Zr_6B)Cl_{14.6}I_{0.4}]$.

It is interesting to note that zirconium cluster iodides only exist in the 6-12 and 6-14 type family. Therefore, this double salt is the first member of a 6-15 type family of compounds in which iodine atoms participate in the inter-cluster connectivity.

Using more chlorine-rich compositions as starting materials, a mixed halide phase is obtained which crystallizes in another structure type than that given above [33]. The structure of $Cs_2[(Zr_6B)Cl_{11.98(2)}I_{3.02}]$ is deduced from that of $Cs[Nb_6Cl_{15}]$, and $K[(Zr_6C)Cl_{15}]$ or $CsK[(Zr_6B)Cl_{15}]$, respectively [31, 32]. As is observed for all members of the 6-15 type family it contains $[(Zr_6Z)X_{12}^i X_{6/2}^{a-a}]$ anions, which are connected through all halogen atoms X^{a-a} thereby forming a

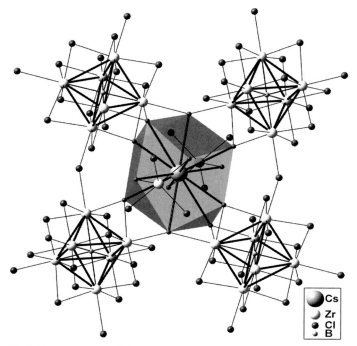

Fig. 5.16 Environment of the Cs cations in crystals of
$Cs[ZrCl_5] \cdot Cs_2[(Zr_6B)X_{15}]$ (X=Cl or Cl + I).

three-dimensional cluster network. A view of the cluster arrangement in the
unit cell of this orthorhombic structure (space group Pmma) is given in
Fig. 5.17.

This structure is built up of two symmetry-independent cluster units, forming
two different types of cluster chains. The first one forms six bent $Zr\text{-}I^{a-a}\text{-}Zr$
bridges (125.57(4)°), whereas the other one has four bent iodine bridges in addition to two linear chlorine bridges. These inter-cluster halogen bridges give two
different types of cluster chains, one zigzag chain running parallel to $\vec{a}$ (horizontally in Fig. 5.17) and in a linear fashion the other one running along $\vec{c}$ (vertically). As we have already seen for other mixed-halide structures (see Fig. 5.17
above) cluster inter-connecting halogen sites in many cases are occupied by only
one type of halogen. Here this holds for the X1-site, which is completely occupied by iodine, and the X3-site, which is completely occupied by chlorine. Similar to the rhombohedral 6-15 mixed-halide structure, discussed above, the X2
site is statistically mixed and occupied by both chlorine and iodine. As usual
this also holds for the inner halogen sites. The Cs cations are distributed and
disordered on three different sites. It is interesting to note that these sites are
arranged along channels and the disorder is observed along these channels (see
Fig. 5.18). This might serve as a structural hint of the possibility that these com-

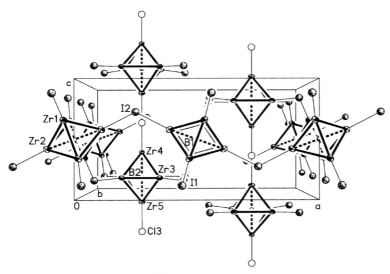

Fig. 5.17 View of the structure of the orthorhombic mixed-halide cluster phase $Cs_2[(Zr_6B)Cl_{11.98}I_{3.02}]$ showing the two types of cluster chains (thermal ellipsoids with 50% probability). Cs cations and inner halides are omitted for clarity.

Fig. 5.18 View of the structure of $Cs_2[(Zr_6B)Cl_{11.98}I_{3.02}]$ along $\vec{a}$ showing the Cs cation sites as located in channels along the view direction. Thermal ellipsoids are draws at the 50% level.

pounds could be good ionic Cs (or Rb) conductors [34]. Powder X-ray investigations conducted in order to find out the phase width of this mixed-halide structure revealed that, with increasing chlorine:iodine ratio, related structures with lower symmetry (monoclinic instead of orthorhombic) also exist [34].

5.7
Mixed Chloride–Iodide Zirconium Cluster Phases with a 6:18 Metal:Halide Ratio – Products from Solid-state Reactions

Increasing the amount of alkaline metal halide in the starting material of the solid-state chemical reactions leads eventually to cluster materials which have more halogen atoms, and alkaline metal cations, respectively, per zirconium octahedron, and which therefore carry fewer inter-cluster halogen bridges. The final end-cluster phases of the 6-18 type family, of which so far only very few members are known, consist of molecular, usually anionic, cluster units [35, 36]. Mixed halide chloride/iodide reactions with sodium as the source for counter-cations and Fe, or C, as the source for interstitial atoms, gave crystalline products, from which single-crystal X-ray studies finally established compositions of the triclinic phases to be $Na_4[(Zr_6Fe)Cl_{14.64(4)}I_{3.36}]$, and $Na_4[(Zr_6C)Cl_{14.41(6)}I_{3.59}]$, respectively [37]. A view of the arrangement of cluster anions and Na cations in the unit cell is given in Fig. 5.19. This compound is isostructural

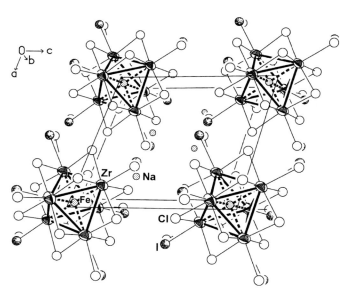

Fig. 5.19 View of the triclinic structure of the mixed-halide $A_4^I[(Zr_6Z)(Cl,I)_{18}]$ phase (here with Z = Fe), which consists of molecular cluster units (thermal ellipsoids with 50% probability).

to $K_4[(Zr_6C)Br_{18}]$ [36]. As can be seen from Fig. 5.19 the cluster units are arranged in layers. So far the substitution of chlorine by iodine atoms, in both mixed-halide structures, is only observed on the outer positions. This resembles observations in mixed-halide niobium [38], as well as molybdenum [39] cluster chemistry. Investigations of the phase widths and the possibility of substituting halides also on the *inner* positions are under way [37].

5.8
Outlook

John D. Corbett once said: "There are *many* wonders still to be discovered!" [4]. This certainly holds generally for all the different areas and niches of early transition cluster chemistry and especially for the mixed-halide systems. The results reported above so far cover a very limited selection of only chloride/iodide systems and basically boron as the interstitial. Because of the very sensitive dependence of the stable structure built in the solid-state reaction type on parameters like optimal bonding electron counts, number of cations present, size and type of cations (bonding requirements for the cations), metal/halide ratio, and type of halide, a much larger mixed-halide cluster chemistry can be expected. Further developments, also in mixed-halide systems, can be expected by using solution chemistry of molecular clusters, excised from solid-state precursors.

Acknowledgments

The research reported here was supported by the *Deutsche Forschungsgemeinschaft* (DFG), the *Fonds der Chemischen Industrie* and the State of North Rhine Westphalia (Bennigsen Foerder award for M.K.). Thanks are due to all the students who did the experimental work and whose names appear in the references. Thanks are also due to Prof. Dr. G. Henkel (University of Paderborn), and especially to Prof. Dr. J.D. Corbett (Ames, IA, USA).

References

1 J.D. Corbett, *J. Alloys Compounds* **1995**, 229, 10.

2 J.D. Corbett, *Modern perspectives in inorganic crystal chemistry*, in: E. Parthé (Ed.), NATO ASI Series C, Kluwer Academic Publishers, Dordrecht, The Netherlands, 1992, p. 27.

3 J.D. Corbett, *J. Chem. Soc. Dalton Trans.* **1996**, 575.

4 J.D. Corbett, *Inorg. Chem.* **2000**, *39*, 5178.

5 H. Schäfer, H.-G. von Schnering, *Angew. Chem.* **1964**, *76*, 833.

6 R.-Y. Qi, J.D. Corbett, *Inorg. Chem.* **1995**, *34*, 1646.

7 R.-Y. Qi, J.D. Corbett, *Inorg. Chem.* **1995**, *34*, 1657.

8 R.-Y. Qi, J.D. Corbett, *Inorg. Chem.* **1997**, *36*, 6039.

9 R.-Y. Qi, J. D. Corbett, *J. Solid State Chem.* **1998**, *139*, 85.

10 T. Hughbanks, G. Rosenthal, J. D. Corbett, *J. Am. Chem. Soc.* **1988**, *110*, 1511.

11 R. P. Ziebarth, J. D. Corbett, *J. Am. Chem. Soc.* **1989**, *111*, 3272.

12 T. Hughbanks, *Prog. Solid State Chem.* **1989**, *19*, 329.

13 M. Sägebarth, A. Simon, *Z. Anorg. Allg. Chem.* **1990**, *587*, 119.

14 A. Yoshiasa, H. Borrmann, A. Simon, *Z. Anorg. Allg. Chem.* **1994**, *620*, 1329.

15 J. D. Smith, J. D. Corbett, *J. Am. Chem. Soc.* **1985**, *707*, 5704.

16 J. D. Smith, J. D. Corbett, *J. Am. Chem. Soc.* **1986**, *108*, 1927.

17 M. Köckerling, R.-Y. Qi, J. D. Corbett, *Inorg. Chem.* **1996**, *55*, 1437.

18 R. P. Ziebarth, J. D. Corbett, *J. Am. Chem. Soc.* **1985**, *707*, 4571.

19 The orthorhombic [Nb_6Cl_{14}] structure is realized in a stuffed version in many zirconium cluster compounds of the general formula $A^I[(Zr_6Z)X_{14}]$, with A^I = Li-Cs, Tl or nothing; Z = B; C, Al, Si, P, Cr, Mn, Fe, Co, or K; X = Cl, Br, or I; see Refs. [1–4], and references cited therein.

20 M. Köckerling, J. B. Willems, P. D. Boyle, *Inorg. Chem.* **2001**, *40*, 1439.

21 M. Köckerling, *J. Solid State Chem.* **2003**, *170*, 273.

22 A. Simon, H.-G. von Schnering, H. Wöhrle, H. Schäfer, *Z. Anorg. Allg. Chem.* **1965**, *339*, 155.

23 D. Bauer, H.-G. von Schnering, H. Schäfer, *J. Less-Common Met.* **1965**, *8*, 388.

24 H. M. Artelt, G. Meyer, *Z. Krist.* **1993**, *206*, 306.

25 M. Köckerling, *Inorg. Chem.* **1998**, *37*, 380.

26 M. Köckerling, *Z. Anorg. Allg. Chem.* **1999**, *625*, 24.

27 M. Sägebarth, A. Simon, *Z. Anorg. Allg. Chem.* **1990**, *587*, 119.

28 C. Kopschütz, M. Köckerling, *so far unpublished research*.

29 J. Zhang, J. D. Corbett, *Inorg. Chem.* **1995**, *34*, 1652.

30 H. W. Rohm, M. Köckerling, *Z. Anorg. Allg. Chem.* **2003**, *629*, 2356.

31 A. Nägele, C. Day, A. Lachgar, H.-J. Meyer, *Z. Naturf.*, **2001**, *56b*, 1238, and references cited therein.

32 R. P. Ziebarth, J. D. Corbett, *J. Am. Chem. Soc.* **1987**, *109*, 4844.

33 M. Köckerling, *manuscript in preparation*.

34 M. Köckerling, *unpublished research*.

35 a) R. P. Ziebarth, J. D. Corbett, *J. Am. Chem. Soc.* **1989**, *111*, 3272; b) J. Zhang, J. D. Corbett, *Z. Anorg. Allg. Chem.* **1991**, *599/600*, 381; c) J. Zhang, R. P. Ziebarth, J. D. Corbett, *Inorg. Chem.* **1992**, *31*, 614.

36 R.-Y. Qi, J. D. Corbett, *J. Solid State Chem.* **1989**, *139*, 85.

37 M. Köckerling, *ongoing, so far unpublished research*.

38 See, for example, B. G. Hughes, J. L. Meyer, P. B. Fleming, R. E. McCarley, *Inorg. Chem.* **1970**, *9*, 1343.

39 W. Preetz, D. Bublitz, H. G. von Schnering, J. Saßmannshausen, *Z. Anorg. Allg. Chem.* **1994**, *620*, 234.

6
Titanium Niobium Oxychlorides:
Ligand Combination Strategy for the Preparation
of Low-dimensional Metal Cluster Materials

Ekaterina Anokhina and Abdessadek Lachgar

Abstract

A substantial part of John's work is dedicated to the synthesis and characterization of metal-rich compounds with focus on those compounds containing metal clusters and, more particularly, octahedral metal clusters. This paper represents a modest contribution to the field of cluster chemistry and aims to describe the use of a combination of ligands to direct the structural characteristics of cluster-based materials. The key idea of this work is to induce directional bonding preferences in cluster units using a combination of ligands with a large difference in charge density. The goal of the work described here was to develop a strategy for the preparation of low-dimensional and open-framework materials using octahedral metal clusters as building blocks. The approach takes its roots from crystal engineering principles where desired framework topologies are achieved through building block design. The niobium oxychloride cluster compounds described here have original structure types with topologies unprecedented in compounds containing octahedral clusters. Comparative analysis of their structural features indicates that the connectivity patterns in these systems result from complex interplay between the effects of anisotropic ligand arrangement and optimization of ligand-counterion electrostatic interactions. The role played by these factors sets the niobium oxychloride system apart from compounds based on clusters containing one ligand type or systems with statistical ligand distribution, where the main structure-determining factor is the total number of ligands.

6.1
Introduction

One of the distinctive aspects of transition-metal and lanthanide chemistry is cluster formation via metal–metal bonding that is characteristic of many of these elements in low oxidation states [1]. The unique structural, chemical, and

Inorganic Chemistry in Focus III.
Edited by G. Meyer, D. Naumann, L. Wesemann
Copyright © 2006 WILEY-VCH Verlag GmbH & Co. KGaA, Weinheim
ISBN: 3-527-31510-1

physical properties of clusters [2] lead to considerable advantages in using metal clusters as nanosized building blocks of materials with extended structures. For example, the large size of the clusters results in larger pore dimensions compared to those found in non-cluster frameworks with the same connectivity [3]. Second, compared to non-cluster frameworks formed by linking coordination polyhedra with typically 4–6 vertices, the cluster frameworks have the potential for a wider variety of topologies due to a larger number of ligands surrounding the metal cores. Third, discrete clusters have remarkable properties such as high catalytic activity, greater electronic flexibility, and strong luminescence [2].

Transition metal clusters exhibit a wide variety of nuclearity and architectures, most of which are based on fragments of closest packing of metal atoms surrounded by a shell of ligands [1, 4]. The extent of metal–metal bonding increases as the metal oxidation state decreases and more valence electrons become available for metal–metal bond formation (Table 6.1).

Metal clusters can be subdivided into two classes: (a) late transition metal clusters which occur preferentially with π-acceptor organic ligands, such as carbonyl [2], and (b) early transition-metal clusters which are typically formed with π-donor ligands (e.g., halogens, oxygen, chalcogens, or alkoxy groups) [6]. While discrete molecular cluster units are common for both classes, extended structures are formed almost exclusively from early transition-metal clusters with π-donor ligands, and are synthesized primarily by high-temperature solid-state chemistry techniques [7]. However, the assembly of soluble cluster units into extended frameworks has recently emerged as a promising new route to cluster-containing frameworks [8].

Solid-state cluster chemistry is dominated by octahedral $(M_6L^i_8)L^a_6$ and $(M_6L^i_{12})L^a_6$ units which are the focus of this paper. These two cluster types are different in the way the metal octahedral core is surrounded by the ligands. In $(M_6L^i_8)L^a_6$-type clusters (Fig. 6.1a), typical for molybdenum and rhenium halides, chalcogenides, and chalcohalides, eight "inner" ligands (L^i) cap the octahedron faces and six "outer" ligands (L^a) are located in the apical positions [9]. For metals with a smaller number of valence electrons, the $(M_6L^i_{12})L^a_6$-type clusters

Table 6.1 Increase in the extent of metal–metal bonding in niobium chlorides as the number of valence electrons per niobium atom (n) increases.

n	0	1	2.33	2.5	2.67
Compound [5]	NbCl$_5$	NbCl$_4$	Nb$_3$Cl$_8$	CsNb$_4$Cl$_{11}$	Nb$_6$Cl$_{14}$
Main structural fragment ($\bullet$ – Nb, $\circ$ – Cl)					

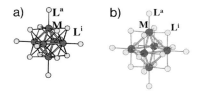

Fig. 6.1 Two types of octahedral cluster units: $(M_6L_8^i)L_6^a$ (a) and $(M_6L_{12}^i)L_6^a$ (b).

(Fig. 6.1b) in which twelve "inner" ligands bridge the edges of the M_6 octahedron, and six "outer" ligands occupy apical positions, predominate. These units are found in reduced zirconium, niobium, tantalum, and rare-earth halides, and niobium, tantalum, molybdenum and tungsten oxides [1a, 6, 10].

6.1.1
Cluster Connectivity and Framework Dimension

In the solid state, metal clusters can connect to each other through outer and/ or inner ligands, or by direct condensation of the metal octahedra (Fig. 6.2), to form a variety of frameworks.

The established strategy for modifying the framework dimensionality in the chemistry of octahedral clusters is based on adjusting the metal-to-ligand ratio [1d, 11], as illustrated in Table 6.2 for M_6L_{18} units linked via outer ligands. When the number of ligands per cluster decreases from 14 to 11, the number of intercluster linkages increases, leading to progressive transformation from structures based on discrete clusters (0D) to 1D, 2D, and 3D frameworks. As the number of ligands decreases, the cluster framework charge decreases, and can be compensated by adjusting the charge of counter-ions or modifying the charge of the cluster unit using a combination of ligands of different charges. Adjusting the metal:ligand ratio is a well-established and successful strategy for

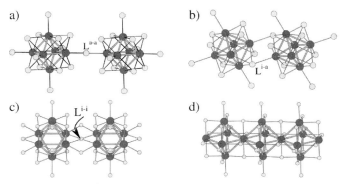

Fig. 6.2 Examples of cluster linkage modes: "outer-outer" (a), "inner-outer" (b), "inner-inner" (c), and condensation through sharing vertices of metal octahedra (d).

Table 6.2 Relationship between framework stoichiometry and dimensionality in compounds based on M_6L_{18}-type clusters.

Framework formula	Framework dimensionality [a]	Connectivity formula and examples
$M_6L_{18}^{n-}$	0D	$(M_6L_{12}^i)L_6^a$ $A_4^INb_6Cl_{18}$ [13], $A_2^IB_2^INb_6Cl_{18}$ [14] $A_2^IB^{II}Nb_6Cl_{18}$ [15], $A^IB^{III}Nb_6Cl_{18}$ [16]
$M_6L_{17}^{(n-1)-}$	1D	$(M_6L_{12}^i)L_{2/2}^{a-a}L_4^a$ $Ba_2(Zr_6B)Cl_{17}$; $Cs_2Nb_6Br_5F_{12}$
$M_6L_{16}^{(n-2)-}$	2D	$(M_6L_{12}^i)L_{4/2}^{a-a}L_2^a$ $Li_2Nb_6Cl_{16}$ [17]
$M_6L_{15}^{(n-3)-}$	3D	$(M_6L_{12}^i)L_{6/2}^{a-a}$ Nb_6F_{15} MNb_6Cl_{15} [18]

a) The figures show schematic representations of the cluster frameworks. The actual structures typically have bent $M\text{-}L^{a-a}\text{-}M$ intercluster bridges.

predictable modification of cluster-framework dimensionality [1d, 9a, 11d,g, 12]. This approach leads, however, to a limited variety of cluster connectivity types, and thus to limited framework topologies.

6.1.2
The Ligand Combination Approach to Creating Anisotropic Frameworks

The major advances in developing strategies for rational framework design in many classes of inorganic, organic, and coordination solids are associated with the use of building blocks with well-defined directional bonding preferences [19]. A similar strategy was investigated with the objective of preparing low-dimensional materials based on octahedral metal clusters. The goal was to induce anisotropic bonding preferences in these highly symmetrical building units, by modifying the charge distribution around the metal core using a combination of ligands with a large difference in their charge density. Investigations of mixed-ligand systems, mostly molybdenum and rhenium chalcohalides, were aimed primarily at changing the overall charge of the cluster unit and in most cases led to statistical ligand distribution

[9a, 20]. One way to decrease the tendency for statistical ligand distribution is to choose a combination of ligands with a large difference in charge densities such as oxide and chloride ligands, which also satisfy three other important criteria:

i) The stability of the same cluster type with L^- and L^{2-} ligands. The octahedral $(M_6L_{12}^i) L_6^a$ type clusters are the most common structural motifs in reduced niobium chlorides and oxides. This factor is essential for the stabilization of mixed-ligand clusters with a wide range of Cl/O ratios.

ii) The differences between inter-cluster connectivity trends in M-Cl and M-O systems. The chlorides are usually characterized by discrete cluster units, while the oxides have dense cluster frameworks in which most of the ligands are shared between clusters.

iii) The combination of oxide and chloride ligands has been shown to favor the formation of low-dimensional structures in the case of ternary oxychlorides of transition and rare-earth metals [21].

6.2
Overview of the Chemistry of Niobium Chloride and Niobium Oxide Cluster Compounds

6.2.1
Synthesis and Chemical Properties

Niobium chloride compounds containing octahedral clusters are generally synthesized by high-temperature reactions between niobium metal, niobium pentachloride, and sources of counter-ions, carried out in sealed silica or niobium containers at 700–900 °C. These reactions occur via a chemical vapor transport mechanism involving volatile niobium penta- and tetrachlorides as transporting agents [22]. The resulting solids are stable in air and dissolve in water and other polar solvents, producing intensely colored olive-green solutions containing octahedral clusters. Their outer chloride ligands are labile and have been the subject of many ligand substitution reactions [23]. The $[(Nb_6Cl_{12}^i)L_6^a]^{n-}$ cluster anions have been crystallized from solutions with a variety of inorganic and organic cations including radical organic cations such as tetrathiafulvaleium (TTF$^+$) which lead to the formation of charge-transfer salts [24].

Niobium oxide cluster compounds (Table 6.11) are prepared by solid-state synthesis techniques at higher temperatures (900–1600 °C), typically starting from niobium metal or NbO and oxides containing niobium in higher oxidation states, such as NbO_2, Nb_2O_5, $ANbO_3$ or A_3NbO_4 (A=alkali metals). These reactions are characterized by slow rates due to high melting points and low volatility of the reagents, and often require the presence of a molten salt or oxide flux (usually an alkali chloride, fluoride, or B_2O_3) [25]. Niobium oxide cluster compounds are chemically inert and can only be dissolved (with decomposition of the cluster) in very strong oxidizing agents.

6.2.2
Electronic Structure, Redox and Magnetic Properties

A number of theoretical studies aimed at determining the electronic structure of transition-metal clusters have been reported [26, 27]. The molecular orbital (MO) diagrams of octahedral $[(M_6Cl^i_{12})L^a_6]$ clusters which are constructed based on symmetry and orbital overlap considerations, consist of a set of eight metal–metal bonding and sixteen metal–metal anti-bonding MOs. Extended Hückel MO calculations have shown that the Nb–Nb bonding states have an Nb–L^i anti-bonding contribution, which is especially significant in the case of the δ-type a_{2u} orbital. This orbital is overall slightly bonding in the case of niobium halides, leading to the preferred electronic configuration of 16 valence electrons per cluster (VEC) found in most compounds obtained at high temperatures. A more recent DFT-based calculation found that the a_{2u} orbital is located in the middle of a large energy gap between the bonding t_{2g} and the anti-bonding e_u levels. This result better explains the observed stability of clusters with VEC=16 and 14, as well as 15 in the case of rare-earth metals. In solution, the 16-electron $[Nb_6Cl_{18}]^{4-}$ cluster is stable in the absence of oxygen. In aqueous solutions the $[Nb_6Cl_{18}]^{4-}$ cluster undergoes slow hydrolytic decomposition, leading to precipitation of hydrated niobium(V) oxide, while in non-aqueous media in acidic conditions, they are oxidized by oxygen or iodine to give red-brown 15-electron clusters [28]. Oxidation of $[(CH_3)_4N]_4Nb_6Cl_{18}$ by chlorine gas in ethanol in the presence of HCl led to the preparation of a 14-electron cluster compound $[(CH_3)_4N]_2Nb_6Cl_{18}$ [29]. Clusters with 13 or 17 electrons were observed in these studies but could not be isolated due to their short lifetime [30]. In agreement with the calculated electronic structures, compounds with VEC=14 or 16 are diamagnetic or exhibit temperature-independent paramagnetism, while 15-electron clusters are paramagnetic, as was demonstrated by magnetic and ESR measurements. The changes in the electronic structure affect the intra-cluster distances, and significantly longer Nb–Nb intra-cluster distances and shorter Nb–Cl^a and Nb–Cl^i bond lengths are observed as the VEC decreases from 16 to 14, and electrons are removed from the Nb–Nb a_{2u} bonding state.

In contrast to chloride compounds, niobium oxides have a VEC of 14 electrons, due to an overall anti-bonding character of the a_{2u} state, caused by a stronger Nb-O anti-bonding contribution. In some cases, the VEC cannot be determined unambiguously due to the uncertainty in the electron distribution between the clusters and additional niobium atoms present in the majority of the structures. The 14-electron compounds exhibit semiconducting properties and weak temperature-independent paramagnetism. Unlike niobium chlorides, the oxides do not exhibit a correlation between the electronic configuration and intra-cluster bond distances.

6.3
Niobium Oxychloride Cluster Compounds

The lack of magnetic interaction between the paramagnetic 15-electron cluster and magnetic cations in $A_xREM_6X_{18}$ (RE = Rare-Earth, A = K, Cs, x = 0, 1, 2) compounds [31] presumably due to long cluster-RE distances led to the preparation and characterization of the first series of rare-earth niobium and tantalum oxyhalide cluster compounds by Cordier and coworkers, namely $Cs_2REM_6X_{18-y}O_y$ (RE = La, Lu, U; y = 1, 3) and $REM_6X_{13}O_3$ [32]. The basis of this work was the substitution of large halogen ligands by oxygen used to decrease cluster size, leading to smaller cluster-RE distances, which would enhance magnetic interactions.

Investigations of the interaction between 3d transition metals and octahedral halide or oxide metal clusters led to the preparation of a number of novel cluster compounds such as the series $A_xB_yNb_6Cl_{18}$ (A = Li, K, Rb, Cs; B = Ti, V, Mn, Cu) [33], and $Ti_2Nb_6O_{12}$ [34].

A number of niobium oxychloride cluster compounds, in which the number of oxygen ligands is 0 to 6, and 12, have been reported [32, 34, 35, 36]. The following sections will briefly describe the overall structural features of some of these oxychlorides that contain titanium as a counter-ion and will highlight the relationship between the topology and the main structure-determining factors, such as the number of oxygen ligands, their arrangements, the role played by the counter-ions, and optimization of electrostatic interactions.

6.3.1
One-dimensional Cluster Frameworks

Examples of recently reported oxychlorides with one- or quasi-one-dimensional frameworks include $A_2Ti_2Nb_6Cl_{14}O_5$ and $Cs_2Ti_4Nb_6Cl_{18}O_6$ in which the octahedral cluster unit has 5 or 6 oxygen ligands, respectively.

6.3.1.1 Frameworks Built from Clusters with Five Oxygen Ligands

The crystal structure of the oxychloride series $A_2Ti_2Nb_6Cl_{14}O_5$ (A = K, Rb, Cs) is based on a one-dimensional framework formed by octahedral niobium oxychloride clusters linked through outer chlorine ligands (Cl^a) and $[TiCl_3O_2]$ square pyramids (Fig. 6.3 a). The chains interact with each other through A^+ cations. The $(Nb_6Cl_7^iO_5^i)Cl_6^a$ cluster unit consists of Nb_6 octahedron in which all edges are bridged by chloride or oxide ligands and six other chloride ligands are in apical positions (Fig. 6.3 b). Four inner ligand positions are fully occupied by oxygen, six by chlorine, and two are statistically occupied by oxygen and chlorine (Fig. 6.3 c). The Nb_6 octahedral cluster core is distorted due to the size and charge difference between the ligands. The intra-cluster Nb–Nb bond lengths are 2804 and 2990 for oxygen- and chlorine-bridged niobiums, respectively. These bond lengths indicate that the number of valence electrons per cluster core is close to 14. Each cluster shares two of its outer chloride ligands with

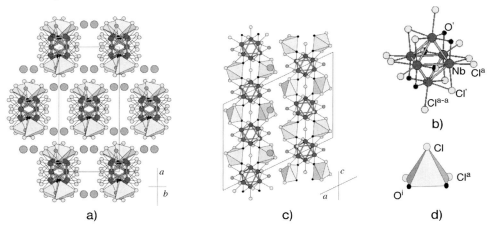

Fig. 6.3 Views of the crystal structure of $A_2Ti_2Nb_6Cl_{14}O_5$ in the [001] (a) and [010] (c) directions. (d) $[TiCl_3O_2]$ square pyramids.

two neighboring units, leading to the formation of zigzag chains (Fig. 6.3 b), with the connectivity formula $(Nb_6L_{12}^i)Cl_{2/2}^{a-a}Cl_4^a$. Titanium atoms provide additional inter-cluster linkages within the chain, with each titanium coordinated by five ligands, two inner oxygens and two outer chlorines (one from each cluster), and one chlorine that belongs to titanium only (Fig. 6.3 d). Bond-valence sum calculations indicate that titanium is present in the +3 oxidation state.

6.3.1.2 Frameworks Built of Clusters with Six Oxygen Ligands

Examples of niobium oxychlorides with clusters containing six oxygen ligands include $Cs_2Ti_4Nb_6Cl_{18}O_6$ and $KLu_3Nb_6Cl_{15}O_6$. The crystal structure of $Cs_2Ti_4Nb_6Cl_{18}O_6$ is based on $(Nb_6Cl_6^iO_6^i)Cl_6^a$ clusters linked through $(TiCl_4O_2)$ octahedra to form linear chains. The chains interact with each other through $(TiCl_6)$ octahedra and cesium ions (Fig. 6.4 a). The $(Nb_6Cl_6^iO_6^i)Cl_6^a$ cluster unit (Fig. 6.4 b) has six oxide ligands which selectively occupy "inner" positions arranged in two sets of three on opposite sides of the Nb_6 octahedron. The ligand arrangement results in an anisotropic cluster unit with symmetry close to D_{3d}, and intra-cluster Nb–Nb bond lengths between 2811 and 2979 Å. Each cluster is connected to two adjacent clusters through titanium ions to form chains. Neighboring clusters are linked by three octahedrally coordinated Ti atoms, which coordinate to one O^i and one Cl^a ligand from each cluster, and two chloride ligands that belong to titanium only. The $(TiCl_4O_2)$ octahedron has two O^i and two Cl ligands in *cis*-positions and two Cl^a ligands in *trans*-positions. The chains are connected by other titanium atoms coordinated by six Cl ligands from three $(TiCl_4O_2)$ octahedra that belong to three different chains. This linkage leads to the formation of tetrameric propeller-like $(Ti_4Cl_{12}O_6)$ units built from four

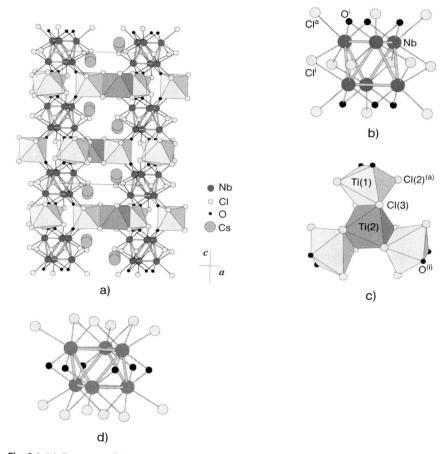

a)

b)

c)

d)

Fig. 6.4 (a) Fragment of the structure of $Cs_2Ti_4Nb_6Cl_{18}O_6$ (light and dark) polyhedra represent $(Ti(1)Cl_4O_2)$ and $(Ti(2)Cl_6)$ octahedra, respectively). (b) Cluster unit in $Cs_2Ti_4Nb_6Cl_{18}O_6$. (c) A tetrameric propeller-like $(Ti_4Cl_{12}O_6)$ unit in $Cs_2Ti_4Nb_6Cl_{18}O_6$ built from four (TiL_6) octahedra that share edges. (d) Cluster unit in $PbLu_3Nb_6Cl_{15}O_6$.

(TiL_6) octahedra sharing edges (Fig. 6.4c). Clusters with the same formulation $(Nb_6Cl_6^iO_6^i)Cl_6^a$ were observed in the compound $PbLu_3Nb_6Cl_{15}O_6$ [36], surprisingly, the units complement each other in terms of the ligand arrangement. The oxide inner ligand positions in the titanium phase are occupied by chloride ligands in the lutetium phase, and vice versa (Fig. 6.4d).

6.3.2
Two-dimensional Cluster Frameworks

When the niobium oxychloride clusters contain four oxygen ligands as inner ligands, different types of 2D cluster frameworks have been obtained. Their structures are briefly described below.

6.3.2.1 2D Oxychloride Frameworks with a Honeycomb-like Structure

A series of niobium oxychloride cluster compounds with the general idealized chemical composition $A_5Ti_8Nb_{18}Cl_{53}O_{12}$ (A = K, In, and Tl) has been reported [35c,h]. The structure of this series of niobium oxychlorides which is better formulated as $[A_5(Ti_2Cl_9)][(Nb_6Cl_{12}O_4)_3(Ti_3Cl_4)_2]$ (Fig. 6.5 a) is composed of honeycomb-like layers built from $(Nb_6Cl_6^iO_4^i)Cl_6^a$ clusters and $Ti_3Cl_7O_6$ trimers. The layers are stacked perfectly on top of each other, generating tunnels in which $Tl_2Cl_9^{3-}$ dimeric anions and A^+ cations are located. Each cluster has four oxygen ligands in inner positions (Fig. 6.5 b), and shares four outer chloride ligands with four adjacent clusters to form a two-dimensional network which generates six- and three-member rings (Fig. 6.5 c) similar to those found in hexagonal tungsten bronzes (HTB) [37]. Additional linkages between clusters within the same layer are provided by $Ti_3Cl_7O_6$ trimers (d(Ti-Ti) = 3.792(3) Å) built from three $TiCl_4O_2$ octahedra that share one vertex and three edges (Fig. 6.5 d). Each trimer connects three clusters through inner oxygen and outer chlorine ligands. Four chloride ligands of the trimer, the central and three apical ligands, labeled as Cl in Fig. 6.6 d, belong to the trimers only. The six-member cluster ring openings contain $[Ti_2Cl_9]^{3-}$ dimers formed by two $TiCl_6$ octahedra that share a face (d(Ti-Ti) = 2.977(8) Å) and are aligned along the [001] direction. Each dimer is surrounded by five A^+ ions statistically distributed over six crystallographic sites forming a distorted trigonal prism (Fig. 6.6 e). Similar $[Ti_2Cl_9]^{3-}$ dimers surrounded by A^+ ions have been observed in the series $A_3Ti_2Cl_9$ (A = Cs, In) [38].

The formation of the unusual HTB-type cluster framework is presumed to be favored by the presence of $[A_5Ti_2Cl_9]^{2+}$ units that play the role of a template around which the six-member cluster ring is assembled. The importance of the structure-directing effect is evidenced by the fact that, in $Ti_2Nb_6Cl_{14}O_4$, where no templating units are present, the same clusters form a framework with different topology. A similar templating effect was observed in other cluster compounds, such as $Cs_3(ZrCl_5)Zr_6Cl_{15}Mn$ [39] in which $[ZrCl_5]^-$ units stabilize a three-dimensional framework formed by octahedral $Zr_6Cl_{18}(Mn)$ clusters, and in $Nb_7S_2I_{19}$ [40] in which NbI_5 molecules act as templates leading to the formation of graphite-like framework formed by triangular $Nb_3I_{10}S$ clusters (Fig. 6.6).

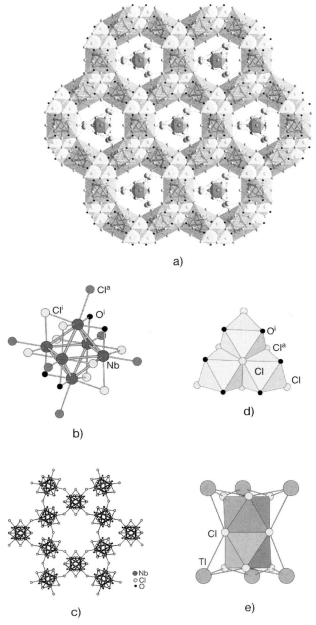

Fig. 6.5 (a) A projection of one layer of $[A_5(Ti_2Cl_9)][(Nb_6Cl_{12}O_4)_3(Ti_3Cl_4)_2]$ (I) structure on the (*ab*) plane (dark octahedra: $(Nb_6Cl_8^iO_4^i)Cl_6^a$ clusters, light: $Ti_3Cl_7O_6$ trimers, centered: Ti_2Cl_9 dimers, large spheres: Tl). (b) A cluster unit in (I). (c) Cluster framework in (I). (d) $Ti_3Cl_7O_6$ trimers and (e) Ti_2Cl_9 dimers surrounded by A^+ ions.

$[A_5(Ti_2Cl_9)][(Nb_6Cl_{12}O_4)_3(Ti_3Cl_4)_2]$ $Nb_7S_2I_{19}$ $Cs_3(ZrCl_5)Zr_6Cl_{15}Mn$

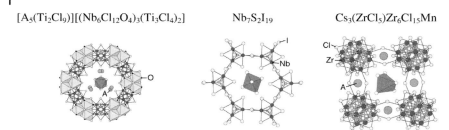

Fig. 6.6 Representations of the crystal structures of $[A_5(Ti_2Cl_9)][(Nb_6Cl_{12}O_4)_3(Ti_3Cl_4)_2]$, $Nb_7S_2I_{19}$, and $Cs_3(ZrCl_5)Zr_6Cl_{15}$ illustrating templating effects. The templating units are shown as dark shaded, the $Ti_3Cl_7O_6$ trimers as light shaded polyhedra.

6.3.2.2 Pillared 2D Oxychloride Frameworks

The clusters $(Nb_6Cl_6^iO_4^i)Cl_6^a$ observed in the HTB-like layered framework is also present in the oxychloride $Ti_2Nb_6Cl_{14}O_4$ featuring a three-dimensional framework formed by layers of octahedral niobium oxychloride clusters pillared by chains of $[TiCl_4O_2]$ octahedral-sharing edges (Fig. 6.7a). This structure represents the only known example of a niobium cluster compound featuring a combination of extended sublattices of different dimensionalities. Each cluster shares four outer chloride ligands with four neighboring clusters, leading to the formation of square-net layers with connectivity pattern $(Nb_6Cl_8^iO_4^i)Cl_{4/2}^{a-a}Cl_2^a$ (Fig. 6.7b). The cluster layers are linked to each other through zigzag chains of edge-sharing $[TiCl_4O_2]$ octahedra (Fig. 6.7c). The presence of octahedral cavities of ca. 2.7 Å in diameter, located within the cluster layers, inside the four-member cluster rings, led to the investigation of the possible insertion of A^+ cations in these cavities. These cavities (one per cluster) have an octahedral environment formed by six chloride ions (two of which do not belong to the clusters) and fit the size of K^+, In^+, Tl^+ or comparable ions. These studies resulted in the formation of the phases $A_xTi_2Nb_6Cl_{14-x}O_{4+x}$ ($A = In$, Tl, $x = 0.1–0.27$) in which the A^+ cations partially occupy the octahedral cavities. The additional charge provided by A^+ ions is compensated by substitution of chlorine by oxygen, keeping the electronic structure of the framework intact.

6.3.2.3 2D Framework with Graphite-like Cluster Connectivity

Clusters containing four oxygen ligands with the oxygen ligand arrangement different from that found in $Ti_2Nb_6Cl_{14}O_4$ and $[A_5(Ti_2Cl_9)][(Nb_6Cl_{12}O_4)_3(Ti_3Cl_4)_2]$ lead to the formation of an unusual layered framework in the compounds $A_2B_3(Nb_6Cl_{12.5}O_4)_2Cl_2$ ($A = Cs$, In; $B = Ti$, Sc) (Fig. 6.8a). The structure is characterized by graphite-like connectivities between clusters through outer chlorine ligands, $[TiCl_2O_3]$ and $[TiCl_4O_2]$ polyhedra. The cluster unit $(Nb_6Cl_8^iO_4^i)Cl_6^a$, has four inner oxide ligands, as is the case in $Ti_2Nb_6Cl_{14}O_4$ and $[A_5(Ti_2Cl_9)]$ $[(Nb_6Cl_{12}O_4)_3(Ti_3Cl_4)_2]$, but differs from the latter in terms of its ligand arrange-

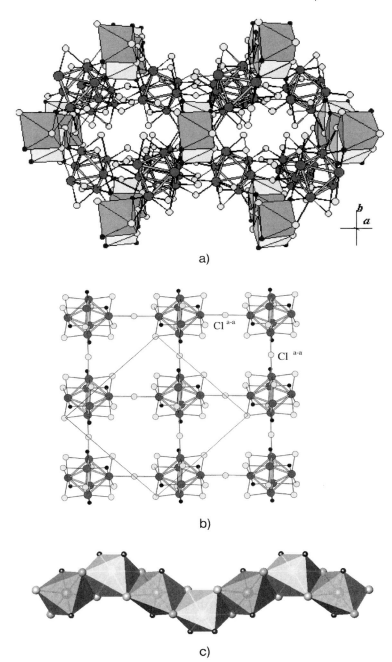

a)

b)

c)

Fig. 6.7 (a) Perspective of $Ti_2Nb_6Cl_{14}O_4$ structure in the [001] direction. (b) Cluster layer in $Ti_2Nb_6Cl_{14}O_4$. (c) Zigzag chain of edge-sharing $[TiCl_4O_2]$ octahedra in $Ti_2Nb_6Cl_{14}O_4$.

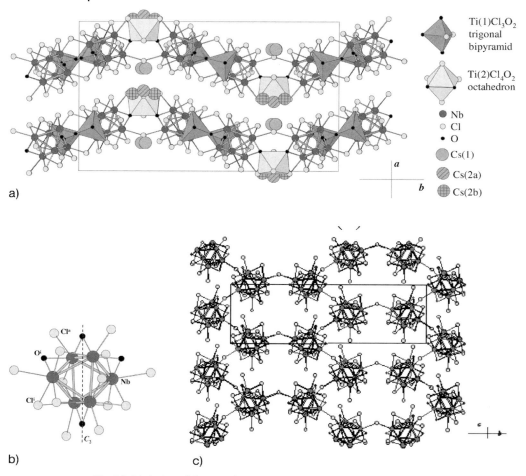

Fig. 6.8 (a) A view of the crystal structure of $Cs_2Ti_3(Nb_6Cl_{12.5}O_4)_2Cl_2$ in the [001] direction. (b) The cluster unit in $Cs_2Ti_3(Nb_6Cl_{12.5}O_4)_2Cl_2$. (c) A cluster layer in $Cs_2Ti_3(Nb_6Cl_{12.5}O_4)_2Cl_2$.

ment. The oxide ligands are distributed in sets of three ("triad") and one on opposite sides of the Nb_6 octahedron (Fig. 6.8 b), leading to an anisotropic chiral cluster unit with symmetry close to C_2. The intra-cluster Nb-Nb, Nb-Cl, and Nb-O indicate the presence of 14 valence electrons per cluster. Each cluster shares three of its six Cl^a ligands with three adjacent clusters to form layers with topology similar to that of graphite (Fig, 6.8 c) with the connectivity formula $(Nb_6Cl_8^iO_4^i)Cl_{3/2}^{a-a}Cl_3^a$ which is unprecedented for compounds containing octahedral clusters. Titanium polyhedra provide additional linkages between the clusters within the layers (Fig. 6.8 d). One titanium has a distorted trigonal-bipyramidal environment formed by three O^i and two Cl^a ligands and connects three clusters. The other titanium polyhedron connects two clusters together and consists of a distorted octahedron formed of two O^i and two Cl^a ligands (*cis*- configuration) and Cl or O that belong to Ti only.

6.4
Summary of Crystallographic Data on Titanium Niobium Oxychlorides

The structural features of most niobium oxychlorides known to-date are summarized in Table 6.1. The use of a combination of chloride and oxide ligands leads to compounds with unique structure types [41], characterized by a remarkable variety of cluster frameworks, ranging from discrete cluster units to chains, layers, and three-dimensional nets, some topologies of which are unprecedented in compounds containing octahedral clusters. Most of the niobium oxychlorides known to date have anisotropic structures (the exceptions are $Cs_2LuNb_6Cl_{17}O$ and $PbLu_3Nb_6Cl_{15}O_6$).

In contrast to the previously investigated mixed-ligand octahedral clusters in molybdenum and rhenium chalcohalides [42], niobium oxychlorides are typically characterized by a selective ligand distribution, which leads to anisotropic cluster units. In these clusters, oxide ligands exclusively occupy inner positions. It is also noteworthy that clusters with the same number of oxide ligands can have different ligand arrangements leading to frameworks with different topologies. Moreover, the same cluster unit can lead to different structure types, indicating the potential for a versatile structural chemistry of metal oxychloride cluster compounds yet to be discovered.

The structural features found in these oxychlorides indicate that inducing anisotropic bonding preferences in the cluster units, using a combination of ligands with large differences in charge density, is a viable strategy for the preparation of low-dimensional and open-framework cluster materials. The comparative analysis of the structural features of the niobium oxychloride allows us to assess the effects on framework topology produced by factors that have significant influence in other systems, such as the total number of ligands, the ligand arrangement, and ligand interactions with counter-ions, as well as to identify structure-determining factors that are unique to the oxychloride family.

6.4.1
Effect of the Total Number of Ligands

The relationship between the number of ligands and cluster connectivity established for metal halides containing M_6L_{18}-type clusters (Table 6.2) is observed only in the case of the oxychlorides $Cs_2LuNb_6Cl_{17}O$ and $Cs_2UNb_6Cl_{15}O_3$, which have structures based on discrete clusters. No direct correlation between the number of ligands and framework connectivity is observed in the other oxychlorides. In oxychlorides, the number of ligands is generally larger than that corresponding

Table 6.3 Ligand arrangement and structural properties in reported niobium oxychloride cluster compounds.

1 oxide ligand	3 oxide ligands	
$Cs_2LuNb_6Cl_{17}O^{32}$	$ScNb_6Cl_{13}O_3^{32}$	$Cs_2UNb_6Cl_{15}O_3^{32}$
CC: discrete units	*CC:* isotropic 3D	*CC:* discrete units
OF: isotropic 3D	*OF:* anis. 3D	*OF:* 2D
4 oxide ligands		

	$Ti_2Nb_6Cl_{14}O_4^{35}$	HLF	$Cs_2Ti_3(Nb_6Cl_{12.5}O_4)_2Cl_2^{35}$	$CsNb_6Cl_{12}O_2^{52}$
CC	2D	2D	*CC:* 2D	*CC:* anis. 3D
	square net	HTB	(graphite topology)	
OF	anis. 3D	2D	*OF:* 2D	*OF:* anis. 3D

5 oxide ligands	6 oxide ligands	
$Rb_2Ti_2Nb_6Cl_{14}O_5^{35}$	$Cs_2Ti_4Nb_6Cl_{18}O_6^{35}$	$PbLu_3Nb_6Cl_{15}O_6^{32}$
CC: 1D	*CC:* discrete units	*CC:* discrete units
OF: 1D	*OF:* quasi 1D	*OF:* isotropic 3D

The table includes structure types only, compounds obtained by isomorphic substitution are not listed. Dark shaded balls: Nb; light shaded balls: Cl; black balls: O; semi-shaded balls: sites partially occupied by Cl and O.

Abbreviations: *CC*=cluster connectivity; *OF*=overall framework; HTB=hexagonal tungsten bronze; HLF=Honeycomb-like layered framework

$[Tl_5(Ti_2Cl_9)][(Nb_6Cl_{12}O_4)_3(Ti_3Cl_4)_2]$.

to the cluster framework dimensionality because most oxychlorides have ligands that do not belong to the clusters. For example, in contrast to niobium chloride compounds with 18 ligands per cluster, which are based on discrete clusters, the compounds $Ti_2Nb_6Cl_{14}O_4$ and the series $A_xTi_2Nb_6Cl_{14-x}O_{4+x}$ (A = In, Tl) have layered cluster frameworks with two chloride ligands coordinating to Ti^{3+} ions only. Furthermore, oxychlorides in which all ligands belong to the cluster, as is the case in $A_xNb_6Cl_{12}O_2$ (A = K, Rb, Cs, In) and $ANb_6Cl_{13}O_3$ (M = Sc and Ti), have framework topologies that do not correspond to those typically found in halide compounds with the same number of ligands. The structure of $Ti_2Nb_6Cl_{14}O_4$ has an overall 3D framework with a pronounced anisotropic character compared to the isotropic structure of halide cluster compounds with 14 ligands [45–52]. In contrast to chloride compounds with 16 ligands ($Li_2Nb_6Cl_{16}$, $Na_4Zr_6Cl_{16}Be$, and $Cs_3Zr_6Cl_{16}C$), which have layered structures with a square-net topology [53, 54], the oxychlorides $ANb_6Cl_{13}O_3$ (M = Sc and Ti) have a unique three-dimensional cluster framework featuring interconnected helices. These examples indicate the existence of factors other than the total number of ligands that determine the framework topology in niobium oxychlorides. A deviation from the typical framework topology for M_6L_{14} composition is also observed in the mixed-ligand compound $Nb_6Cl_{12}I_2$ which has a cubic framework formed via I^{a-a-a} linkages [55].

6.4.2
Cluster Configuration

6.4.2.1 Relationships Between Ligand Arrangement and Direct Inter-cluster Linkages

Ligand arrangement has an important affect on cluster connectivities as can be illustrated by comparing the crystal structures of $Ti_2Nb_6Cl_{14}O_4$ and $[A_5(Ti_2Cl_9)][(Nb_6Cl_{12}O_4)_3(Ti_3Cl_4)_2]$. These compounds are based on the same cluster unit in which the arrangement of oxide and chloride ligands leads to two sets of niobium atoms with different coordination environments, and thus, different effective charges (Fig. 6.9). The first set, Nb1, coordinate to one oxide and four chloride ligands and are located in the highlighted plane, while the second set, Nb2, coordinate to two oxide and three chloride ligands and are located axial positions.

Density functional calculations [56] show that Nb1 set has a lower effective charge than that of Nb2, and that the differentiation between niobium atoms leads to different effective charges of the corresponding outer chloride ligands. The charge difference between these ligands presumably determines their coordination preferences between the niobium atoms from neighboring clusters ($Nb^{2.67+}$) and Ti^{3+} ions. The equatorial outer ligands have lower effective charge and are expected to coordinate preferentially to niobium, leading to the formation of two-dimensional frameworks, as is indeed observed in the compounds $Ti_2Nb_6Cl_{14}O_4$ and $[A_5(Ti_2Cl_9)][(Nb_6Cl_{12}O_4)_3(Ti_3Cl_4)_2]$. The correlations between the ligand arrangement and cluster connectivity are further supported by the structural features of $Cs_2UNb_6Cl_{15}O_3$, $Cs_2Ti_4Nb_6Cl_{18}O_6$, and $PbLu_3Nb_6Cl_{15}O_6$.

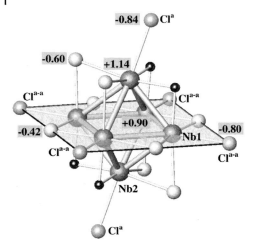

Fig. 6.9 Differentiation of niobium atoms and outer chloride ligands in the cluster unit of $Ti_2Nb_6Cl_{14}O_4$ and $[Tl_5(Ti_2Cl_9)][(Nb_6Cl_{12}O_4)_3(Ti_3Cl_4)_2]$. The numbers in the boxes are electrostatic charges calculated by the DFT method.

The ligand arrangement in these oxychlorides leads to equivalent coordination of all niobium atoms, which correlates with the formation of structures containing discrete clusters.

6.4.2.2 Relationships Between the Ligand Arrangement and Inter-cluster Linkages via Counter-ions

The ligand arrangement in the oxychloride clusters affects not only the direct intercluster linkages, but also the linkages via the counter-ions, and therefore plays an important role in determining the dimensionality of the overall structure. The highly charged "hard" counter-ions (Ti^{3+}, Sc^{3+}, Lu^{3+}, U^{3+}) preferentially coordinate to oxide ligands, forming strong intercluster $Nb-O-M^{3+}-O-Nb$ linkages that affect the overall framework connectivity. The directions of these linkages are defined by the arrangement of oxide ligands in the cluster unit, as illustrated in Fig. 6.10.

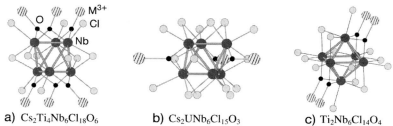

a) $Cs_2Ti_4Nb_6Cl_{18}O_6$ b) $Cs_2UNb_6Cl_{15}O_3$ c) $Ti_2Nb_6Cl_{14}O_4$

Fig. 6.10 Examples of the directional effect of oxide ligand arrangement on the location of "hard" cations M^{3+} and orientation of $M-O^i$ linkages.

The anisotropic ligand distribution in $Cs_2Ti_4Nb_6Cl_{18}O_6$ and $Cs_2UNb_6Cl_{15}O_3$ leads to the formation of chains and layers respectively, which indicates that an overall low-dimensional framework may be obtained even when the ligand distribution does not induce anisotropic direct inter-cluster linkages. Figure 6.10c shows the effect of the ligand arrangement on the location of titanium in $Ti_2Nb_6Cl_{14}O_4$, which are located between the cluster layers, rather than within the layers as is the case for Li^+ ions in $Li_2Nb_6Cl_{16}$ leading to a pillared structure.

6.4.3
Anion Segregation

The conclusion that the cluster configuration is the major factor determining the framework topology, implies that the symmetry of the framework must reflect that of the cluster unit. However, this is not the case, for example, in $Cs_2Ti_3(Nb_6Cl_{12.5}O_4)_2Cl_2$ where the C_2 symmetry of the cluster units is not preserved in the framework connectivity, which indicates the presence of other structure-determining factors that take precedence over charge distribution in the cluster unit. Analysis of the structures of $Cs_2Ti_3(Nb_6Cl_{12.5}O_4)_2Cl_2$ and other oxychlorides indicates that optimization of electrostatic interactions between the ligands and M^{3+} counter-ions are also important.

In the presence of highly charged counter-ions, oxide and chloride ligands tend to segregate to form aggregates of oxide ions surrounded by chlorides, which allows one to maximize cation–anion attractive interactions and minimize anion–anion repulsion (Fig. 6.11) which may explain some of the unusual structural features of oxychlorides, such as the formation of a framework with graphite-type topology in $Cs_2Ti_3(Nb_6Cl_{12.5}O_4)_2Cl_2$.

The structures of niobium oxychlorides exhibit not only anion segregation, but also segregation between "hard" and "soft" counter-ions (Fig. 6.11). In contrast to "hard" cations that are located within or on the surface of the oxide cores, "soft" cations are surrounded exclusively by chloride ions. This indicates that simultaneous presence of "hard" and "soft" cations leads to the formation of an overall low-dimensional structure compared to compositions with "hard" counter-ions only, such as $Ti_2Nb_6Cl_{14}O_4$ and $ScNb_6Cl_{13}O_3$. The non-uniform distribution of "hard" and "soft" ions in oxychlorides is analogous to the organization phenomena in concentrated surfactant solutions resulting in segregation between polar and non-polar components, which frequently leads to the formation of anisotropic structures [57].

6.4.4
Structure-determining Factors in the Absence of "Hard" Cations

Trivalent cations are present in most of niobium oxychloride cluster compounds known to date and play an important role in their structural chemistry. The series $A_xNb_6Cl_{12}O_2$ (A = K, Rb, Cs, In), on the other hand, represents an example of an oxychloride structure stabilized without a trivalent counter-ion (Fig. 6.12).

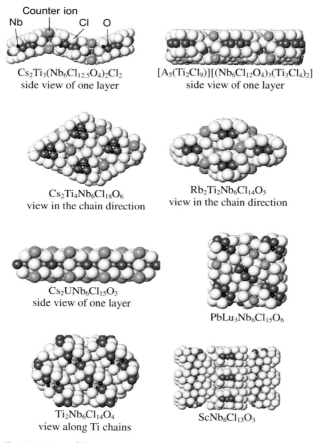

Counter ion

Nb   Cl   O

$Cs_2Ti_3(Nb_6Cl_{12.5}O_4)_2Cl_2$
side view of one layer

$[A_5(Ti_2Cl_9)][(Nb_6Cl_{12}O_4)_3(Ti_3Cl_4)_2]$
side view of one layer

$Cs_2Ti_4Nb_6Cl_{18}O_6$
view in the chain direction

$Rb_2Ti_2Nb_6Cl_{14}O_5$
view in the chain direction

$Cs_2UNb_6Cl_{15}O_3$
side view of one layer

$PbLu_3Nb_6Cl_{15}O_6$

$Ti_2Nb_6Cl_{14}O_4$
view along Ti chains

$ScNb_6Cl_{13}O_3$

Fig. 6.11 Space-filling views of the structures of niobium oxychlorides showing the trend of anion segregation.

This series also stands apart from all other oxychlorides in terms of the arrangement of oxide ligands. Unlike most oxychloride clusters, where oxide ligands occupy exclusively inner positions, the cluster unit in $A_xNb_6Cl_{12}O_2$ has two inner and two outer oxide ligands. Comparison between oxide ligand environments in the latter series and the other oxychlorides indicates that this difference is probably related to the absence of trivalent cations. The trivalent cations coordinate to all inner oxide ions, increasing the coordination number of the O^i ligands to three. In the series $A_xNb_6Cl_{12}O_2$, the three-fold coordination of O^i ligands is achieved by the formation of an additional linkage to a niobium atom from another cluster, which can be formally considered as a cation $Nb^{2.5+}$. Similar O^{i-a} linkages, leading to three-coordinated oxide ligands, are present in the compound $Nb_3Cl_5O_2$ [58, 59] which contains triangular clusters. Thus, the choice

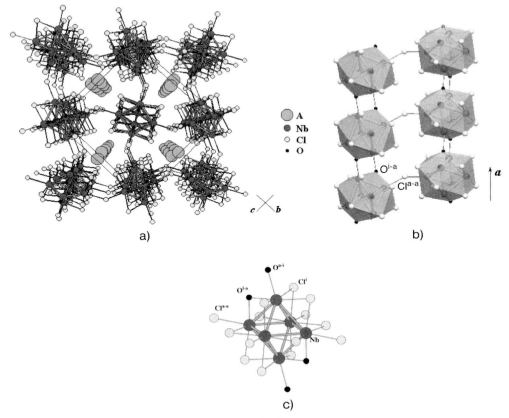

Fig. 6.12 (a) A perspective of $A_xNb_6Cl_{12}O_2$ structure in the $\vec{a}$ direction. (b) A view of the cluster framework in $A_xNb_6Cl_{12}O_2$ emphasizing the difference in intercluster linkages in the $\vec{a}$ direction and within the bc plane. (c) Cluster unit in $A_xNb_6Cl_{12}O_2$.

between inner or outer position for oxide ions seems to be determined by their preference to occupy sites with larger coordination number.

6.5
Electronic Configuration of Niobium Oxychloride Clusters

The number of valence electrons in niobium oxychloride clusters decreases as the number of oxide ligands increases (Table 6.4). The compound $Cs_2LuNb_6Cl_{17}O$ has the VEC of 16 as found in most chloride clusters, suggesting that the presence of one oxide ligand per cluster does not tip the balance between the Nb–Nb bonding and Nb–L^i anti-bonding contributions to the "a_{2u}" state. The VEC of most clusters

Table 6.4 Number of valence electrons per cluster in niobium oxychlorides.

	Number of O^i ligands	VEC
$Cs_2LuNb_6Cl_{17}O$	1	16
$A_xNb_6Cl_{12}O_2$ (A = K, Rb, Cs, In, x = 0.63–0.94)	2	$14.63 \leq VEC \leq 14.94$
$ScNb_6Cl_{13}O_3$	3	14
$Cs_2UNb_6Cl_{15}O_3$	3	14
$Ti_2Nb_6Cl_{14}O_4$	4	14
$A_xTi_2Nb_6Cl_{14-x}O_{4+x}$ (A = In, Tl, x = 0–0.27)	4 + x	14
$[Tl_5(Ti_2Cl_9)][(Nb_6Cl_{12}O_4)_3(Ti_3Cl_4)_2]$	4	14
$Cs_2Ti_3(Nb_6Cl_{12.5}O_4)_2Cl_2$	4	14
$Cs_2Ti_4Nb_6Cl_{18}O_6$, $PbLu_3Nb_6Cl_{15}O_6$	6	14
$KLu_3Nb_6Cl_{15}O_6$ [36a]	6	13 [a]

a) no data of physical properties are available.

with 3–6 oxide ligands is 14, indicating the dominating influence of oxide ligands. The $A_xNb_6Cl_{12}O_2$ series where the clusters have two inner and two outer oxide ligands seems to correspond to an intermediate case, since the outer ligands do not contribute to the "a_{2u}" state. In these clusters, the addition of electrons to the "a_{2u}" state neither decreases nor increases the stability of the cluster, which leads to partial population of this state with a variable number of electrons. Among oxychlorides with six oxide ligands, a 13-electron compound, $KLu_3Nb_6Cl_{15}O_6$was reported; however, its electronic configuration has not yet been confirmed by measurements of the physical properties.

6.6
Conclusion and Outlook

The investigation of niobium oxychloride cluster compounds has demonstrated that inducing anisotropic bonding preferences in the cluster units, using a combination of ligands with a large difference in charge density, leads to the formation of low-dimensional cluster compounds. All but two of the oxychlorides reported exhibit novel and diverse anisotropic frameworks with structural features that indicate the potential for the applications of these materials in ion exchange and intercalation reactions.

Investigation of these properties should be one of the significant directions of future development of these compounds, especially worthwhile are the exploration of the possibility of removing the templates from the channels of the honeycomb-like material $[Tl_5(Ti_2Cl_9)][(Nb_6Cl_{12}O_4)_3(Ti_3Cl_4)_2]$ and the studies of this and other oxychloride materials as hosts in redox intercalation processes.

A broader and more important implication of the oxychlorides is the potential of expanding the ligand combination to other transition-metal cluster systems. The advances in soft-chemistry techniques open up new possibilities for the stabilization of metal and ligand combinations that are not accessible by conventional high-temperature techniques, and one can envision that the niobium oxychlorides described here represent only a few examples of a large class of mixed-ligand materials with novel structure types yet to be discovered. Recently, Long et al. reported the synthesis of tungsten oxychloride clusters via a soft-chemistry route [60], which confirms the potential for wide applications of the ligand combination strategy in cluster chemistry.

References

1 (a) Simon, A. *Angew. Chem. Int. Ed. Engl.* **1988**, *27*, 159–183; (b) Mingos, D.P.M., Wales, D.J. *Introduction to Cluster Chemistry*, Prentice Hall, Englewood Cliffs, NJ, **1990**; (c) Köhler, J., Svensson, G., Simon, A. *Angew. Chem. Int. Ed. Engl.* **1992**, *31*, 1437–1456; (d) Corbett, J.D. in Parthé, E. (ed.), *Modern Perspectives in Inorganic Crystal Chemistry*, Kluwer Academic Publishers, Netherlands, **1992**; (e) González-Moraga, G. *Cluster Chemistry*, Springer, New York, **1993**; (f) Corbett, J.D. *J. Alloys Compd.* **1995**, *229*, 10–23; (g) Cotton, F.A. *Dalton Trans.* **2000**, 1961–1968 and references therein.

2 (a) Gates, B.C., Guczi, L., Knozinger, H. (eds.), *Metal Clusters in Catalysis*, Elsevier, Amsterdam, **1986**; (b) Thimmappa, B.H.S. *Coord. Chem. Rev.* **1995**, *143*, 1–34; (c) Goddard, C.A., Long, J.R., Holm, R.H. *Inorg. Chem.* **1996**, *35*, 4347–4354; (d) Gray, T.G., Rudzinski, C.M., Nocera, D.G., Holm, R.H. *Inorg. Chem.* **1999**, *38*, 5932–5933.

3 Shores, M.P., Beauvais, L.G., Long, J.R. *J. Am. Chem. Soc.* **1999**, *121*, 775–779.

4 Hughbanks, T. *J. Alloys Compd.* **1995**, *229*, 40–53.

5 (a) Hönle, W., von Schnering, H.G. *Z. Kristallogr.* **1990**, *191*, 139; (b) Kepert, D.L., Mandyczewsky, R. *Inorg. Chem.* **1968**, *7*, 2091; (c) von Schnering, H.G., Wöhrle, H., Schäfer, H. *Naturwissenschaften* **1961**, *48*, 159; (d) Broll, A., Simon, A., von Schnering, H.-G., Schäfer, H. *Z. Allg. Anorg. Chem.* **1969**, *367*, 1–18; (f) Simon, A., von Schnering, H.-G., Wöhrle, H., Schäfer, H. *Z. Allg. Anorg. Chem.* **1965**, *339*, 155–170.

6 McCarley, R.E. in Chisholm, M.H. (ed.), *Early Transition Metal Clusters with π-Donor Ligands*, VCH Publishers, New York, **1995**.

7 West, R.A. *Basic Solid State Chemistry*, 2nd edition, John Wiley **1999**.

8 (a) Beauvais, L.G., Shores, M.P., Long, J.R. *J. Am. Chem. Soc.* **2000**, *122*, 2763–2772; (b) Naumov, N.G., Virovets, A.V., Fedorov, V.E. *J. Struct. Chem.* **2000**, *41*, 499–520 and references therein; (c) Naumov, N.G., Soldatov, D.V., Ripmeester, J.A., Artemkina, S.B., Fedorov, V.E. *Chem. Commun.* **2001**, 571–572; (d) Y. Kim, S.-M. Park, W. Nam, and S.-J. Kim, *Chem. Commun.* **2001**, 1470; (e) Y. Kim, S.-M. Park, and S.-J. Kim, *Inorg. Chem. Commun.* **2002**, *5*, 592; (f) S. Jin, F.J. DiSalvo, *Chem. Mater.* **2002**, *14*, 3448; (f) Yan, M.B., Zhou A., Lachgar, A. *Inorg. Chem.* **2003**, *42*, 8818–8822. (g) Yan, Z., Day, C.S., Lachgar, A. *Inorg. Chem.* **2005**, *44(13)*, 4499–4505; (h) Yan, B., Day, C.S., Lachgar, A. *Chem. Comm.* **2004**, *21*, 2390–2391; (i) Zhou, H., Day, C.S., Lachgar, A. *Chem. Mater.* **2004** *16(24)*, 4870–4877.

9 The notation employed here was initially defined in Schäfer, H., von Schnering, H.G. *Angew. Chem.* **1964**, *76*, 833–849.

10 (a) Hibble, S.J., Cooper, S.P., Hannon, A.C., Patat, S., McCarroll, W.H. *Inorg. Chem.* **1998**, *37*, 6839–6846; (b) Ritter, A., Lyddsan, T., Harbrecht, B., Z. *Allg. Anorg. Chem.* **1998**, *684*, 1791–1795.

11 (a) Schäfer, H., von Schnering, H.G., Tillack, J., Kuhnen, F., Wöhrle, H., Baumann, H. *Z. Allg. Anorg. Chem.* **1967**, *353*; (b) Leduc, L., Perrin, A., Sergent, M., Le Traon, F., Pilet, J.C., Le Traon, A.

Mater. Lett. **1985**, *3*, 209–215; (c) Potel, M., Perrin, C., Perrin, A., Sergent, M. *Mat. Res. Bull.* **1986**, *21*, 1239–1245; (d) Perrin, A., Leduc, L., Sergent, M. *Eur. J. Solid State Inorg. Chem.* **1991**, *28*, 919–931; (e) Boeschen, S., Keller, H.L. *Z. Kristallogr.* **1992**, *200*, 305–315; (f) von Schnering, H.G., May, W., Peters, K. *Z. Kristallogr.* **1993**, *208*, 368–369; (g) Perrin, C. *J. Alloys Compd.* **1997**, *262*, 10–21 and references therein; (h) Tulsky, E.G., Long, J.R. *Chem. Mater.* **2001**, *13*, 1149–1166 and references therein.

12 (a) Perrin, C., Sergent, M. *J. Less-Common Metals* **1986**, *123*, 117–133; (b) Ihmaïne, S., Perrin, C., Sergent, M., El Ghadraoui, E.H. *Ann. Chim.-Sci. Mat.* **1998**, *23*, 187–190 and references therein; (c) Perricone, A., Slougui, A., Perrin, A. *Solid State Sci.* **1999**, *1*, 657–666 and references therein; (d) Perrin, A., Sergent, M. *New J. Chem.* **1988**, *12*, 337–356; (e) Long, J.R., McCarty, L.S., Holm, R.H. *J. Am. Chem. Soc.* **1996**, *118*, 4603–4616; (f) Long, J.R., Williamson, A.S., Holm, R.H. *Angew. Chem. Int. Ed. Engl.* **1995**, *34*, 226–229; (g) Mironov, Y.V., Cody, J.A., Ibers, J.A. *Acta Crystallogr. Sect. C* **1996**, *52*, 281–283.

13 (a) Simon, A., von Schnering, H.G., Schäfer, H. *Z. Allg. Anorg. Chem.* **1968**, *361*, 235–248; (b) Huges, B.G., Meyer, J.L., Fleming, P.B., McCarley, R.E. *Inorg. Chem.* **1970**, *9*, 1343–1346; (c) Broll, A., Juza, D., Schäfer, H. *Z. Allg. Anorg. Chem.* **1971**, *382*, 69–79; (d) Reckeweg, O., Meyer, H.-J. *Z. Kristallogr.* **1996**, *211*, 396; (e) Baján, B., Meyer, H.-J. *Z. Naturforsch.* **1995**, *B50*, 1373–1376.

14 (a) Lachgar, A., Meyer, H.-J. *J. Solid State Chem.* **1994**, *110*, 15–19; (b) Sitar, J., Lachgar, A., Meyer, H.-J. *Z. Kristallogr.* **1996**, *211*, 395; (c) Sitar, J. *Master Thesis*, Wake Forest University, **1995**.

15 Sitar, J., Lachgar, A., Womelsdorf, H., Meyer, H.-J. *J. Solid State Chem.* **1996**, *122*, 428–431.

16 (a) Ihmaïne, S., Perrin, C., Peña, O., Sergent, M. *J. Less-Common Metals* **1988**, *137*, 323–332; (b) Nägele, A., Anokhina, E., Sitar, J., Meyer, H.-J., Lachgar, A., *Z. Naturforsch. B* **2000**, *55 (2)*, 139–144.

17 Baján, B., Meyer, H.-J. *Z. Allg. Anorg. Chem.* **1997**, *623*, 791–795.

18 (a) Sägebarth, M.E., Simon, A., Imoto, H., Wepper, W., Kliche, G. *Z. Allg. Anorg. Chem.* **1995**, *621*, 1589–1596; (b) Womelsdorf, H., Meyer, H.-J., Lachgar, A. *Z. Allg. Anorg. Chem.* **1997**, *623*, 908–912; (c) Baján, B., Balzer, G., Meyer, H.-J. *Z. Allg. Anorg. Chem.* **1997**, *623*, 1723–1728; (d) Nägele, A. *Ph. D. Thesis*, University of Tübingen, Germany, **2001**; (e) Nägele, A., Day, C.S., Meyer, H.-J., Lachgar, A. *Z. Naturforsch B*, **2001**, *56*, 1238–1240.

19 (a) Desiraju, G.R. *Angew. Chem. Int. Ed. Engl.* **1995**, *34*, 2311–2327; (b) Aakeroy, C.B. *Acta Crystallogr. Sect. B* **1997**, *53*, 569–586; (c) Yaghi, O.M., Li, H.L., Davis, C., Richardson, D., Groy, T.L. *Acc. Chem. Res.* **1998**, *31*, 474–484; (d) Li, H.L., Eddaoudi, M., Laine, A., O'Keeffe, M., Yaghi, O.M. *J. Am. Chem. Soc.* **1999**, *121*, 6096–6097 and references therein; (e) O'Keeffe, M., Eddaoudi, M., Li, H.L., Reineke, T., Yaghi, O.M. *J. Solid State Chem.* **2000**, *152*, 3–20; (f) Sharma, C.V.K. *J. Chem. Ed.* **2001**, *78*, 617–622.

20 Gabriel, J.-C.P., Boubekeur, K., Uriel, S., Batail, P. *Chem. Rev.* **2001**, *101*, 2037–2066.

21 (a) Lind, M.D. *Acta Crystallogr. B* **1970**, *26*, 1058–1062; (b) Snegireva, E.M., Troyanov, S.I., Rybakov, V.B. *Zh. Neorg. Khim.* **1990**, *35*, 1945–1946; (c) Haase, A., Brauer, G. *Acta Crystallogr.* **1975**, *31*, 2521–2522; (d) Forsberg, H.E. *Acta Chem. Scand.* **1962**, *16*, 777–778; (e) Meyer, G., Staffel, T. *Z. Allg. Anorg. Chem.* **1986**, *532*, 31–36; (f) Brandt, G., Diehl, R. *Mat. Res. Bull.* **1974**, *9*, 411–420; (g) Seifert, H.J., Uebach, J. *Z. Allg. Anorg. Chem.* **1981**, *479*, 32–40; (h) Sands, D.E., Zalkin, A., Elson, R.E. *Acta Crystallogr.* **1959**, *12*, 21–23; (i) Corbett, J.D. in Whittingham, M.S., Jacobson A.J. (eds.), *Intercalation Chemistry*, Academic Press, New York, **1982** and references therein; (j) Song, K., Kauzlarich, S.M. *Chem. Mater.* **1994**, *6*, 386–394; (k) Takehara, Z., Sakaebem, H., Kanamura, K. *J. Power Sources* **1993**, *44*, 627–634.

22 (a) Schäfer, H. *Chemical Transport Reactions*, Academic Press, New York, **1964**; (b) Gruehn, R., Schweizer, H.-J. *Angew. Chem. Int. Ed. Engl.* **1983**, *22*, 82–95; (c) Miller, G. J. *J. Alloys Compd.* **1995**, *229*, 93–106.

23 Prokopuk, N., Shriver, D. F. *Adv. Inorg. Chem.* **1999**, *46*, 1–49.

24 (a) Imoto, H., Hayakawa, S., Morita, N., Saito, T. *Inorg. Chem.* **1990**, *29*, 2007–2014; (b) Pénicaud, A., Batail, P., Davidson, P., Levelut, A.-M., Coulon, C., Canadell, E., Perrin, C. *Chem. Mater.* **1990**, *2*, 117–123; (c) Beck, U., Simon, A., Brnicevic, N., Sirac, S. *Croatica Chemica Acta* **1995**, *68*, 837–848; (d) Reckeweg, O., Meyer, H.-J. *Z. Naturforsch. B* **1995**, *50*, 1377–1381; (e) Reckeweg, O., Meyer, H.-J. *Z. Allg. Anorg. Chem.* **1996**, *622*, 411–416; (f) Sirac, S., Planinic, P., Maric, L., Brnicevic, N., McCarley, R. E. *Inorg. Chimica Acta* **1998**, *271*, 239–242; (g) Prokopuk, N., Weinert, C. S., Kennedy, V. O., Siska, D. P., Jeon, H. J., Stern, C. L., Shriver, D. F. *Inorg. Chimica Acta* **2000**, *300*, 951–957; (h) Batail, P., Boubekeur, K., Fourmigué, M., Gabriel, J.-C. P. *Chem. Mater.* **1998**, *10*, 3005–3015.

25 (a) Köhler, J., Simon, A., Hibble, S. J., Cheetham, A. K. *J. Less-Common Metals* **1988**, *142*, 123–133; (b) Köhler, J., Tischtau, R., Simon, A. *J. Chem. Soc., Dalton Trans.* **1991**, 829–832; (c) Köhler, J., Simon, A. *Z. Allg. Anorg. Chem.* **1987**, *553*, 106–122; (d) Köhler, J., Simon, A. *Z. Allg. Anorg. Chem.* **1989**, *572*, 7–17; (e) Geselbracht, M. J., Stacy, A. M. *J. Solid State Chem.* **1994**, *110*, 1–5.

26 (a) Cotton, F. A., Hass, R. E. *Inorg. Chem.* **1964**, *3*, 10.

27 (a) Hughbanks, T. *Prog. Solid State Chem.* **1989**, *19*, 329–372; (b) Lin, Z., Williams, I. D. *Polyhedron* **1996**, *15*, 3277; (c) Lin, Z., Fan, M.-F. *Struct. Bonding* **1997**, *87*, 35; (d) Ogliaro, F., Cordier, S., Halet, J.-F., Perrin, C., Saillard, J.-Y., Sergent, M. *Inorg. Chem.* **1998**, *37*, 6199–6207.

28 (a) Fleming, P. B., Dougherty, T. A., McCarley, R. E. *J. Am. Chem. Soc.* **1967**, *89*, 159–160; (b) Koknat, F. W., McCarley, R. E. *Inorg. Chem.* **1974**, *13*, 295–300.

29 Koknat, F. W., McCarley, R. E. *Inorg. Chem.* **1972**, *11*, 812–816.

30 (a) Klendworth, D. D., Walton, R. A. *Inorg. Chem.* **1981**, *20*, 1151–1155; (b) Quigley, R., Barnard, P. A., Hussey, C. L., Seddon, K. R. *Inorg. Chem.* **1992**, *31*, 1255–1261.

31 (a) Ihmaïne, S., Perrin, C., Sergent, M. *Acta Crystallogr.* **1987**, *C43*, 813; (b) Ihmaïne, S., Perrin, C., Sergent, M. *Acta Crystallogr.* **1989**, *C45*, 705; (c) Cordier, S., Perrin, C., Sergent, M. *Z. Anorg. Allg. Chem.* **1993**, *619*, 621; (d) Cordier, S., Perrin, C., Sergent, M. *J. Solid State Chem.* **1995**, *118*, 274.

32 (a) Cordier, S., Perrin, C., Sergent, M. *Eur. J. Solid State Inorg. Chem.* **1994**, *31*, 1049–1060; (b) Cordier, S., Perrin, C., Sergent, M. *J. Solid State Chem.* **1995**, *120*, 43; (c) Cordier, S., Perrin, C., Sergent, M. *Croat. Chem. Acta* **1995**, *68*, 781; (d) Cordier, S., Perrin, C., Sergent, M. *Mater. Res. Bull.* **1996**, *31*, 683; (e) Cordier, S., Perrin, C., Sergent, M. *Mater. Res. Bull.* **1997**, *32*, 85.

33 Duraisamy, T., Qualls, J., Lachgar, A. *J. Solid State Chem.* **2003**, *170(2)*, 227–231.

34 Anokhina, E. V., Essig, M. W., Day, C. S., Lachgar, A. *J. Am. Chem. Soc.* **1999**, *121*, 6827–6833.

35 (a) Anokhina, E. V. *Ph. D. Thesis,* **2000**; (b) Anokhina, E. V., Essig, M. W., Lachgar, A. *Angew. Chem. Int. Ed. Engl.* **1998**, *37*, 522–525; (c) Anokhina, E. V., Essig, M. W., Day, C. S., Lachgar, A. *Inorg. Chem.* **2000**, *39 (10)*, 2185–2188; (d) Anokhina, E. V., Day, C. S., Lachgar, A. *Angew. Chem. Int. Edit.* **2000**, *39 (6)*, 1047; (e) Anokhina, E. V., Day, C. S., Lachgar, A. *Chem. Comm.* **2000**, *16*, 1491–1492; (f) Anokhina, E. V., Day, C. S., Lachgar, A. *Inorg. Chem.,* **2001**, *40*, 5072–5076; (g) Anokhina, E. V., Day, C. S., Meyer, H.-J., Ströbele, M., Kauzlarich, S. M., Kim, H., Whangbo, M.-H., Lachgar, A. *Journal of Alloys and Compounds* **2002**, *338*, 218–228; (h) Anokhina, E. V., Duraisamy, T., Lachgar, A. *Chem. Mater.* **2002**, *14*, 4111–4117; (i) Duraisamy, T., Yan, Z., Anokhina, E. V., Choi, C. S., Day, C. S., Lachgar, A. *J. Solid State Chem.* **2003**, *173*, 46–51.

36 (a) Cordier, S., Gulo, F., Perrin, C. *Solid State Sci.* **1999**, *1*, 637–646; (b) Gulo, F., Roisnel, T., Perrin, C. *J. Mater. Chem.* **2001**, *11*, 1237–1241.

37 Dickens, P. G., Whittingham, M. S. *Chem. Soc. Quart. Rev.* **1968**, *22*, 30–44.

38 (a) Bajan, B., Meyer, H.-J. *Z. Kristallogr.* **1996**, *211*, 817; (b) Gloger, T., Hinz, D., Meyer, G., Lachgar, A. *Z. Kristallogr.* **1996**, *211*, 821.

39 Zhang, J., Corbett, J. D. *Inorg. Chem.* **1995**, *34*, 1652–1656.

40 (a) Miller, G. J. *J. Alloys Comp.* **1995**, *229*, 93–106; (b) Miller, G. J., Lin, J. *Angew. Chem. Int. Ed. Engl.* **1994**, *33*, 334–336.

41 With the exception of $Cs_2LuNb_6Cl_{17}O$ which is isotypic with $Cs_2EuNb_6Cl_{18}$: Perrin, C., Ihmaïne, S., Sergent, M. *New J. Chem.* **1988**, *12*, 321–332.

42 (a) Perrin, A., Perrin, C., Sergent, M. *J. Less-Common Metals* **1988**, *137*, 241–265; (b) Perricone, A., Slougui, A., Perrin, A. *Solid State Sci.* **1999**, *1*, 657–666 and references therein.

43 Corbett, J. D. *Alloys Compd.* **1995**, *229*, 10–23 and references therein.

44 Perrin, C. *Alloys Compd.* **1997**, *262*, 10–21 and references therein.

45 Baján, B., Meyer, H.-J. *Z. Kristallogr.* **1995**, *210*, 607.

46 Artelt, H. M., Meyer, G. *Z. Kristallogr.* **1993**, *206*, 306–307.

47 Ziebarth, R. P., Corbett, J. D. *J. Solid State Chem.* **1989**, *80*, 56–67.

48 Smith, J. D., Corbett, J. D. *J. Am. Chem. Soc.* **1985**, *107*, 5704–5711.

49 Zhang, J., Corbett, J. D. *J. Solid State Chem.* **1994**, *109*, 265–271.

50 Smith, J. D., Corbett, J. D. *J. Am. Chem. Soc.* **1986**, *108*, 1927–1934.

51 Hughbanks, T., Rosenthal, G., Corbett, J. D. *J. Am. Chem. Soc.* **1988**, *110*, 1511–1516.

52 Hinz, D. J., Meyer, G. *Chem. Commun.* **1994**, 125–126.

53 Baján, B., Meyer, H.-J. *Z. Allg. Anorg. Chem.* **1997**, *623*, 791–795.

54 Ziebarth, R. P., Corbett, J. D. *Inorg. Chem.* **1989**, *28*, 626–631.

55 Sägebarth, M., Simon, A. *Z. Allg. Anorg. Chem.* **1990**, *587*, 119–128.

56 Geometry optimization was performed on the discrete cluster unit to eliminate the effects of crystal packing and interactions with titanium ions. The calculations were carried out in local spin density approximation using a SPARTAN 5.0.3 package (Wavefunction, Inc., Irvine, CA 92612 USA).

57 Luisi, P. L., Straub, B. E. (eds.) *Reverse Micelles: Biological and Technological Relevance of Amphiphilic Structures in Apolar Media*, Plenum, **1984**.

58 Gulo, F., Perrin, C. *Mater. Res. Bull.* **2000**, *35*, 253–262.

59 Lachgar, A., Anokhina, E., Day, C. *219th National Meeting of American Chemical Society*, San Francisco, CA, March **2000**, Abstract INOR-742.

60 Crawford, N. R. M., Long, J. R. *Inorg. Chem.* **2001**, *40(14)*, 3456–3462.

7

Trinuclear Molybdenum and Tungsten Cluster Chalcogenides: From Solid State to Molecular Materials

Rosa Llusar and Cristian Vicent

7.1
Introduction

The chemistry of transition metal halides and chalcogenides is often dominated by the presence of metallic clusters, both in the solid state and in solution. In the solid state these cluster units can be condensed (apex, edge or face-sharing) forming extended solids, or they can be discrete with only bridging interactions between adjacent cluster motifs in the solid. Many metal-halide and chalcogenide clusters are easily available via high-temperature solid-state synthesis from the elements. Solution strategies for the preparation of cluster complexes include self-assembly procedures, which very much resemble a designed synthesis in the solid state consisting of trial and error experiments, with only limited rational planning.

The exploratory solid-state synthetic work of John Corbett has illustrated the diversity, beauty and richness of this chemistry with a large variety of new phases and structures [1–3]. John Corbett was also the pioneer who recognized the potential of these cluster polymers in the development of a versatile solution chemistry [4]. Once the cluster unit has been identified in the solid state, the excision of this motif appears as the most rational method for accessing these cluster complexes in solution [5].

The clusters of interest here possess a central Mo_3 or W_3 triangular unit with one capping and three bridging chalcogenides (type I) or dichalcogenides (type II) groups, as represented in Fig. 7.1. The general formula of clusters with type I structure is $M_3(\mu\text{-}Q)(\mu\text{-}Q)_3L_9$ where the outer positions are occupied by different L ligands in a pseudooctahedral environment, without considering the metal-metal interactions. Complexes with type II structures correspond to the $M_3(\mu\text{-}Q)(\mu\text{-}Q_2)_3L_6$ general formula. If one neglects the metal-metal bonding, the metal atoms can be viewed as being seven coordinated. The bridging dichalcogenides in this structure are almost perpendicular to the metal plane with three chalcogen atoms occupying an equatorial position (Q_{eq}) essentially in the M_3 plane and with other axial chalcogen atoms (Q_{ax}) located out of the trimetallic

Inorganic Chemistry in Focus III.
Edited by G. Meyer, D. Naumann, L. Wesemann
Copyright © 2006 WILEY-VCH Verlag GmbH & Co. KGaA, Weinheim
ISBN: 3-527-31510-1

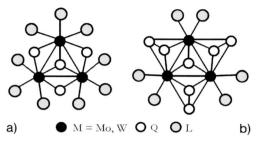

a)   ● M = Mo, W   ○ Q   ○ L   b)

Fig. 7.1 Idealized structures of the trinuclear cluster
chalcogenides $M_3Q_4L_9$ (type I, (a)) and $M_3Q_7L_6$ (type II, (b)).

plane. The M_3Q_4 unit in clusters with type I structures can be considered as incomplete M_4Q_4 cubes, in which M and Q occupy adjacent vertices with a missing metal atom. Selective removal of the chalcogen atoms in equatorial positions in clusters with type II structures results in the formation of type I complexes.

A simplified description of the metal-to-metal bonding in these trinuclear
complexes showed that the molecular orbitals concerned are as follows: three
bonding $1a_1$ and $1e$ orbitals, one $2a_1$ orbital which is essentially M–M non-bonding and five anti-bonding ($2e$, $3e$ and $1a_2$) orbitals [6]. According to this scheme,
these trinuclear clusters should be stable when there are six metal "d" electrons
available to enter the low-energy $1a_1$ and $1e$ metal-based orbitals, which correspond to three metal–metal bonds. Compounds with seven or eight metal electrons may also be stable, since the additional electrons occupy the non-bonding
$2a_1$ orbital. A topological analysis of the Electron Localization Function (ELF) in
these complexes shows that the M_3 unit behaves as a specific entity where the
bonding arises from the presence of a three-center-bond associated with a group
of basins involving three disynaptic V(M,M) and one trisynaptic V(M, M, M) basins. The larger orbital contributions to these basins come from the $1a_1$ and $1e$
metal orbitals [7].

This chapter is organized as follows, the first section describes the preparation of trinuclear complexes with M_3Q_4 and M_3Q_7 units using dimensional reduction and excision from solid-state phases. The second section presents a general overview of the ligand substitution reactions on these molecular M_3Q_4 and
M_3Q_7 compounds and finally some recent advances regarding the use of these
trinuclear cluster complexes in the development of molecular conductors and
supramolecular aggregates are discussed.

7.2
Synthesis and Structure of Molecular M_3Q_4 and M_3Q_7 Cluster Complexes

The presence of an M_3Q_7 cluster unit was first identified in 1968 in the inorganic polymers of crystal formula $[Mo_3(\mu_3\text{-}S)(\mu\text{-}S_2)_3X_2X_{4/2}]$, also represented as $\{Mo_3S_7X_4\}_n$ [8]. The structure consists of a zigzag chain of $Mo_3S_7X_2$ units bridged by four halogen atoms as shown in Fig. 7.2 [9]. These $\{Mo_3S_7X_4\}_n$ (X=Cl, Br) phases were prepared from Mo, S_2Cl_2 and sulfur (X = Cl) or from the elements (X=Br) at 350–400 °C in evacuated sealed borosilicate ampoules and reactivity studies were hindered by their low reactivity.

A systematic study of this class of compounds did not start until twenty years later and led to the preparation of a series of $\{M_3Q_7X_4\}_n$ (M=Mo, W; Q=S, Se and X=Cl, Br) inorganic polymers by high-temperature reactions (ca. 350 °C) of the elements in a sealed tube [10–14]. The interest on these cluster phases was mainly motivated by their excellent role as synthons for the preparation of molecular M_3Q_7 and M_3Q_4 cluster complexes, as will be presented in this section.

In contrast with the other $\{M_3Q_7X_4\}_n$ phases, the inorganic polymer $\{W_3S_7Cl_4\}_n$ reacts at temperatures ranging between 320 and 380 °C, liberating sulphur to produce a hexagonal phase of formula $W_3S_4Cl_4$ characterized by single-crystal X-ray diffraction studies [15]. $W_3S_4Cl_4$ is a layered tungsten thiohalide with a basic structure similar to that of TiS_2 and no cluster motif could be identified in this solid phase.

The chemistry of metal telluride clusters has received less attention than the rest of the chalcogenide cluster family. Reactions of Mo, Te and I_2 lead, depending on the reagents ratios, either to the coordination polymer $\{Mo_3Te_7I_4\}_n$ or to the ionic compound $[Mo_3Te_7(TeI_3)]I$ [16, 17]. On the other hand, the reaction of WTe_2, Te and Br_2 in a 3:2:1 molar ratio, presumably affords the $\{W_3Te_7Br_4\}_n$ phase, based on its reactivity. The poor crystallinity of these tellurium solid phases prevents any definite structural elucidation, but it is reasonable to sup-

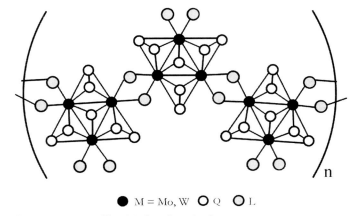

● M = Mo, W   ○ Q   ○ L

Fig. 7.2 Structure of $\{M_3Q_7X_4\}_n$ polymeric phases.

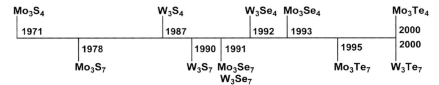

Scheme 7.1 Chronological evolution of the first structural data for M_3Q_4 and M_3Q_7 complexes.

pose that they are isostructural with the $\{Mo_3S_7Cl_4\}_n$ and $\{W_3S_7Br_4\}_n$ polymers structurally characterized by single crystal X-ray diffraction experiments.

While pioneering synthetic methods for the preparation of molecular M_3Q_7 and M_3Q_4 complexes were based on self-assembly procedures, most efficient approaches rely on dimensional reduction or excision methods that use polymeric $\{M_3Q_7X_4\}_n$ cluster phases as precursors. Scheme 7.1 presents a chronological evolution of the structural reports on these molecular clusters from the characterization of the first Mo_3S_4 and Mo_3S_7 clusters in the seventies until the systematic extension of this chemistry to tungsten and other chalcogenides (Se, Te) in the nineties starting from the 1D $\{M_3Q_7X_4\}_n$ polymers [18]. Details on the synthetic procedures employed are described next.

7.2.1
Solid-state Synthesis: Dimensional Reduction

Dimensional reduction implies the rupture of covalent bonds with the concomitant generation of molecular species and has proved to be a general method for dismantling trinuclear clusters from 1D $\{M_3Q_7X_4\}_n$ phases. Halide salts have been used as efficient dimensional reduction agents. High-temperature reactions at ca. $300\,^{\circ}C$ of polymeric $\{M_3S_7X_4\}_n$ (M = Mo, W; X = Cl, Br) with Ph_4PX afford the molecular anionic clusters of formula $[M_3S_7X_6]^{2-}$ according to Eq. (1).

$$\{M_3S_7X_4\}_n + Ph_4PX \rightarrow (Ph_4P)_2[M_3S_7X_6] \tag{1}$$

Dimensional reduction can also be achieved by mechanochemical activation of solid $\{M_3S_7X_4\}_n$ in the presence of Et_4NX (X = Cl, Br) in a vibration mill. A mechanochemical reaction is considered to occur during deformation, friction and fraction of the reacting solids. Molecular $[Mo_3S_7Br_6]^{2-}$ is an excellent starting material for the preparation of other Mo_3S_7 derivatives by ligand substitution, as will be later illustrated. A high yield route to $[Mo_3S_7Br_6]^{2-}$ is to treat $[Mo_3S_{13}]^{2-}$, the first molecular complex reported to have a Mo_3S_7 core, with HX acids [19].

Reaction of $\{M_3Se_7X_4\}_n$ with molten Ph_4PX or with Et_4NX under mechanochemical conditions also yields $[M_3Se_7X_6]^{2-}$ although in this case the resulting salts are contaminated and can not be recrystallized without decomposition [13, 20]. The stability of the selenohalide complexes of molybdenum and tungsten is lower than that of the sulfides and the selenocomplexes are often stabilized by substitution of the terminal halides, i.e., with dithiocarbamate ligands. No dimensional reduction of the telluride $\{M_3Te_7X_4\}_n$ polymers with halides has been reported. The reactivity of these telluride clusters is much lower than that of their sulphur and selenium analogues and strong nucleophiles are required to overcome their inertness.

Other ligand melts, different from halide salts, have been successfully used to obtain molecular clusters. Melting $\{Mo_3Q_7X_4\}_n$ (Q=S, Se) with 1,10-phenanthroline gives trinuclear $[Mo_3Q_7(phen)_3]^{4+}$ complexes where the six terminal positions, two on each metal atom, are occupied by three phenanthroline ligands [20]. On the other hand, mechanochemical activation of $\{M_3Q_7X_4\}_n$ (Q=S, Se) in the presence of potassium oxalate produces, after water extraction, a series of molecular $[M_3Q_7(C_2O_4)_3]^{2-}$ complexes [21]. As for the tellurides, the reaction of polymeric $\{Mo_3Te_7I_4\}_n$ or $\{W_3Te_7Br_4\}_n$ with KNCSe melt gives the molecular $[M_3Se_7(CN)_6]^{2-}$ (M=Mo, W) clusters as final products, where the tellurium atoms in the starting cluster core have been replaced by selenium atoms, while preserving the type-II structure. It is interesting to point out that this substitution proceeds sequentially according to the different labilities of the three distinct chalcogen atoms present in the cluster core, that is $Q_{eq} > Q_{ax} > \mu_3$-Q and mixed Te/Se intermediates have been isolated, which gives access to one of the few examples of complexes containing the $SeTe^{2-}$ group as a ligand [22].

In some cases, dimensional reduction of the $\{M_3Q_7X_4\}_n$ phases is accompanied by a cluster core change from M_3Q_7 to M_3Q_4 or, less often, to M_4Q_4. For example, the $\{M_3Se_7X_4\}_n$ selenides react in molten diphosphanes such as diphenylphosphanoethane (dppe) to produce cationic $[M_3S_4X_3(dppe)_3]^+$ clusters [23]. The $\{M_3Q_7Br_4\}_n$ (M=Mo, W; Q=S, Se) or $\{Mo_3Te_7I_4\}_n$ cluster polymers react in molten KCN at ca. 450 °C to form the molecular cubane-type cyano complexes of formula $[M_4Q_4(CN)_{12}]^{6-}$ or $[Mo_4Te_4(CN)_{12}]^{7-}$ [24]. The tungsten analogue $[W_4Te_4(CN)_{12}]^{6-}$ is made by heating WTe_2 with KCN.

7.2.2
Solution Routes: Excision

Excision is a solution-based procedure that consists in transferring into solution the molecular form of a certain cluster unit, in this case a trinuclear unit [5]. Indeed, this approach is quite similar to dimensional reduction except that, here, the rupture of the bridging-halide in the 1D $\{M_3Q_7X_4\}_n$ parent solid may be accompanied by replacement of the peripheral ligands by solvent molecules. This is illustrated by the reaction of $\{M_3Q_7X_4\}_n$ (Q=S, Se) polymeric phases with H_3PO_2 in aqueous media, which represent the most efficient synthetic entry to the aqua complexes of the general formula $[M_3Q_4(H_2O)_9]^{4+}$ (M=Mo, W; Q=S, Se) [25]. This process implies the reduction of the dichalcogenide bridges present in the parent solid to chalcogenide and this transformation is considered to be the driving force for these excision reactions. Scheme 7.2 presents a summary of some illustrative examples of the reactivity of these polymeric 1D phases, both in the solid state and in solution.

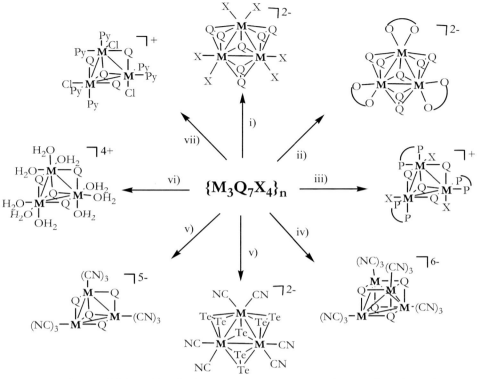

Scheme 7.2 Reactivity of polymeric $\{M_3Q_7X_4\}_n$ phases.
i) Et$_4$NX (mechanochemical activation) or molten PPh$_4$X;
ii) K$_2$ oxalate (mechanochemical activation);
iii) diphos/CH$_3$CN or molten dppe;
iv) molten KCN; v) aqueous KCN;
vi) aqueous H$_3$PO$_2$; vii) PPh$_3$/pyridine.

Excision of polymeric $\{M_3Q_7X_4\}_n$ (Q = S, Se) phases can also be achieved with other chalcogen-abstracting reagents such as CN$^-$, mono- or diphosphanes to invariably obtain the corresponding molecular M$_3$Q$_4$ complexes with a type-I structure in moderate to high yields. Remarkably, the less energetic conditions achieved by excision routes, as compared with dimensional reduction methods with cyanide salts, results in the formation of completely different products. While the reaction of $\{M_3Q_7X_4\}_n$ (Q = S, Se) with KCN in boiling water affords molecular [M$_3$Q$_4$(CN)$_9$]$^{5-}$ clusters, dimensional reduction with molten KCN gives tetranuclear [M$_4$Q$_4$(CN)$_{12}$]$^{6-}$ complexes, as presented in the previous section [24]. On the other hand, reaction of $\{Mo_3Te_7I_4\}_n$ or $\{W_3Te_7Br_4\}_n$ in aqueous KCN gives [M$_3$Te$_7$(CN)$_6$]$^{2-}$ with a type-II structure [16, 22]. These complexes are inert to tellurium abstraction unlike the S and Se analogues and reduction of bridging ditellurides cannot be invoked in this case as the driving force for the excision reaction.

The excision of polymeric $\{M_3Q_7X_4\}_n$ (Q=S, Se) phases with phosphanes as chalcogen abstracting ligands has proved to be a general method for the preparation of type-I M_3Q_4 incomplete cuboidal clusters that may have, i.e., $[Mo_3S_4Cl_4(PEt_3)_4(CH_3OH)]$ [26] or may not have, i.e. $[Mo_3S_4Cl_4(C_5H_5N)_5]$ [27], phosphanes as ancillary ligands. The yield of an excision reaction depends on the nature of the solvent employed. In particular, excision of $\{M_3Q_7X_4\}_n$ (Q = S, Se) polymers with diphosphanes in refluxing acetonitrile affords molecular species of general formula $[M_3Q_4X_3(diphos)_3]^+$ in high yields [28, 29]. It is worth mentioning that the specific coordination of the diphosphane ligand, with one phosphorus atom *trans* to the capping chalcogen and the other *trans* to the bridging chalcogen atom, results in cubane-type chalcogenide clusters with backbone chirality as shown in Fig. 7.3.

An unprecedented stereoselective procedure to obtain enantiomerically pure transition cluster M_3Q_4 complexes consists of the direct excision of the $\{M_3Q_7X_4\}_n$ polymers using chiral diphosphanes, namely (+)-1,2-bis[(2R,5R)-2,5-(dimethylphospholano)]ethane [(R,R)-Me-BPE] and its respective enantiomer [(S,S)-Me-BPE] to afford the trinuclear complexes (P)-$[Mo_3S_4Cl_3(R,R$-Me-BPE$)_3]^+$ and (M)-$[Mo_3S_4Cl_3(S,S$-Me-BPE$)_3]^+$, respectively [30]. The structures of both enantiomers are shown in Fig. 7.3. The symbols (P) and (M) refer to the rotation of the chlorine atoms around the C_3 axis, with the capping sulfur pointing towards the viewer.

These trinuclear complexes can act as metalloligands to afford a whole series of heterobimetallic $M_3M'Q_4$ cubane-type complexes and several review articles on the topic have been published in the last decade [31, 32].

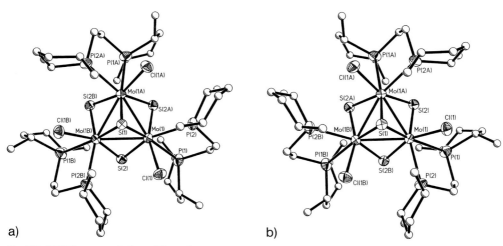

a)

b)

Fig. 7.3 ORTEP representation of the cations (P)-$[Mo_3S_4Cl_3(R,R$-Me-BPE$)_3]^+$ (a) and (M)-$[Mo_3S_4Cl_3(S,S$-Me-BPE$)_3]^+$ (b).

7.2.3
Ligand Exchange Reactions

The outer ligands in $M_3Q_4L_9$ (type-I) and $M_3Q_7L_6$ (type-II) clusters can be easily replaced through nucleophilic substitution reactions. These reactions are discussed separately for each cluster type.

7.2.3.1 M_3Q_4 Cluster Complexes

Reactivity studies on these systems have been mainly focused on the $M–H_2O$ bonds in the $[M_3Q_4(H_2O)_9]^{4+}$ aqua ions and the $M–Y$ bonds in the $[M_3Q_4Y_3(di$-phos$)_3]^+$ (Y=Cl, Br, OH, H) diphosphano complexes. The chemistry of the trinuclear aqua clusters is restricted to acidic media due to stability reasons. The $[M_3Q_4(H_2O)_9]^{4+}$ aqua ion can be easily derivatized by substitution of the water molecules with ligands such as nitrilotriacetate, oxalate, dithiophosphates, thiocyanate, etc. In some cases ligand substitution results in complexes having an enhanced stability, a fact that has allowed the crystallization and further characterization of certain cluster cores. Kinetic studies on the substitution of coordinated water in these aqua complexes revealed the existence of two types of kinetically different water molecules, the one *trans* to the capping chalcogen being less labile than the other two [33]. Replacement of the capping sulfur by selenium in the molybdenum trimer results in a retardation of the substitution of the more labile water molecules, the opposite effect being observed upon substitution of the bridging sulfide by selenide. In all cases, there is an acceleration of the substitution reaction when the acid concentration is decreased, which is interpreted as a consequence of the formation of more labile hydroxo complexes. Chloride coordination to $[M_3Q_4(H_2O)_9]^{4+}$ is weak with a formation constant of 3 M^{-1}. $[Mo_3S_4(H_2O)_9]^{4+}$ aqua ions and the $[W_3S_4(NCS)_9]^{5-}$ complex react with acetylene derivatives in acidic media to afford new alkenedithiolate cluster complexes [34].

The halide ligand in $[W_3Q_4X_3(diphos)_3]^+$ can be substituted by a hydroxo group or a hydride ligand through treatment with NaOH in acetonitrile-water solutions or with borohydride in methanol, respectively. Complexes $[W_3Q_4(OH)_3(diphos)_3]^+$ and $[W_3Q_4H_3(diphos)_3]^+$ react with HX acids under mild conditions in a variety of solvents to form the corresponding halide complexes while no reaction is observed in the presence of halide salts. These observations indicate that protons play an important role in the process. Protonation of the hydride cubane-type $[W_3Q_4H_3(dmpe)_3]^+$ (Q = S, Se) with acids is represented in Eq. (2).

$$[W_3Q_4H_3(dmpe)_3]^+ + 3\ HX \rightarrow [W_3Q_4X_3(dmpe)_3]^+ + 3\ H_2 \qquad (2)$$

The reaction mechanism in acetonitrile or water/acetonitrile mixtures occurs with three kinetically distinctive steps according to Eqs. (3)–(6), a kinetics significantly more complicated than that observed for the substitution of the aqua ligands in $[M_3Q_4(H_2O)_9]^{4+}$ complexes [35, 36].

$$[W_3Q_4H_3(dmpe)_3]^+ + HX \rightarrow [W_3Q_4H_2(H_2)(dmpe)_3]^{2+} + X^-; \; 3 \, k_1 \qquad (3)$$

$$[W_3Q_4H_2(H_2)(dmpe)_3]^{2+} \rightarrow [W_3Q_4H_2(dmpe)_3]^{2+} + H_2; \; \text{fast} \qquad (4)$$

$$[W_3Q_4H_2(dmpe)_3]^{2+} + H_2O \rightarrow [W_3Q_4H_2(H_2O)(dmpe)_3]^{2+}; \; 3 \, k_2 \qquad (5)$$

$$[W_3Q_4H_2(H_2O)(dmpe)_3]^{2+} + X^- \rightarrow [W_3Q_4H_2X(dmpe)_3]^+ + H_2O; \; 3 \, k_3 \qquad (6)$$

The first step consists in the attack of a proton on the W–H bond to yield a labile dihydrogen intermediate (Eq. (3)) that rapidly releases H_2 to form a coordinatively unsaturated complex (Eq. (4)). This complex adds water in the next step to form an aqua complex (Eq. (5)) that completes the reaction by substituting the coordinated water by the X^- anion (Eq. (6)). Steps (3)–(6) are repeated for each W–H bond and the factor of 3 in the rate constants appears as a consequence of the statistical kinetics at the three metal centers. The rate constants for both the initial attack by the acid (k_1) and water attack to the coordinatively unsaturated intermediate (k_2) are faster in the sulfur complex, whereas the substitution of coordinated water (k_3) is faster for the selenium compound.

Kinetic studies of the hydride cluster $[W_3S_4H_3(dmpe)_3]^+$ with acids in a non-coordinating solvent, i.e., dichloromethane, under the pseudo-first-order condition of acid excess, show a completely different mechanism with three kinetically distinguishable steps associated to the successive formal substitution of the coordinated hydrides by the anion of the acid, i.e., Cl^- in HCl [37]. The first two kinetic steps show a second-order dependence with the acid concentration. This is the first example of a proton transfer process to a hydride complex with a second-order dependence. Theoretical calculations indicate that the role of the HX molecules is the formation of W-H$\cdots$H-Cl$\cdots$H-Cl adducts that convert into W-Cl, H_2 and HCl_2^- in the rate-determining state through hydrogen complexes as transition states.

7.2.3.2 M_3Q_7 Cluster Complexes

The lability of the bromine ligands in $[Mo_3Q_7Br_6]^{2-}$ can be conveniently used to prepare a large variety of derivatives. Examples include various non-reducing ligands, such us thiocyanate, oxalate, dithiocarbamate, imidodiphosphanechalcogenide, aniline, dithiophosphates, catecholate, 2-thiopyridine, 8-hydroxyquinoline and several 1,2-bis-dithiolates [38]. A characteristic feature of these compounds is the electrophilic character of the axial chalcogen atoms, those out of the Mo_3 plane, that provides them with the ability to bind mono-anions such as halogens or to form dimers through chalcogen–chalcogen interactions with the peripheral ligands of a neighbor cluster. This is clearly illustrated in Fig. 7.4 with the solid-state structure of the $\{[Mo_3S_7(C_2O_4)_3]Br\}^{3-}$ and $\{[Mo_3S_7(mnt)_3]_2\}^{4-}$ (mnt = maleonitriledithiolate) aggregates. In the case of the oxalate cluster, the presence of this triply charged adduct is also detected in solution by mass spec-

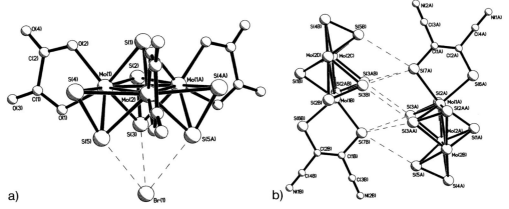

Fig. 7.4 Structure of the $\{[Mo_3S_7(C_2O_4)_3]Br\}^{3-}$ (a) and $\{[Mo_3S_7(mnt)_3]_2\}^{4-}$ (b) adducts where the maleonitrile ligands on two molybdenum sites are omitted for clarity.

trometry experiments using electrospray as an ionization source. It is interesting to point out that some of these clusters have been shown to be good optical limiters, namely the bromide, thiocyanate and maleonitriledithiolate derivatives.

The coordination of redox-active ligands such as 1,2-bis-dithiolates, to the Mo_3Q_7 cluster unit, results in oxidation-active complexes in sharp contrast with the electrochemical behavior found for the $[Mo_3S_7Br_6]^{2-}$ di-anion for which no oxidation process is observed by cyclic voltammetry in acetonitrile within the allowed solvent window [38]. The oxidation potentials are easily accessible and this property can be used to obtain a new family of single-component molecular conductors as will be presented in the next section. Upon reduction, $[Mo_3S_7$ (dithiolate)$_3]^{2-}$ type-II complexes transform into $[Mo_3S_4(dithiolate)_3]^{2-}$ type-I di-anions, as represented in Eq. (7).

$$[Mo_3S_7(dithiolate)_3]^{2-} + 3\ PPh_3\ \rightarrow\ [Mo_3S_4(dithiolate)_3]^{2-} + 3\ S{=}PPh_3 \qquad (7)$$

The structure of $[Mo_3S_4(dmit)_3]^{2-}$ (dmit = 1,3-dithiole-2-thione-4,5-dithiolate) represents one of the rare examples of M_3S_4 clusters where each metal atom appears as pentacoordinate instead of its more common type-I structure octahedral environment [39]. Complexes $[M_3Q_4(dmit)_3]^{2-}$ (M = Mo, W; Q = S, Se) degrade in air with an almost quantitative yield and afford a series of M(V) dimers of formula $[M_2O_2(\mu\text{-}Q)_2(dmit)_2]^{2-}$ where the oxygen atoms are in a *syn* configuration.

7.3
Trinuclear Clusters as Building Units

The potential of the cluster units described here to participate in intermolecular chalcogen–chalcogen interactions combined with the easy modification of their outer coordination sphere with ligands of different nature, i.e., redox active, hydrogen donors, bi-functional, etc., make these systems useful blocks for the construction of supramolecular materials with multi-physical properties.

7.3.1
Molecular Conductors Based on M_3Q_7 Cluster Complexes

Carrier generators in molecular conductors have been associated for a long time to a partial charge transfer between the HOMO (or LUMO) electronic band and other chemical species. These systems are known as two-component molecular conductors. Tetrathiofulvalene derivatives are versatile systems for the formation of molecular organic conductors due to their electron donor capacity by transferring one π-electron from the HOMO orbital, and to their planar shape that promotes their stacking as a consequence of the π-π orbital overlap. The electronic properties of these salts are essentially determined by the packing pattern of the donor molecules which, in turn, depends on the counter-ion.

The $[Mo_3S_7Br_6]^{2-}$ cluster has been used as a charge-compensating anion for the preparation of ET [*bis*(ethylenedithio)-tetrathiofulvalene] charge transfer salts by electrochemical oxidation of the ET donor molecules. Phases with different stoichiometries, namely $(ET)_3[Mo_3S_7Br_6]_2$, $(ET)_2[Mo_3S_7Br_6] \cdot 1.1\,CH_2Br_2$ and $(ET)_3\{[-Mo_3S_7Br_6]Br\} \cdot 0.5\,C_2H_4Cl_2$ have been isolated depending on the solvent employed [40]. Except for $(ET)_3[Mo_3S_7Br_6]_2$, where one-third of the donor molecules are incorporated into the structure as ET^{2+}, the remaining organic molecules have a $+1$ charge. The role of the anion in this salts is merely structural, due to the absence of redox activity upon oxidation and the anion participates in various $S \cdots S$ and $S \cdots Br$ contacts with the ET oxidized donor molecules that combined with interactions with the solvent give rise to unique arrangements of the organic molecules. A general feature of these structures is the presence of alternating layers of dimerized organic donor molecules $(ET^+:ET^+)$ and of inorganic clusters, where the long axis of the donor dimers runs almost parallel to the cluster layer, as represented in Fig. 7.5. The $(ET)_2[Mo_3S_7Br_6] \cdot 1.1\,CH_2Br_2$ phase has semiconducting properties. There is a strong tendency of the combination $\{[Mo_3S_7Br_6]:ET\}$ to accomodate a third bulky component as tetrabutylammonium in the (ET) $(Bu_4N)[Mo_3S_7Br_6]$ phase, that also presents semiconducting properties.

At the turn of this century it was realized that carrier generation was also possible between the HOMO and LUMO band even in neutral single-component materials assuming that there was a small HOMO-LUMO gap and conduction paths have been associated with the presence of large transverse intermolecular interactions. The most relevant examples of single-component molecular conductors are the mononuclear $M(dithiolate)_2$ (M = Co, Ni, Cu, Au) complexes with

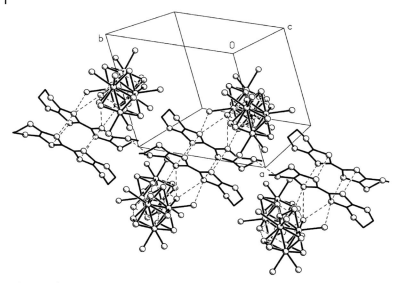

Fig. 7.5 Characteristic cluster$_{aggregate}$: (ET$^+$: ET$^+$) : cluster$_{aggregate}$ packing for the system {[Mo$_3$S$_7$Br$_6$] : ET}.

extended TTF ligands prepared by oxidation of their corresponding anions [41]. Substitution of the Br atoms in [Mo$_3$S$_7$Br$_6$]$^{2-}$ by redox-active ligands such as dithiolates gives the [Mo$_3$S$_7$(dithiolate)$_3$]$^{2-}$ di-anions which can be oxidized at easily accessible potentials. The facile oxidation upon ligand substitution results in a change from a $1a_1^2 1e^4$ to a $1a_1^2 1e^2$ ground-state configuration, suggesting the presence of a significant contribution of the dithiolate ligand to the HOMO "e" orbital, which leads to a partial occupation of the degenerate HOMO e-type orbitals with the concomitant production of radicals, a prerequisite in the formation of single-component molecular conductors.

Recently we have found that chemical or electrochemical oxidation of the [Mo$_3$S$_7$(dmit)$_3$]$^{2-}$ di-anion provides neutral paramagnetic Mo$_3$S$_7$(dmit)$_3$ species which are semiconducting in the solid state with small activation energies (11–22 meV) [42]. Neutral Mo$_3$S$_7$(dmit)$_3$ crystallizes in a trigonal space group with the cluster units oriented along the c direction and connected through various S$\cdots$S contacts of ca. 3.6 Å as represented in Fig. 7.6, to produce infinite chains. Parallel chains are related by an inversion center and connected through short contacts (3.1–4.0 Å) across the ab plane between the dmit and bridging sulfur atoms of one chain, with those of the adjacent chain, as represented in Fig. 7.6, which results in an extended hexagonal network.

The room temperature electrical conductivity of single crystals of Mo$_3$S$_7$ (dmit)$_3$ along the c axis is 25 S cm^{-1}, which is very high for a neutral molecular crystal. The variation of the magnetic susceptibility (χ) with the temperature does not follow a Curie-Weis law with a continuous decrease of χT vs. T upon cooling

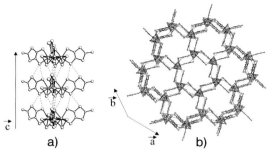

Fig. 7.6 Crystal packing of neutral $Mo_3S_7(dmit)_3$ cluster along the c direction (a) and view of the hexagonal packing across the ab plane (b).

and a χT value of 0.80 emu K mol^{-1} at room temperature, clearly below the expected value for a triplet state. These observations indicate the presence of antiferromagnetic exchange interactions between the unpaired electrons of the neutral $Mo_3S_7(dmit)_3$ molecules. First-principles spin-polarized DFT-type calculations estimate the antiferromagnetic (AFM) state to be very close in energy to the ferromagnetic (FM) state, the AFM state being only 0.02 eV/molecule below the FM state in agreement with the susceptibility data. Band structure calculations on the AFM state show the existence of predominant electronic interactions along the c direction for the electrons near the Fermi level so that $Mo_3S_7(dmit)_3$ is a one-dimensional magnetic semiconductor with a small but non-negligible dispersion along c. In addition there is a small energy gap at the Fermi level (ca. 200 meV) in harmony with the activated, although high, conductivity of the system. The hypothetical metallic state for $Mo_3S_7(dmit)_3$ was calculated to be only 50 meV higher in energy than the AFM state. Chemical modifications aimed to stabilize this metallic state are now in progress.

7.3.2
Formation of Supramolecular Adducts

The construction of supramolecular aggregates based on organic macrocycles such as cucurbit[n]uril relies heavily upon complementary hydrogen bond formation [43]. Triangular thio and seleno M_3Q_4 aquo complexes yield supramolecular compounds with cucurbit[6]uril (cuc) [44, 45]. This organic molecule has a barrel-like shape having a hollow core with a diameter of about 5.5 Å and two identical portals surrounded by carbonyl groups. The driving force for the crystallization process is the formation of several complementary hydrogen bonds between the C=O groups and the six aqua ligands located in *cis* positions with respect to the μ_3-capping chalcogen atom. This arrangement results in the formation of very stable hybrid organic–inorganic supramolecular compounds in which the portals of cucurbit[6]uril are efficiently closed by one or two cluster cations as represented in Fig. 7.7.

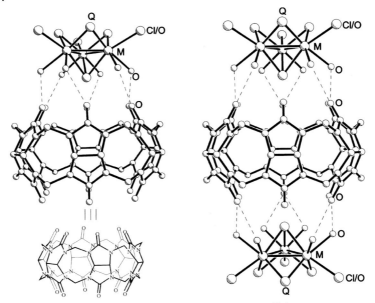

Fig. 7.7 General packing diagram of the $[M_3Q_4Cl_x(H_2O)_{9-x}]^{(4-x)+}$:cuc and $[M_3Q_4Cl_x(H_2O)_{9-x}]^{(4-x)+}$:cuc:$[M_3Q_4Cl_x(H_2O)_{9-x}]^{(4-x)+}$ supramolecular aggregates.

This supramolecular approach has been used as a way to facilitate the crystallization of cuboidal aqua ions and to capture single intermediates present in complex solution mixtures. For example, $[M_3Q_4Cl_x(H_2O)_{9-x}]^{4-x}$ (Q=S, Se) supramolecular adducts have been isolated for x =1–5 from aqueous HCl solutions by varying the acid concentrations where the aggregates contain in each case only one of all possible isomers [46]. In addition to the hydrogen bonds, other kinds of interactions, namely $Cl \cdots H_2O$, $Cl \cdots Cl$ and $Q \cdots Q$ contacts, are also involved in the network propagation. This supramolecular approach has also been efficiently employed for the crystallization of a large number of aqua complexes within the $[M_3M'Q_4]$ cubane-type family, where M' is a transition or post-transition metal [47].

This entry to supramolecular chemistry allows one, at the same time, to develop a host–guest chemistry because the space inside the cucurbituril barrel is sufficient to confine small "guest" molecules. This is illustrated with the crystallization of the supramolecular adduct $\{[W_3S_4(H_2O)_8Cl](pyH^+\subset cuc)\}Cl_4a \cdot 5.5\,H_2O$ with a pyridinium cation inside the cucurbituril cavity [48]. The introduction of guest molecules may vary the net charge on the assembly and consequently the whole packing in the solid state.

Acknowledgments

The authors express their sincere thanks to their co-workers whose contribution to the work presented in this chapter has been essential; their names are quoted in the references. Current work is funded by the "Ministerio de Ciencia y Tecnología" (Grant BQU2002-00313), "Generalitat Valenciana" (Grants GV04B-029, IIARCO/2004/161, IIARCO/2004/163) and "Fundació Bancaixa-UJI" (Grant P1 1B2004-19).

References

1 J. D. Corbett, *J. Chem. Soc., Dalton Trans.* (1996), 575.

2 J. D. Corbett, *Structural and Electronic Paradigms in Cluster Chemistry,* 87 (1997), 157.

3 J. D. Corbett, *Inorg. Chem.,* 39 (2000), 5178.

4 F. Rogel, J. D. Corbett, *J. Am. Chem. Soc.,* 112 (1990), 8198.

5 S. C. Lee, R. H. Holm, *Angew. Chem., Int. Ed. Engl.,* 29 (1990), 840.

6 A. Müller, R. Jostes, F. A. Cotton, *Angew. Chem. Int. Engl.,* 19 (1980), 875.

7 M. Feliz, R. Llusar, J. Andres, S. Berski, B. Silvi, *New J. Chem.,* 26 (2002), 844.

8 A. A. Opalowskii, V. Y. Fedorov, K. A. Khaldoyanidi, *Akad. Nauk. SSSR,* 182 (1968), 1095.

9 J. Marcoll, A. Rabenau, D. Mootz, H. Wunderlich, *Rev. Chim. Miner.,* 11 (1974), 607.

10 F. A. Cotton, P. A. Kibala, M. Matusz, C. S. McCaleb, R. B. W. Sandor, *Inorg. Chem.,* 28 (1989), 2623.

11 T. Saito, A. Yoshikawa, T. Yamagata, H. Imoto, K. Unoura, *Inorg. Chem.,* 28 (1989), 3588.

12 V. P. Fedin, M. N. Sokolov, O. A. Gerasko, B. A. Kolesov, V. Y. Fedorov, A. V. Mironov, D. S. Yufit, Y. L. Slovohotov, Y. T. Struchkov, *Inorg. Chim. Acta,* 175 (1990), 217.

13 V. P. Fedin, M. N. Sokolov, K. G. Myakishev, O. A. Gerasko, V. Y. Fedorov, *Polyhedron,* 10 (1991), 1311.

14 V. P. Fedin, M. N. Sokolov, V. Y. Fedorov, D. S. Yufit, Y. T. Struchkov, *Inorg. Chim. Acta,* 179 (1991), 35.

15 P. E. Rauch, F. J. DiSalvo, W. Zhou, D. Tang, P. P. Edwards, *J. Alloys Compd.,* 182 (1992), 253.

16 V. P. Fedin, H. Imoto, T. Saito, W. McFarlane, A. G. Sykes, *Inorg. Chem.,* 34 (1995), 5097.

17 V. P. Fedin, H. Imoto, T. Saito, *J. Chem. Soc. Chem. Commun.,* (1995), 1559.

18 *Cambridge Crystallographic Data Center CCDC,* 12 Union Road, Cambridge, CB2 1EZ UK.

19 V. P. Fedin, M. N. Sokolov, Y. V. Mironov, B. A. Kolesov, S. V. Tkachev, V. Y. Fedorov, *Inorg. Chim. Acta,* 167 (1990), 39.

20 V. P. Fedin, M. N. Sokolov, O. A. Gerasko, A. V. Virovets, N. V. Podberezskaya, V. Y. Fedorov, *Inorg. Chim. Acta,* 187 (1991), 81.

21 M. N. Sokolov, A. L. Gushchin, D. Y. Naumov, O. A. Gerasko, V. P. Fedin, *Inorg. Chem.,* 44 (2005), 2431.

22 M. N. Sokolov, P. A. Abramov, A. L. Gushchin, I. V. Kalinina, D. Y. Naumov, A. V. Virovets, E. V. Peresypkina, C. Vicent, R. Llusar, V. P. Fedin, *Inorg. Chem.,* 44 (2005), 8116.

23 M. Feliz, R. Llusar, S. Uriel, C. Vicent, M. G. Humphrey, N. T. Lucas, M. Samoc, B. Luther-Davies, *Inorg. Chim. Acta,* 349 (2003), 69.

24 V. P. Fedin, I. V. Kalinina, D. G. Samsonenko, Y. V. Mironov, M. N. Sokolov, S. V. Tkachev, A. V. Virovets, N. V. Podberezskaya, M. R. J. Elsegood, W. Clegg, A. G. Sykes, *Inorg. Chem.,* 38 (1999), 1956.

25 V. P. Fedin, A. G. Sykes, *Inorg. Synth.,* 33 (2002), 162.

26 T. Saito, N. Yamamoto, T. Yamagata, H. Imoto, *Chem. Lett.* (1987), 2025.

27 J. Mizutani, S. Yajima, H. Imoto, T. Saito, *Bull. Chem. Soc. Jpn.,* 71 (1998), 631.

28 F. Estevan, M. Feliz, R. Llusar, J. A. Mata, S. Uriel, *Polyhedron,* 20 (2001), 527.

29 R. Llusar, S. Uriel, C. Vicent, *J. Chem. Soc., Dalton Trans.* (2001), 2813.

30 M. Feliz, E. Guillamon, R. Llusar, S. E. Stiriba, J. Prieto, M. Barberis, C. Vicent, *Chem. Eur. J.,* 12 (2005), 1486.

31 R. Hernandez-Molina, M. N. Sokolov, A. G. Sykes, *Acc. Chem. Res.,* 34 (2001), 223.

32 R. Llusar, S. Uriel, *Eur. J. Inorg. Chem.* (2003), 1271.

33 R. Hernandez-Molina, A. G. Sykes, *J. Chem. Soc., Dalton Trans.* (1999), 3137.

34 Y. Ide, M. Sasaki, M. Maeyama, T. Shibahara, *Inorg. Chem.,* 43 (2004), 602.

35 M. G. Basallote, M. Feliz, M. J. Fernandez-Trujillo, R. Llusar, V. S. Safont, S. Uriel, *Chem. Eur. J.,* 10 (2004), 1463.

36 M. G. Basallote, F. Estevan, M. Feliz, M. J. Fernandez-Trujillo, D. A. Hoyos, R. Llusar, S. Uriel, C. Vicent, *Dalton Trans.* (2004), 530.

37 A. G. Algarra, M. G. Basallote, M. Feliz, M. J. Fernandez-Trujillo, R. Llusar, V. S. Safont, *Chem. Eur. J.,* 12 (2006), 1413.

38 J. M. Garriga, R. Llusar, S. Uriel, C. Vicent, A. J. Usher, N. T. Lucas, M. G. Humphrey, M. Samoc, *Dalton Trans.* (2003), 4546, and references therein.

39 R. Llusar, S. Triguero, C. Vicent, M. Sokolov, B. Domercq, M. Fourmigué, *Inorg. Chem.,* 44 (2005), 8937.

40 R. Llusar, S. Triguero, S. Uriel, C. Vicent, E. Coronado, C. J. Gomez-Garcia, *Inorg. Chem.,* 44 (2005), 1563.

41 A. Kobayashi, E. Fujiwara, H. Kobayashi, *Chem. Rev.,* 104 (2004), 5243.

42 R. Llusar, S. Uriel, C. Vicent, J. M. Clemente-Juan, E. Coronado, C. J. Gomez-Garcia, B. Braida, E. Canadell, *J. Am. Chem. Soc.* 126 (2004), 12076.

43 J. Lagona, P. Mukhopadhyay, S. Chakrabarti, L. Isaacs, *Angew. Chem., Int. Ed. Engl.,* 44 (2005), 4844.

44 M. N. Sokolov, A. V. Virovets, D. N. Dybtsev, O. Gerasko, V. P. Fedin, R. Hernandez-Molina, W. Clegg, A. G. Sykes, *Angew. Chem., Int. Ed. Engl.,* 112 (2000), 1725.

45 V. P. Fedin, *Russ. J. Coord. Chem.,* 30 (2004), 163.

46 M. N. Sokolov, O. A. Gerasko, D. N. Dybtsev, E. V. Chubarova, A. V. Virovets, C. Vicent, R. Llusar, D. Fenske, V. P. Fedin, *Eur. J. Inorg. Chem.* (2004), 63.

47 R. Hernandez-Molina, M. Sokolov, P. Esparza, C. Vicent, R. Llusar, *Dalton Trans.* (2004), 847.

48 D. N. Dybtsev, O. A. Gerasko, A. V. Virovets, M. N. Sokolov, V. P. Fedin, *Inorg. Chem. Commun.* 3 (2000), 345.

8

Current State on (B,C,N) Compounds of Calcium and Lanthanum

H.-Jürgen Meyer

8.1
Introduction

A remarkable variety of compounds in the Ca-(B,C,N) system has opened a window for research in related fields. With the elements boron, carbon and nitrogen, substance classes such as borocarbides, boronitrides, and carbonitrides can be considered to contain anionic derivatives of binary compounds B_4C, BN, and C_3N_4. Until now, most compounds in these substance classes have been considered to contain alkali, alkaline-earth, or lanthanide elements. Lanthanide borocarbides are known from the work of Bauer [1]. Lanthanide boronitrides represent a younger family of compounds, also assigned as nitridoborates [2] following the nomenclature of oxoborates.

The compounds that are being reported here are mostly those with combined anions, such as BN_x anions in nitridoborates. On the other hand, we know of compounds with mixed anions, like the well-known Ti(N,C) in which anions share the same lattice site, or NbBN with separate B and N sublattices. How can these compounds be explained in comparison with compounds having C–N or B–N bonding? Are there separate strategies for their synthesis, what are the differences in stability, what are their properties, and what are their structures like? These and some other questions will be discussed, mainly for Ca-(B,C,N) and La-B-N systems.

8.2
Problems and Pitfalls of some Calcium Compounds with (mixed) B,C,N Anions

In the past, there was quite a lot of confusion regarding the true nature of some alkaline-earth compounds. Several boride, nitride, and carbide compounds, as well as some mixed species, especially of calcium, were in doubt or at least subject to some ongoing discussion. For example, the phase relationships and structures of CaC_2 phases I–IV were not well understood. The three nitride

Inorganic Chemistry in Focus III.
Edited by G. Meyer, D. Naumann, L. Wesemann
Copyright © 2006 WILEY-VCH Verlag GmbH & Co. KGaA, Weinheim
ISBN: 3-527-31510-1

phases reported as (a-,β-,γ-) Ca_3N_2, and the unusual (electron-deficient) nitride $Ca_{11}N_8$ could not be confirmed experimentally. The reported calcium mono-chloride (CaCl) is considered not to exist – at least not as a solid. And the symmetries of $Ca_3(BN_2)_2$ and $Sr_3(BN_2)_2$ structures were in question.

Recently, some light was shed on these and other compounds as well as on their structures and properties. Today it may be considered that some of the mistakes made earlier could have been avoided by the availability of elements or compounds of higher purity, and thereby have prevented the incorporation of unexpected impurities. In this context it is important to note some experimental principles of solid-state synthesis that should be considered: In solid-state synthesis, great attention should be always dedicated to the *purity of the starting materials*, the *purity of the container material (and gas atmosphere)* used in the reaction, and the *control and adjustment of the reaction conditions* in order to finally obtain a *high yield* product.

A surprising variety of calcium compounds with combinations of mixed B, C or N anions is known, with some of them listed in Table 8.1. Their structures may contain anionic networks, or very often, triatomic anions with 16 valence electrons, and other anionic species. Some examples of Ca-(B,C,N) compounds are displayed in Fig. 8.1, and will be briefly discussed here.

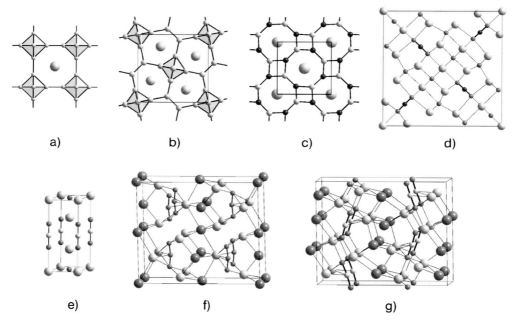

a)　　　　　b)　　　　　c)　　　　　d)

e)　　　　　f)　　　　　g)

Fig. 8.1 Crystal structures of some calcium-(B,C,N) compounds: (a) CaB_6 (LaB_6 type); (b) $CaB_{4-x}C_x$ (LaB_4 type); (c) CaB_2C_2 (corresponding with LaB_2C_2); (d) $Ca_{11}(CN_2)_2N_6$; (e) $Ca_3(BN_2)N$; (f) $Ca_3Cl_2C_3$; (g) $Ca_3Cl_2(CBN)$.

Table 8.1 Some Ca-(B,C,N) and Ca-Cl-(B,C,N) compounds.

	Ca	Ca, CaCl$_2$
B	CaB$_6$	
C	CaC$_2$ (I–IV)	Ca$_3$Cl$_2$C$_3$
N	Ca$_2$N	
	Ca$_3$N$_2$	Ca$_2$ClN
B,C	CaB$_2$C$_2$	Ca$_9$Cl$_8$(BC$_2$)$_2$
	CaB$_{4-x}$C$_x$	Ca$_5$Cl$_3$(BC$_2$)C$_2$
B,N	Ca$_3$(BN$_2$)N	Ca$_2$Cl(BN$_2$)
	Ca$_3$(BN$_2$)$_2$ ($\alpha+\beta$)	
C,N	Ca(CN$_2$)	
	Ca$_4$(CN$_2$)N$_2$	
	Ca$_{11}$(CN$_2$)$_2$N$_6$	
B,C,N	Ca$_{10}$(CBN)$_4$(C$_2$)$_2$	Ca$_3$Cl$_2$(CBN)

8.2.1
Borides of Calcium and Lanthanum

A new carbon-stabilized calcium tetraborate was recently found as CaB$_{4-x}$C$_x$ (with x $\approx$ 0,3), isostructural with the well-known (tetragonal) lanthanide tetraborides LnB$_4$. In spite of the unsuccessful isolation of a pure calcium tetraboride, it is interesting to note an isotypic relation between carbon-stabilized CaB$_4$ and LaB$_4$, which is also known for the couples CaB$_6$/LaB$_6$, or CaC$_2$/LaC$_2$. However, compounds in each single couple clearly exhibit distinct electronic properties, as can be derived from the respective band structure. Alkaline-earth hexaborides like CaB$_6$ with Ca^{2+} are small gap semiconductors, and LaB$_6$ compounds with La^{3+} are metal-like, as a result of one extra electron (per formula unit) being present in the conduction band.

The reported ferromagnetism in Ca$_{1-x}$La$_x$B$_6$ has attracted great interest, since the apparent magnetism was obviously induced by a compound that is composed of non-magnetic elements [3]. Systematic investigations with doped and non-doped samples have, however, not always revealed ferromagnetism. In addition, ferromagnetic Ca$_{1-x}$La$_x$B$_6$ crystals have shown a *wash away magnetism*, meaning that, after washing the crystals carefully with HCl, no ferromagnetism was retained. It was therefore assumed that the obtained ferromagnetism arose from iron impurities on the surface of crystals. But the magnetic transition temperature was not consistent with that of iron. A possible explanation could be that the magnetic transition temperature of iron particles was altered through partial bonding with the surfaces of the crystals. A similar contamination was made responsible for the reported ferromagnetism [4] obtained for CaB$_2$C$_2$ [5] that is, in fact, diamagnetic [6].

It can therefore be concluded that the unexpected ferromagnetism obtained independently for CaB$_2$C$_2$ and for lanthanum-doped CaB$_6$ does not reflect an intrinsic property of these compounds [7].

8.2.2
The CaC$_2$ Problem and Ca$_3$Cl$_2$C$_3$

Calcium carbide is probably thought of as one of the best known compounds – but is it really? Its application in the carbide lamp in households or its technical application in the process of generating acetylene is legend, but the so-called technical CaC$_2$ is still the same: a mixture of compounds and phases. Four phases are known for CaC$_2$, of which up to three can coexist at room temperature. These phases are conventionally assigned by roman numerals as tetragonal CaC$_2$-I, the two monoclinic phases CaC$_2$-II and -III, and the cubic high-temperature phase CaC$_2$-IV. Their structures can all be derived from the distorted NaCl-type structure with different alignments of C$_2$ units (Fig. 8.2).

For example, the difference between the monoclinic CaC$_2$ phases is the presence of only one type of acetylide ion in phase II and two distinct C$_2$ species in phase III, as concluded from their crystal structures and from ^{13}C-NMR studies (Fig. 8.3) [8]. The transformation between phases II and III is induced by heating phase II above 150 °C until the metastable phase, III, is formed. Phase III remains stable even when being cooled down to room temperature. However, when the metastable phase III is ground in a mortar at room temperature, it transforms back into phase II.

$$CaC_2\text{-II} \underset{\text{grinding}}{\overset{>150°C}{\rightleftarrows}} CaC_2\text{-III}$$

The phase relationship between the most commonly known tetragonal phase CaC$_2$-I and phase(s) II (or III) is still in question. CaC$_2$-I transforms reversibly into the cubic phase IV, as does the phase (II →) III. It is interesting to note that the cubic phase IV seems to have a memory for its respective precursor phase (III or I) that is regained after cooling down phase IV. The synthesis of

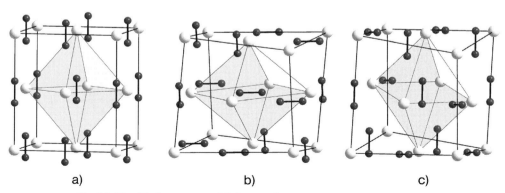

a)   b)   c)

Fig. 8.2 Simplified structures of CaC$_2$ I–III (from (a) to (c)), displayed as derivatives of the NaCl type structure, neglecting the relation of their true crystallographic unit cells.

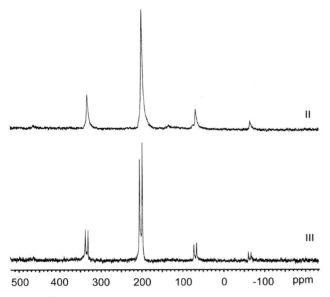

Fig. 8.3 ^{13}C-NMR spectra of CaC_2-II and CaC_2-III.

phase I or phase II was found to be influenced by the relative amounts of calcium and graphite used in reactions. The reason for this and the influence of lattice defects on the relative stabilities of the different CaC_2 phases are not yet well understood.

With the presence of $CaCl_2$, another carbide is formed as $Ca_3Cl_2C_3$ [9]. Its crystal structure contains an allene-like $[C=C=C]^{4-}$ unit – the first one being refined by single crystal X-ray diffraction. $Ca_3Cl_2C_3$ occurs as a deep red-colored crystalline solid from the reactions of Ca, $CaCl_2$, and graphite (1) above 800 °C (in Ta tubes). Its synthesis was the result of systematic attempts to reproduce results on the so-called calcium monochloride. This compound was reported a long time ago as a product from solid state reactions of (carbon contaminated) Ca and $CaCl_2$, or as a byproduct in the electrochemical reduction of calcium on graphite electrodes from a $CaCl_2$ flux.

$$2\,Ca + CaCl_2 + 3\,C \rightarrow Ca_3Cl_2C_3 \tag{1}$$

The $[C=C=C]^{4-}$ unit in $Ca_3Cl_2C_3$ is slightly bent with C–C distances of 134 pm. This less-common anion is also known from structures of Mg_2C_3 as well as Sc_3C_4, and has been studied by spectroscopic means (IR, RA), ^{13}C-NMR, and by MO and band-structure calculations [10]. The ^{13}C-NMR of $[C=C=C]^{4-}$ in $Ca_3Cl_2C_3$ exhibits signals at 173 ppm and 122 ppm in a 1:2 ratio. $Ca_3Cl_2C_3$ is closely related with $Ca_3Cl_2(CBN)$ [11], which was the first compound compromising a $[C=B=N]^{4-}$ ion.

8.2.3
Calcium Nitride and Calcium Carbodiimides

Out of three phases reported for Ca_3N_2 [12], to date only the cubic (a) nitride phase is confirmed, crystallizing in the anti-bixbyite type structure. In addition, the carbodiimide nitride $Ca_4(CN_2)N_2$ [13] may hold for another unconfirmed calcium nitride phase, and $Ca_{11}(CN_2)_2N_6$ stands for the ill-defined $Ca_{11}N_8$ [14]. Here we note again that carbon impurities may produce significant difficulties in reactions.

8.2.4
Calcium Nitridoborates

$Ca_3(BN_2)_2$ is readily formed when (distilled) calcium metal is melted in the presence of (layer-type) boron nitride. This reaction provides some insight on how alkaline-earth metals like calcium may act as a *catalyst* in the phase transformation of layered a-BN into its cubic modification. Instead of metals, nowadays alkaline-earth (Ca, Sr, Ba) nitridoborates can be used as a flux catalyst in high-pressure and high-temperature transformation reactions to produce cubic boron nitride [15].

$$10\,Ca + 12\,BN \rightarrow 3\,Ca_3(BN_2)_2 + CaB_6 \qquad (2)$$

Since a separation process for $Ca_3(BN_2)_2$ and CaB_6 is unknown, Ca_3N_2 is better employed as a starting material for the synthesis of calcium nitridoborate nitride [16], and calcium nitridoborate:

$$Ca_3N_2 + BN \rightarrow Ca_3(BN_2)N \qquad (3)$$

$$Ca_3N_2 + 2\,BN \rightarrow Ca_3(BN_2)_2 \qquad (4)$$

Reactions suggest that $Ca_3(BN_2)N$ can be viewed as an intermediate in the reaction (4). The substitutional effect of nitride by nitridoborate ions reflects that the linear $[N=B=N]^{3-}$ ion (with a typical B–N distance around 135 pm) can be viewed as a pseudo-nitride.

8.2.5
A Comparison of $Ca_3(BN_2)_2$ and $Sr_3(BN_2)_2$ Structures

A long time passed between the first discovery of $AE_3(BN_2)_2$ phases [17] (AE = Ca, Sr) and their structural characterizations [18, 19]. Two distinct phases are known to exist for both calcium and strontium nitridoborate, denoted as low-temperature β-$AE_3(BN_2)_2$ and high-temperature a-$AE_3(BN_2)_2$ [20]. Their phase transitions have been studied by temperature-dependent XRD, thermoanalysis, space group theory, and their crystal structures have been established [21]. The high-temperature phases crystallize cubic ($Im\overline{3}m$) containing three formula units $AE_3(BN_2)_2$ per unit cell. In this space group the two special cation positions $2a$ (0, 0, 0; 1/2, 1/2, 1/2) and $8f$ (1/4, 1/4, 1/4; 3/4, 1/4, 1/4; ...), are occupied with a total of nine out of ten possible AE sites of the nominal $AE_2[AE_8](BN_2)_6$. Interestingly, for AE = Ca and Sr, two distinct patterns of cation occupations are obtained, most likely as a result of their different ionic radii. In a-$Ca_3(BN_2)_2$ calcium ions occupy 7/8 of $8f$ positions, corresponding to $Ca_2[Ca_7](BN_2)_6$; and in a-$Sr_3(BN_2)_2$ the strontium ions occupy 1/2 of $2a$ positions, corresponding to $Sr_1[Sr_8](BN_2)_6$ (Fig. 8.4).

Following the high-temperature syntheses used for these compounds according to the reaction (4), a phase transition into the low-temperature modification may occur on slow cooling. The high-temperature phase, on the other hand, may be frozen out at room temperature through fast quenching (Fig. 8.5).

The low-temperature (β-)$AE_3(BN_2)_2$ phases exhibit two distinct structures for AE = Ca and Sr that can be derived from the cation disordering in their respective high-temperature phases. For β-$Ca_3(BN_2)_2$ an orthorhombic (Cmca) superstructure of the cubic cell with $a_o \approx \sqrt{2} \cdot a_c$, $b_o \approx a_c$, $c_o \approx 2\sqrt{2} \cdot a_c$ was obtained, in which the former $8f$ sites are occupied by seven calcium ions in an ordered fashion. In contrast, the structure of β-$Sr_3(BN_2)_2$ is simply the result of a transition from a cubic body-centered ($Im\overline{3}m$) into a primitive structure ($Pm\overline{3}m$), in which the former $2a$ position (0, 0, 0; 1/2, 1/2, 1/2) is split into two independent positions, of which only one is occupied by strontium (Fig. 8.6).

The observation of phase transitions may be prominent for structures containing stick-packings resulting from diatomic or triatomic units. Well known

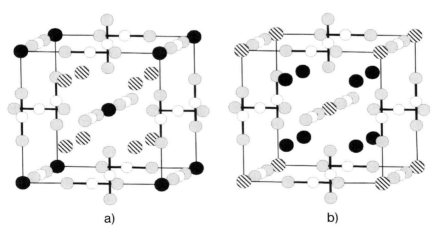

a)   b)

Fig. 8.4 Crystal structures of a-$Ca_3(BN_2)_2$ (a) and a-$Sr_3(BN_2)_2$ (b) high-temperature phases. Striped atoms indicate partial occupations on special positions $8f$ (occupied by 7/8) and $2a$ (occupied by 1/2).

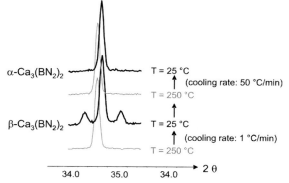

Fig. 8.5 Sections of (*in situ*) X-ray powder patterns of α-Ca$_3$(BN$_2$)$_2$ with the 220 reflection and β-Ca$_3$(BN$_2$)$_2$ with 200, 224, and 400 reflections. (Low-temperature) β-Ca$_3$(BN$_2$)$_2$ is obtained on slow cooling (1°C/min) from 250°C. Heating to 250°C again produces (high-temperature) α-Ca$_3$(BN$_2$)$_2$ that can be frozen out at room temperature on fast cooling (50°C/min).

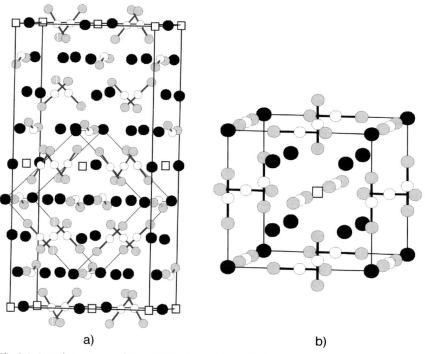

a)   b)

Fig. 8.6 Crystal structures of β-Ca$_3$(BN$_2$)$_2$ (a) and β-Sr$_3$(BN$_2$)$_2$ (b). Squares indicate cation vacancies. The related cubic cell of α-Ca$_3$(BN$_2$)$_2$ is highlighted in the structure of β-Ca$_3$(BN$_2$)$_2$.

examples are CaC_2, $Li_3(BN_2)$, $EA_3(BN_2)_2$, and others. For $(\beta\text{-})Eu_3(BN_2)_2$ only one phase was reported consistent with the primitive ($Pm\bar{3}m$) structure. Due to the similarities of ionic radii for Eu^{2+} and Sr^{2+}, a corresponding high-temperature phase should exist as $\alpha\text{-}Eu_3(BN_2)_2$. For $Sr_3(BN_2)_2$ the transition temperature is, however, quite high (around 820 °C).

8.3
Metal-nitridoborates

As nitridoborates are known as $A_3(BN_2)$ for alkaline elements (A=Li, Na) and as $AE_3(BN_2)_2$ for alkaline-earth elements (AE=Ca, Sr, Ba) it would be interesting to find methods for the synthesis of nitridoborates of transition or lanthanide elements. Can they be made straightforward like $Li_3(BN_2)$ and $AE_3(BN_2)_2$ from metal nitrides and layer-like $\alpha\text{-}BN$, or do they require new preparative strategies – if they can be made at all?

Mixed compounds, e.g., in the M-B-N system, where M is a transition metal, are well known with separate B and N sublattices or atoms. Prominent examples are NbBN [22] containing kinked B-chains and isolated nitride ions; or Ti(N,C) with a NaCl-like structure containing disordered C and N ions sharing equivalent positions.

8.3.1
Electronic Considerations

A simple electronic picture may help to understand the differences in bonding of nitridoborate ions with different metal atoms. By comparing the relative energies of (B–N) bonding levels of BN_x anions and (H_{ii}) energies of the metals valence electrons, we may distinguish interactions with alkaline-earth (AE), lanthanides (Ln), or 3d-elements (M), respectively. The discrete energy levels of metal atoms are unperturbed and broaden in energy if the various bonding and anti-bonding interactions are taken into consideration. A comparison of the energy blocks of B–N bonding states of BN_x with alkaline-earth, lanthanides and 3d-metals is shown in Fig. 8.7. We note that the valence orbitals of alkaline-earth and lanthanide elements are well above the energy of bonding BN_x levels, such as to allow salt-like interactions between metal atoms such as Ca or La and (BN_x). However, lanthanide nitridoborates with a smaller band gap can occur as metal-like compounds when excess electrons are present. Coming to the 3d elements, their interactions with BN_x seem rather difficult to establish. The conditions indicated here remind us of the electronic conditions of Zintl phases or metal borides, with a small band gap or a zero band gap, respectively. If we consider more electronegative 3d elements like nickel, shown in Fig. 8.7, their 3d energy states tend to fall as low in energy as the upper B–N bonding states of BN_x, suggesting covalent interactions between M and BN_x, or even a reversed charge transfer, as described for anionic gold in the anti-perovskite structured $(Ca^{2+})_3Au^{(-)}N^{(3-)}(e^-)_2$ [23].

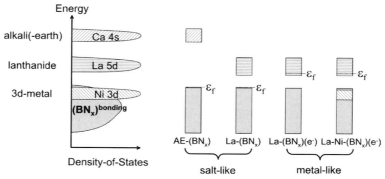

Fig. 8.7 Superposition of density-of-states for B–N bonding BN_x states with corresponding metal states, reflected by their valence states from alkaline-earth (as Ca), lanthanide (as La), and 3d-metal (as Ni), and corresponding block schemes for salt-like Ca-(BN_x), salt-like La-(BN_x), metal-like La-$(BN_x)(e^-)$, and metal-like La-Ni-$(BN_x)(e^-)$. The region of BN_x bonding levels is shown in grey. The (e^-) denotes an electron-rich situation; and ε_f indicates the fermi level.

Indeed, solid state reactions between some transition-metal halides and $Li_3(BN_2)$ have already demonstrated the decomposition of $(BN_2)^{3-}$ ions into boron nitride. The remaining nitride can lead to the formation of a binary metal nitride or reduce the transition metal ion under the formation of N_2. Both mechanisms have been obtained experimentally, depending on the stability of the metal nitride. For instance niobium pentachloride forms NbN, titanium trichloride forms TiN, and nickel dichloride forms Ni, plus BN and nitrogen, respectively, in reactions with $Li_3(BN)_2$ (at 300–600 °C) [24].

$$M^{3+} + (BN_2)^{3-} \rightarrow MN + BN \tag{5}$$

$$M^{3+} + (BN_2)^{3-} \rightarrow M + BN + 1/2\,N_2 \tag{6}$$

We note that the valence orbitals of metal atoms order in energy as AE > Ln > M. The d-levels of transition elements (M) range the lowest, and are therefore most sensitive for reduction, or to form a stable binary metal nitride. This may also explain the virtual absence of d-element compounds with 16 (valence) electron species, such as $[N=N=N]^-$, $[N=C=N]^{2-}$, $[N=B=N]^{3-}$, $[C=C=C]^{4-}$, or $[C=B=C]^{5-}$, at least through high-temperature syntheses.

The same reasoning may hold for the virtual absence of transition metal dicarbides with $(C_2)^{n-}$ anions. On the other hand, stable multinary compounds in A-M-(BN_x), AE-M-(BN_x), or Ln-M-(BN_x) systems can exist with an electron-rich transition metal (M), similar to the known ternary dicarbides $AM(C_2)$ with A = alkali and M = Pd, Pt [25]. In these cases the valence electrons provided from A or AE metals fill the bonding $(BN_x)^{n-}$ or $(C_2)^{m-}$ levels, and the transition metal retains an approximate d^{10} configuration.

8.4
Lanthanum Nitridoborates

Nitridoborates of lanthanum and the lanthanides were obtained from reactions of lanthanide metal or lanthanide metal nitride with layer-like (a-)BN at elevated temperatures ($\gg$1200 °C). These reactions require elaborated techniques in the inert gas sample-handling and the use of efficient heating sources, such as induction heating. Only some compounds remain stable in this high-temperature segment, and the yields of such reactions are often limited due to the competing stability of binary phases, allowing only the most (thermodynamically) stable compounds to exist.

The chemistry of lanthanide nitridoborates was developed by the more flexible and more efficient solid-state metathesis route by using nitridoborate salts and a lanthanide trichloride, at reaction temperatures as low as 600 °C. This type of reaction has been previously established and studied in some detail for reactions, such as the synthesis of lanthanide nitrides (LnN). Following this concept, lanthanum nitridoborates are obtained from reactions of lanthanum trichloride and lithium nitridoborate (or calcium nitridoborate), performed in a salt-balanced manner with respect to the formation of the co-produced LiCl:

$$3\,LaCl_3 + 3\,Li_3(BN_2) \rightarrow La_3(B_3N_6) + 9\,LiCl \tag{7}$$

In this exothermic reaction (Fig. 8.8) the $[BN_2]^{3-}$ ions undergo a trimerization to form a cyclic $[B_3N_6]^{9-}$ ion with three exocyclic nitrogen atoms (Fig. 8.9). Afterwards $La_3(B_3N_6)$ is washed with water in order to remove LiCl.

For the generation of more nitrogen-rich compounds, a nitride is included in the reaction. This synthesis allows the formation of different nitridoborate (8) ions as well as nitridoborate nitrides (9) [26]:

$$5\,LaCl_3 + 4\,Li_3(BN_2) + Li_3N \rightarrow La_5(B_3N_6)(BN_3) + 15\,LiCl \tag{8}$$

$$6\,LaCl_3 + 4\,Li_3(BN_2) + 2\,Li_3N \rightarrow La_6(B_3N_6)(BN_3)N + 18\,LiCl \tag{9}$$

Metal-rich compounds are readily obtained when a metallothermic reduction is included into the metathesis reaction by using an electropositive metal (10). In addition, metal-rich and nitrogen-rich compounds are obtained when a metal and a metal nitride are employed in reactions (11, 12):

$$3\,LaCl_3 + 2\,Li_3(BN_2) + 3\,Li \rightarrow La_3(B_2N_4) + 9\,LiCl \tag{10}$$

$$4\,LaCl_3 + 2\,Li_3(BN_2) + 3\,Li + Li_3N \rightarrow La_4(B_2N_4)N + 12\,LiCl \tag{11}$$

$$5\,LaCl_3 + 2\,Li_3(BN_2) + 3\,Li + 2\,Li_3N \rightarrow La_5(B_2N_4)N_2 + 15\,LiCl \tag{12}$$

Previously, all these variations led to the discovery of lanthanide nitridoborates. Products were obtained as powders or single crystals, and crystal structures

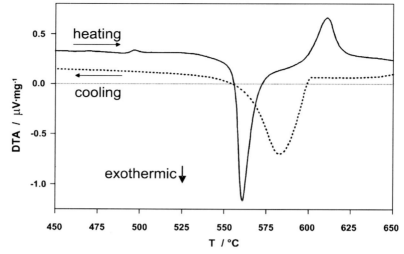

Fig. 8.8 Thermal effects on the formation of La$_3$(B$_3$N$_6$) by metathesis reaction from equimolar amounts of LaCl$_3$ and Li$_3$(BN$_2$). The strong exothermic effect in the heating curve (solid line) between 550 and 575 °C indicates the product formation. The following endothermic effect above 600 °C is due to the melting of LiCl. The exothermic effect on cooling (dotted line) represents the crystallization of LiCl.

were solved by X-ray diffraction techniques. A collection of known nitridoborates of calcium and lanthanum is listed here.

Compounds in Ca-B-N and La-B-N systems

a-Ca$_3$(BN$_2$)$_2$
β-Ca$_3$(BN$_2$)$_2$
Ca$_3$(BN$_2$)N
La$_3$(B$_3$N$_6$)
La$_5$(B$_3$N$_6$)(BN$_3$)
La$_6$(B$_3$N$_6$)(BN$_3$)N
La$_3$(BN$_3$)N
La$_{15}$(BN$_3$)$_8$N
La$_3$(BN$_3$)O$_6$

La$_3$(B$_2$N$_4$)
La$_4$(B$_2$N$_4$)N
La$_5$(B$_2$N$_4$)N$_2$

CaNi(BN)
CaPd(BN)
LaNi(BN)
La$_3$Ni$_2$(BN)$_2$N

8.4.1
Nitridoborate Ions

The prominent features of nitridoborate ions are their close relationships to other known anions, and their unique environments in structures with lanthanide ions. Nitridoborate anions involve the examples $[BN]^{n-}$, $[BN_2]^{3-}$, $[B_2N_4]^{8-}$, $[B_3N_6]^{9-}$, and $[BN_3]^{6-}$ (see Fig. 8.9). Their geometries and electronic structures can be easily derived from related anions and by the octet rule. Three electronic configurations can be considered for the $[BN]^{n-}$ ion, isoelectronic with the $[C_2]^{n-}$ ion with $n = 2$, 4, 6. Due to the rather covalent interactions of $[BN]^{n-}$ in compounds, no clear charge can be assigned to this anion. Considering the B–N bond distance of $[BN]^{n-}$ at 138–143 pm, a double bond slightly longer than in $[NBN]^{3-}$ at 132–137 pm may be addressed. The $[NBN]^{3-}$ ion may be well compared with the carbodiimide $[N=C=N]^{2-}$ or the cyanamide $[N-C\equiv N]^{2-}$ anion. So far, known examples of the dinitridoborate ions are mostly symmetrical as $[N=B=N]^{3-}$, like the carbodiimide ion. A reductive dimerization of two $[NBN]^{3-}$ ions via boron atoms (at B–B distances of 177–182 pm) yields $[N_2B-BN_2]^{8-}$, which is isoelectronic with the oxalate ion, with idealized D_{2h} symmetry. The B–N distances in $[B_2N_4]^{8-}$ are typically at 147–150 pm. The trimerization of $[BN_2]^{3-}$ yields the cyclic $[B_3N_6]^{9-}$ ion with three exocyclic B–N bonds. Structures are known with planar (D_{3h}) and chair-like (C_{2v}) conformations of $[B_3N_6]^{9-}$, with typical B–N distances at 144–151 pm. The trinitridoborate ion $[BN_3]^{6-}$ is isoelectronic with the carbonate ion, showing typical B–N bond lengths of 145–149 pm and idealized D_{3h} symmetry.

In this younger field of chemistry, some examples of anions may be still missing. One possible candidate is the $[BN_4]^{9-}$ ion with a tetrahedral coordination of boron as in cubic BN, others may be adopted from oxoborates. Another interesting feature is the existence of structures with condensed nitridoborate anions derived from portions of BN structures. Until now the only example with a condensed anion structure is U(BN), containing kinked B–B bonded chains of $(BN)_x$ with B–B distances near 188 pm, slightly longer than in $[B_2N_4]^{8-}$.

Nitridoborate anions are surrounded by metal atoms in a typical pattern that allows the prediction of new structures, to some degree. In any case, this typical pattern helps one to understand the construction scheme of structures and to memorize the particular structures.

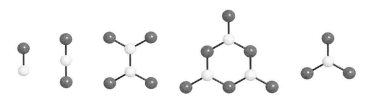

Fig. 8.9 Structures of [BN], [BN$_2$], [B$_2$N$_4$], [B$_3$N$_6$], and [BN$_3$] anions.

8.4.2
Structures of Lanthanum Nitridoborates

Structures of the lanthanide nitridoborates appear as layered structures with approximate hexagonal arrangements of metal atoms, and typical coordination preferences of anions. As in many metal nitrides, the nitride ion prefers an octahedral environment such as in lanthanum nitride (LaN). As a terminal constituent of a BN_x anion, the nitrogen atom prefers a six-fold environment, such as B-N $\cdots$ Ln_5, where Ln atoms form a square pyramid around N. Boron is typically surrounded by a trigonal prismatic arrangement of lanthanide atoms, as in many metal borides (Fig. 8.10). All known structures of lanthanide nitridoborates compromise these coordination patterns.

Lanthanide nitridoborates can be divided into three classes: salt-like compounds, semiconductors, and conductors or superconductors, as already shown in Fig. 8.7. Salt-like structures are usually transparent materials, marked by the typical color of the lanthanide ion. Here we discuss only nitridoborate compounds of lanthanum. The compounds $La_3(B_3N_6)$ [27], $La_5(B_3N_6)(BN_3)$ [28], $La_6(B_3N_6)(BN_3)N$ [29], and $La_3(BN_3)N$ all count as salt-like materials, with La^{3+}, $[B_3N_6]^{9-}$, $[BN_3]^{6-}$ and N^{3-} (Fig. 8.11). Band-structure calculations performed for $La_3(B_3N_6)$ revealed a band gap in the order of 4 eV. The corresponding nitridoborate oxide $La_6(BN_3)O_6$ [30] is also salt-like, owing the typical nitridoborate structure pattern regarding the environment of the $[BN_3]^{6-}$ ion with lanthanum

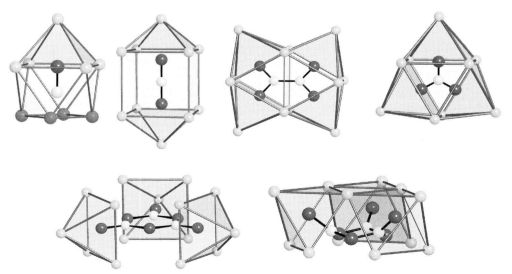

Fig. 8.10 Metal surroundings of [BN], [BN₂], [B₂N₄], and [BN₃] anions, with the two distinct [B₃N₆] configurations shown below.

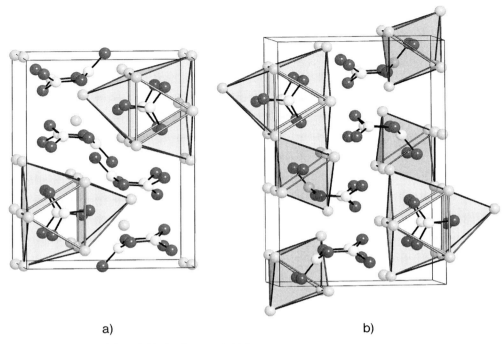

a)   b)

Fig. 8.11 Structures of the salt-like lanthanum nitridoborates (a) $La_5(B_3N_6)(BN_3)$ and (b) $La_6(B_3N_6)(BN_3)N$. Surroundings of $[BN_3]$ and N anions with lanthanum atoms are highlighted.

ions. The reported $La_{15}(BN_3)_8N$ [31] cannot be explained with the ionic charges, and its identity seems to be still in question.

Three different semiconducting lanthanum nitridoborates are known as $La_3(B_2N_4)$, $La_4(B_2N_4)N$ [32], and $La_5(B_2N_4)N_2$ [33] with the oxalate-like $[B_2N_4]^{8-}$ ions, that occur as dark, almost black, solids. Their structures are closely related to each other, being constructed of stacks of $[B_2N_4]^{8-}$ ions as shown for $La_3(B_2N_4)$ in Fig. 8.12, and nitride ions.

Their electronic structures are similar and could be satisfied by the fully occupied B–B bonding energy band shown for $La_3(B_2N_4)$ near –11 eV. However, one additional electron is present per $La_3(B_2N_4)$ formula unit. The alignment of $[B_2N_4]^{8-}$ ions allows B–B p_x-interactions between adjacent $[B_2N_4]$ anions along the short a-axis stacking direction (Fig. 8.12). The corresponding energy band contains one extra electron yielding a semi-metal-like band structure situation of $(La^{3+})_3(B_2N_4)^{8-}(e^-)$ [34]. As in a one-dimensional chain of H-atoms this band is half-filled by one electron. But a Peierls distortion is not likely to occur in this case.

Lanthanide compounds with $[BN]^{n-}$ ions can be considered as metal-like. In spite of our speculations regarding a hypothetical Ca(BN), the $[BN]^{n-}$ ions are known only from ternary compounds until now. Prominent examples are CaNi(BN)

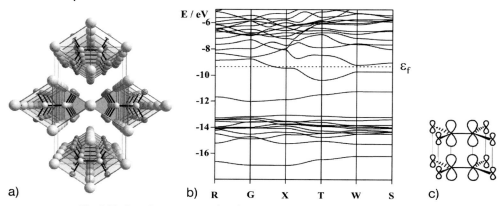

a) b) R G X T W S c)

Fig. 8.12 Crystal structure (a), band structure of La$_3$(B$_2$N$_4$) (b), and orbital interactions along [B$_2$N$_4$] stacks (c) (interactions with lanthanum orbitals are omitted for clarity).

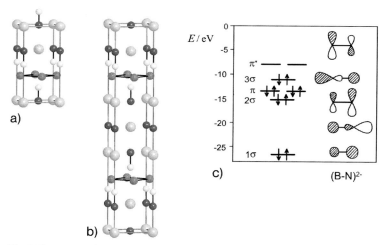

a)

b)

c)

(B-N)$^{2-}$

Fig. 8.13 Tetragonal crystal structures of CaNi(BN) or LaNi(BN) (a) and La$_3$Ni$_2$(BN)$_2$N (b), and MO scheme of [BN]$^{2-}$ (c). Nickel atoms (dark grey) are arranged in plane layers.

[35], LaNi(BN), and La$_3$Ni$_2$(BN)$_2$N [36] (Fig. 8.13). CaNi(BN) forms crystals with a metallic cluster, and its crystal structure is isotypic with that of LaNi(BN).

Calculated band structures of all these compounds feature the fermi level above a density-of-state peak that is consistent with the 3d bands for nickel. The [BN]$^{n-}$ anion in CaNi(BN) compromises an electronic situation with a filled 3σ (HOMO) level that is B–N anti-bonding (Fig. 8.13). Any additional electron will

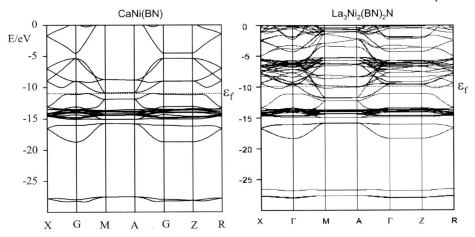

Fig. 8.14 Comparison of band structures of CaNi(BN) and $La_3Ni_2(BN)_2N$.

occupy anti-bonding π^* orbitals. The bonding situation of the $[BN]^{n-}$ anion may be well compared with that of a $[CC]^{n-}$ anion. Considering the tetrahedral environment of boron (of $[BN]^{n-}$) around nickel, this fragment reminds us of the structure of the $Ni(CO)_4$ molecule. The calculated band structure reveals a small indirect band gap for CaNi(BN) shown in Fig. 8.14. A very similar band structure, however, with a zero band gap is obtained for LaNi(BN), due to an additional electron in anti-bonding π^* orbitals of $[BN]^{n-}$.

The crystal structure of superconducting $La_3Ni_2(BN)_2N$ ($T_c = 14$ K) contains two blocks of LaNi(BN) alternating with one LaN layer along the tetragonal c-axis (Fig. 8.13). The band structure of $La_3Ni_2(BN)_2N$ reveals steep bands at the fermi level (ε_f) along certain directions that indicate high electronic mobilities (Fig. 8.14).

Based on the results of our band-structure calculations we assume that the metal-like properties of lanthanum nitridoborates are related by B–B interactions between adjacent BN_x units in structures.

8.5
Outlook

Mixed B-C-N compounds of lanthanum may be subdivided into La-(BN_x), La-(BC_x), and La-(CN_x) compounds (Fig. 8.15). The chemistry of lanthanide nitridoborates has been developed in some detail and some properties were studied. But still more work is necessary, especially in the field of quaternary Ln-metal-(BN_x) compounds.

Borocarbides of the lanthanides are also well-known with the examples LaB_2C_2 [37], $La_5B_2C_6$ [38], $La_5B_4C_{5-x}$ [39], $La_{10}B_9C_{12}$ [40], $La_{15}B_{14}C_{19}$ [41], and the interesting $LnM(BC_2)$ [42] compounds amongst several others.

Fig. 8.15 Concentration simplex of the La-B-C-N system. Combinations include one quaternary system, four ternary systems, and six binary systems.

In contrast, less is known about La-(CN_x) compounds. The composition $La_2(CN_2)_3$ was reported many years ago [43], without any structural information. Solid-state metathesis reactions of lanthanum chloride with $Li_2(CN_2)$ or $Zn(CN_2)$ have recently brought up three series of the lanthanide compounds $Ln_2(CN_2)_3$ [44], $LnCl(CN_2)$ [45], and $Ln_2Cl(CN_2)N$ [46]. Syntheses routes for Ln-(CN_x) compounds containing new anions such as $[C_2N_4]^{4-}$ are to be developed, as well as for compounds in the La-B-C-N system (Fig. 8.15).

As a defined ternary B-C-N compound is not known, other starting materials are to be considered in order to synthesize one of the, yet unknown, La-B-C-N compounds. Solution chemistry routes were successfully performed with tetra-cyanoborates to synthesize the compounds $A[B(CN)_4]$ with monovalent A=Li, Ag, Cu cations, and $M[B(CN)_4]_2$ with divalent M=Hg, Cu, Zn, Mn cations [47].

Acknowledgments

Many skillful and insightful coworkers have contributed to this research and their names are listed in the references. Their discoveries have created the progress and results which we have presented here. Continuous financial support for this research was provided by the Deutsche Forschungsgemeinschaft.

References

1 J. Bauer, J.-F. Halet, J.-Y. Saillard, *Coord. Chem. Rev.* **1998**, *178–180*, 723.

2 B. Blaschkowski, H. Jing, H.-J. Meyer, *Angew. Chem. Int. Ed.* **2002**, *41*, 3322.

3 D. P. Young, D. Hall, M. E. Torelli, Z. Fisk, J. L. Sarrao, J. D. Thompson, H. R. Ott, S. B. Oseroff, R. G. Goodrich, R. Zysler, *Nature* **1999**, *397*, 412.

4 J. Akimitsu, K. Takenawa, K. Suzuki, H. Harima, Y. Kuramoto, *Science* **2001**, *293*, 1125.

5 B. Albert, K. Schmitt, *Inorg. Chem.* **1999**, *38*, 6159; T. Breant, D. Pensec, J. Bauer, J. Debuigne, *Acta Crystallogr.* **1980**, *B36*, 1540.

6 T. Mori, S. Otani, *J. Phys. Soc. Japan* **2002**, *71*, 1789.

7 T. Mori, S. Otani, *Solid State Comm.* **2002**, *123*, 287.

8 M. Knapp, U. Ruschewitz, *Chem. Eur. J.* **2001**, *7*, 874; J. Glaser, S. Dill, M. Marzini, H. A. Mayer, H.-J. Meyer, *Z. Anorg. Allg. Chem.* **2000**, *627*, 1; O. Reckeweg, A. Baumann, H. A. Mayer J. Glaser, H.-J. Meyer, *Z. Anorg. Allg. Chem.* **1999**, *625*, 1686; V. Vohn, W. Kockelmann, U. Ruschewitz, *J. Alloys Compds.* **1999**, *284*, 132.

9 H.-J. Meyer, *Z. Anorg. Allg. Chem.* **1991**, *593*, 185.

10 R. Hoffmann, H.-J. Meyer, *Z. Anorg. Allg. Chem.* **1992**, *607*, 57.

11 H.-J. Meyer, *Z. Anorg. Allg. Chem.* **1991**, *594*, 113.

12 P. Y. Laurent, J. Lang, M.-T. Bihan, *Acta Cryst.* **1968**, *B24*, 494; Y. Laurent, J. David, J. Lang, *C.R. Acad. Sci. Paris* **1964**, *C259*, 1132.

13 O. Reckeweg, F. J. DiSalvo, *Angew. Chem. Int. Ed.* **2000**, *39*, 412.

14 Y. Laurent, J. Lang, M. T. L. Bihan, *Acta Cryst.* **1969**, *B25*, 199.

15 O. Fukunaga, S. Nakano, T. Taniguchi, *Diamond Related Mat.* **2004**, *13*, 1709; T. Endo, O. Fukunga, M. Iwata, *J. Mat. Sci.* **1981**, *16*, 2227.

16 M. Häberlen, J. Glaser, H.-J. Meyer, *Z. Anorg. Allg. Chem.* **2002**, *628*, 1959.

17 J. Goubeau, W. Anselment, *Z. Anorg. Allg. Chem.*, **1961**, *310*, 248.

18 H. Womelsdorf, H.-J. Meyer, *Z. Anorg. Allg. Chem.* **1994**, *620*, 265.

19 M. Wörle, H. Meyer zu Altenschilde, R. Nesper, *J. Alloys Compds.*, **1998**, *264*, 107.

20 M. Häberlen, J. Glaser, H.-J. Meyer, *J. Solid State Chem.* **2005**, *178*, 1478.

21 Only a powder XRD structure refinement is available for β-Ca$_3$(BN$_2$)$_2$.

22 P. Rogl, H. Klesnar, P. Fischer, *J. Am. Ceramic Soc.* **1988**, *71*, 450.

23 J. Jäger, D. Stahl, P. C. Schmidt, R. Kniep, *Angew. Chem. Int. Ed.* **1993**, *32*, 709.

24 K. Gibson, M. Ströbele, B. Blaschkowski, J. Glaser, M. Weisser, R. Srinivasan, H.-J. Kolb, H.-J. Meyer, *Z. Anorg. Allg. Chem.* **2003**, *629*, 1863.

25 U. Ruschewitz, *Z. Anorg. Allg. Chem.* **2001**, *627*, 1231; M. Weiss, U. Ruschewitz, *Z. Anorg. Allg. Chem.* **1997**, *523*, 1208.

26 H. Jing, B. Blaschkowski, H.-J. Meyer, *Z. Anorg. Allg. Chem.* **2002**, *628*, 1955.

27 O. Reckeweg, H.-J. Meyer, *Angew. Chem. Int. Ed.* **1999**, *38*, 1607.

28 O. Reckeweg, H.-J. Meyer, *Z. Anorg. Allg. Chem.* **1999**, *625*, 866.

29 H. Jing, J. Pickardt, H.-J. Meyer, *Z. Anorg. Allg. Chem.* **2001**, *627*, 2070.

30 H. Jing, H.-J. Meyer, *Z. Anorg. Allg. Chem.* **2002**, *628*, 1548.

31 J. Gaude, P. L'Haridon, J. Guyada, J. Lang, *J. Solid State Chem.* **1985**, *59*, 143.

32 H. Jing, O. Reckeweg, B. Blaschkowski, H.-J. Meyer, *Z. Anorg. Allg. Chem.* **2001**, *627*, 774.

33 H. Jing, H.-J. Meyer, *Z. Anorg. Allg. Chem.* **2000**, *626*, 514.

34 R. Schmitt, H.-J. Meyer, *J. Solid State Chem.* **2003**, *176*, 306.

35 B. Blaschkowski, H.-J. Meyer, *Z. Anorg. Allg. Chem.* **2002**, *628*, 1249.

36 B. Blaschkowski, H.-J. Meyer, *Z. Anorg. Allg. Chem.* **2003**, *629*, 129.

37 K. Ohoyama, K. Kaneko, K. Indoh, H. Yamauchi, A. Tobo, H. Onodera, Y. Yamaguchi, *J. Phys. Soc. Japan* **2001**, *70*, 3291; K. Cenzual, L. M. Gelato, M. Penzo, E. Parthe, *Acta Cryst.* **1991**, *47*, 433; J. Bauer, O. Bars, *Acta Cryst.* **1980**, *B36*, 1540.

38 O. Oeckler, J. Bauer, H.-J. Mattausch, A. Simon, *Z. Anorg. Allg. Chem.* **2001**, *627*, 779; J. Bauer, O. Bars, *J. Less-Comm. Met.* **1983**, *95*, 267.

39 V. S. Babizhetskyy, Hj. Mattausch, A. Simon, *Z. Krist.* **2003**, *218*, 417.

40 V. Babizhetskyy, Hj. Mattausch, A. Simon, *Z. Krist.* **2004**, *219*, 11.

41 P. Gougeon, J.-F. Halet, D. Ansel, J. Bauer, *Z. Krist.* **1996**, *211*, 823.

42 E. Tominez, E. Alleno, P. Berger, M. Bohn, C. Mazumdar, C. Godart, *J. Solid State Chem.* **2000**, *154*, 114.

43 H. Hartmann, W. Eckelmann, *Z. Anorg. Chem.* **1948**, *257*, 13.

44 M. Neukirch, S. Tragl, H.-J. Meyer, *in preparation.*

45 R. Srinivasan, M. Ströbele, H.-J. Meyer, *Inorg. Chem.* **2003**, *42*, 3406.

46 R. Srinivasan, J. Glaser, S. Tragl, H.-J. Meyer, *Z. Anorg. Allg. Chem.* **2005**, *631*, 479.

47 T. Küppers, E. Bernhardt, H. Willner, H. W. Rohm, M. Köckerling, *Inorg. Chem.* **2005**, *44*, 1015; M. Neukirch, S. Tragl, H.-J. Meyer, T. Küppers, H. Willner, *Z. Anorg. Allg. Chem.* **2006**, *632*, 939.

9

Compositional, Structural and Bonding Variations in Ternary Phases of Lithium with Main-group and Late-transition Elements

Claude H. E. Belin, Monique Tillard

9.1
Introduction

The rapid increase in literature reports on Zintl phases reflects their promotion to one of the main fields in solid state and material chemistry. Zintl phases, formed between electropositive (metal) and moderately electronegative elements (non-metal, semi-metal), were formerly depicted as polar intermediates between salts and metallic phases [1]. Their most fascinating feature is that simple electron transfer from the electropositive to the electronegative element produces partial anionic structures, the topology of which evolves from simple ions to more complex and multidimensional frameworks. This has been discussed within the Zintl-Klemm concept, now recognized as one of the most successful concepts in solid state chemistry. It has provided generations of researchers with a powerful tool for interpreting chemical bonding and connectivities in electron-rich, electron-precise and electron-deficient anionic networks within the framework of the adaptive generalized valence theory. Zintl phase structures can be tuned by varying either the size of constituting elements or valence electron concentrations (size and electron tuning). The pioneering work of Corbett (naked Zintl anions) and Von Schnering and Schäfer (main group intermetallics) raised strong emulation inside the international scientific community [2–6]. The worldwide spread of the "Zintl Phase Attitude" has nourished the reciprocal exchange of ideas and injected knowledge into many fields. The richness and originality of Zintl compounds have fostered theoretical efforts, spanning simple electron counting approaches [7–10] to semi-empirical and more sophisticated DFT or first-principle methods. The never-concluded exploration in the area of Zintl science will remain a continual source of amazement and satisfaction for all of us. Studies on Zintl phases have unveiled properties of interest in the domain of physics, particularly in semiconductor, superconductor, magnetic and thermoelectric material fields. Unfortunately many of these products will never be of practical use due to their high sensitivity to air and moisture. Nevertheless, soluble Zintl anions have recently been used as precursors for the fabri-

Inorganic Chemistry in Focus III.
Edited by G. Meyer, D. Naumann, L. Wesemann
Copyright © 2006 WILEY-VCH Verlag GmbH & Co. KGaA, Weinheim
ISBN: 3-527-31510-1

cation of ordered mesostructures via surfactant-driven self-organization methods [11–14]. A class of Zintl phase materials containing large amounts of lithium has also gained noticeable consideration with regard to their electrochemical properties and possible utilization as high-capacity negative electrodes in lithium-ion batteries [15–18].

9.2
Tuning Structures and Properties in Lithium Binary and Ternary Systems

The Zintl-phase family covers numbers of intermetallic compounds exhibiting multiple and versatile anionic frameworks formed of electronegative elements. They range from isolated single atoms, oligomeric species, chains, rings and cages to three-dimensional structures built of clusters and aggregates. The estimated Valence Electron Concentration (VEC) is a good guide to evaluating structural and bonding properties in intermetallic compounds. In the spirit of the Zintl-Klemm concept, alkali metals are taken as one-electron donors and the VEC is calculated as the total number of valence electrons divided by the number of the electronegative atoms. For example, the lithium-rich phase $Li_{21.2}Ge_5$ (VEC=8.24) can be formally regarded as containing Ge^{4-} single anions [19], while in the sodium-poor $NaZn_{13}$ phase (2.08) the zinc atoms are highly coordinated within a tetrahedrally close-packed structure of Zn-centered, edge-to-edge interconnected icosahedra (stella quadrangulae) [20]. In the binary system Li-Ga, subtle anionic structures are designed through valence electron variations within the gallium-poor to gallium-rich ends [21, 22]. In Li_2Ga (VEC=5), the anionic gallium sub-lattice consists of zig-zag chains. Puckered layers are found in Li_3Ga_2 (4.5), corrugated twinned layers in Li_5Ga_4 (4.25) and a classical tetrahedral network of four-coordinated Ga^- ions in LiGa (VEC=4). In the electron-poorer phases, gallium aggregates into clusters to maximize electron sharing (Fig. 9.1). The complex structure of Li_5Ga_9 (VEC=3.55) contains, in addition to four-coordinated Ga atoms, three kinds of anionic units: icosahedra, open-face 11-vertex *nido* polyhedra and atom-deficient 17-vertex icosioctahedra (3 square and 25 deltahedral faces). Li_2Ga_7 (3.28) is built of *closo*, apex-to-apex interconnected icosahedra linked to four-coordinated Ga atoms [23].

Main group element anionic networks can be modeled, not only by varying the valence electron contents but also by adapting atomic size proportions. For this, hetero-elements can be partially substituted for some of the electronegative elements. Generally there are trends for these substituting elements to occupy strategic positions in the structure and this will be exemplified in the following with the compound $Li_{13}Cu_6Ga_{21}$ and some other structurally related phases.

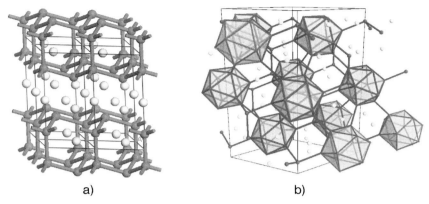

a)   b)

Fig. 9.1 Atomic packing in the intermetallic phases Li_5Ga_4 (a) and Li_2Ga_7 (b).

9.3
Clustering in Condensed Lithium Ternary Phases: A Way Towards Quasicrystals

Lithium has been alloyed with gallium and small amounts of valence-electron poorer elements Cu, Ag, Zn and Cd. Like the early p-block elements (especially group 13), these elements are *icosogen*, a term which was coined by King for elements that can form icosahedron-based clusters [24]. In these combinations, the valence electron concentrations are reduced to such a degree that low-coordinated Ga atoms are no longer present, and icosahedral clustering prevails [25]. Periodic 3-D networks are formed from an icosahedron kernel and the icosahedral symmetry is extended within the boundary of a few shells.

With a VEC reduced to 2.77, $Li_{13}Cu_6Ga_{21}$ (cubic $Im\bar{3}$) appears to be one of the most condensed gallium phases [26]. As in Bergman-type packing, the icosahedral symmetry expands from a unique repeating Ga_{12} icosahedron (1st shell) to the dual Li_{20} pentagonal dodecahedron (2nd shell), then to the enlarged copper icosahedron (3rd shell). Note that the 2nd and 3rd shells constitute the Bergman 32-atom rhombic triacontahedron. Icosahedral symmetry also survives in the fourth shell (mean radius of 6.54 Å) with the fullerene-like Ga_{60} polyhedron (12 pentagonal and 20 hexagonal faces). In preference to fullerene, the term fullerane was used for the first time by Nesper to designate such M_{60} carbon-free units in which there are no double bonds [27]. In the compound $Li_{13}Cu_6Ga_{21}$ [26], the successive onion-like shells of electronegative and electropositive elements form the so-called 104-atom Samson polyhedron referred to as $Ga_{12}@Li_{20}@Cu_{12}@Ga_{60}$ in the endohedral nomenclature of Smalley [28]. In the cubic unit cell, there are two such polyhedra that are fused by edge and hexagonal face-sharing (Fig. 9.2).

Interestingly, no disorder or atomic mixing has been found in $Li_{13}Cu_6Ga_{21}$, which is isostructural with the quasicrystal approximant r-Li_3CuAl_5 ($Li_{52}Cu_{19.3}$-$Al_{88.7}$). Crystals of r-Li_3CuAl_5 were obtained from a melt of composition $Li_{30.5}$-$Cu_{10.5}Al_{59}$ and were surrounded by a continuous layer of the i-Li_3CuAl_6 quasi-

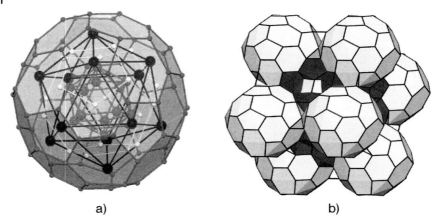

a)   b)

Fig. 9.2 (a) The extended icosahedral symmetry
in $Ga_{12}@Li_{20}@Cu_{12}@Ga_{60}$ and (b) packing of such units
in the cubic cell ($Im\bar{3}$) of $Li_{13}Cu_6Ga_{21}$.

crystalline phase [29]. In the approximant aluminum phase there is some Cu/Al
mixing and although the unit cell of both $Li_{13}Cu_6Ga_{21}$ and r-Li_3CuAl_5 contains
160 atoms, the latter has slightly a lower VEC (2.64).

With increased electron concentrations (3.35, 3.29 and 3.04), the structures of
$Li_{68}Zn_{16}Ga_{133}$ [30], $Li_{29}Cd_3Ga_{64}$ [31] and $Li_{38}(Zn_{0.337}Ga_{0.663})_{101}$ [32] are more complex and display lower symmetries (trigonal and hexagonal). In addition to Ga_{12}
icosahedra, they contain disordered and defective variants of M_{16} deltahedral ico-
sioctahedron. In these phases, unlike in $Li_{13}Cu_6Ga_{21}$, the 104-atom Samson polyhedra interpenetrate in such a manner that an icosahedron which is found at the
center of a fullerane cage also contributes, through pentagonal cuttings, to the surface of other fulleranes. Structures containing non-defective M_{16} polyhedra have
been found for the Fd3m cubic $Na_{35}Cd_{24}Ga_{56}$ and $Na_{144}Ag_{33}Ga_{287}$ compounds
[33, 34]. Samson polyhedra also exist in potassium–thallium intermetallic phases
[26] while other condensed fullerides, stuffed with partially disordered $In_{10}Z$
(Z=Ni, Pd, Pt) clusters, were found in the sodium–indium system [36].

9.4
Exploration of New Lithium Ternary Systems Containing Ag, Zn, Al, Si, Ge

9.4.1
Background

The need for better performing and secure anodic materials in lithium-ion batteries compared with those based on carbon, has boosted research in various do-
mains. Lithium/post-transition element (Al, Si, Sn, Sb...) binary systems have
been widely investigated. Owing to the numerous intermetallic compounds that

can form, especially with Sn and Si, lithium electrochemical insertion proceeds through several plateaus with pronounced potentials. They are accompanied by drastic volume and structurally damaging modifications that lead to rapid fading on cycling and loss in cell capacities. It has been shown that these problems can be partially palliated by the use of ternary lithium alloys. Our work on lithium/transition or late-transition/post-transition element systems was motivated by the good electrochemical properties found for some lithium-transition metal pnictides [37–39], lithium-aluminum silicides and germanides.

9.4.2
The System Li-Al-Ag

Before this work was undertaken, only Pauly's data and Pearson's structure report on Li_2AlAg were available [40, 41]. The original paper of Pauly, based on X-ray powder pattern interpretations, reports Li_2AlAg to crystallize as two cubic polymorphs ($F\bar{4}3m$ and $Fd3m$, a = 6.350 Å). We have recently discovered two further compounds in this system: $LiAlAg_2$ and Li_7Al_4Ag [42].

Seemingly, atomic substitutions in this system have little influence on the structural arrangement and do not modify the electronic properties significantly. This is because Li, Ag and Al have similar atomic radii (1.55, 1.44, 1.43 Å respectively) and Li, Ag are monovalent. Both Li_2AlAg and $LiAlAg_2$ display an ordered cF16 cubic structure ($F\bar{4}3m$, Z = 4, a = 6.350 and 6.312 Å, respectively) which can be seen as deriving from that of LiAl ($Fd3m$, Z = 8, a = 6.367 Å) after substitution of silver for either Al (Li_2AlAg) or both Al and Li ($LiAlAg_2$). Despite the one unit difference in their conventional VEC, Li_2AlAg and $LiAlAg_2$ are

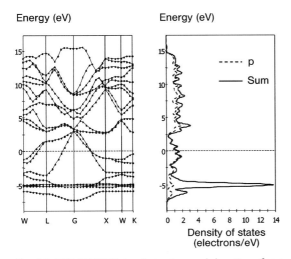

Fig. 9.3 DFT (CASTEP) band structure and densities of states calculated for the cF16-type compound Li_2AlAg.

open-shell compounds. They have quite identical band structures (Fig. 9.3) which, inasmuch as silver d electrons are shown to play no important role in bonding, can accommodate the same number of valence electrons (24 per unit cell) [42]. Population analysis from DFT calculations indicate that silver is the most reduced element in both phases, in agreement with the fact that silver is the most electronegative component ($\chi = 4.44$, against 3.01 and 3.23 for Li and Al) [43]. In Li_2AlAg, the anionic network is based on tetrahedral Al-Ag (2.752 Å) bonding, while in $LiAlAg_2$, Al is 8-coordinated by Ag (2.734 Å). The newly discovered compound Li_7Al_4Ag displays a disordered cF12-type structure ($F\bar{4}3m$, a = 6.344 Å) with only 20 valence electrons in the unit cell.

9.4.3
Compositional and Structural Variations in the System Li-Al-Si

Before 2001, several phases were reported in this system: Li-Al-Si [44, 45], $Li_{12}Al_3Si_4$ [46], $Li_8Al_3Si_5$ [47, 48], $Li_{5.3}Al_{0.7}Si_2$ [49] and $Li_{4+x}Al_{2-x}Si_2$, $x \sim 1$ [50], some of which had controversial compositions and structures. We have since confirmed the exact stoichiometries and determined the structures of LiAlSi, $Li_7Al_3Si_4$, Li_9AlSi_3 and Li_5AlSi_2 by single-crystal X-ray diffraction [51, 52]. First-principle calculations have helped in understanding the subtle variations of these structures with respect to the electron content. Contrary to the system Li-Zn-Ge for which Li_2ZnGe was reported to exist in cubic and trigonal polymorphic forms [53–56], no phase with such a stoichiometry has been found in the system Li-Al-Si. Any attempt to electrochemically add an additional lithium atom at the vacant tetrahedral site in the cF12 structure of LiAlSi ($F\bar{4}3m$, a = 5.928 Å) remained unsuccessful. Why Li_2AlSi cannot form was explained from DFT electronic structure calculations [57]. With a VEC of 4.5 (36 valence electrons per unit cell), Li_2AlSi would be an open shell compound with partially filled and strongly anti-bonding Al-Si crystal orbitals. The preparation of an alloy at this nominal composition by direct melting of the elements actually led to the formation of the compound $Li_7Al_3Si_4$ in association with some unreacted silicon. The cubic $F\bar{4}3m$ (a = 6.115 Å) structure of $Li_7Al_3Si_4$ is disordered with a 75/25% sharing of site 4a by Al and Li, and a 50% vacancy of lithium at 4b. The band structure calculated for the ordered supercell is characterized by a gap opening, and such reversion to a closed shell configuration makes $Li_7Al_3Si_4$ a stable alternative to the unlikely Li_2AlSi.

In the lithium-rich part of the Li-Al-Si system, two new compounds, Li_5AlSi_2 and Li_9AlSi_3, were isolated and their structures determined unambiguously (Fig. 9.4). Li_5AlSi_2 ($P6_3/m$, a = 7.549, c = 8.097 Å) exhibits a disordered pseudo-graphitic anionic network $[Li_3Al_3Si_6]^{12-}$ mingled with a tetrahedral-like Li cationic sublattice [51]. Al/Li atomic mixing occurs at sites 2a (47/53%) and 2c (91/9%). In a first interpretation ($15Li^+$, Al^{3+}, $2(AlSi_3^{9-})$), the anionic layer would appear to contain discrete trigonal $AlSi_3^{9-}$ units with Al–Si bond lengths of 2.467, 2.516 and 2.567 Å (Pauling's Al–Si single bond is 2.421 Å). The band structure calculated with the one-electron extended Hückel tight-binding (EHT) method

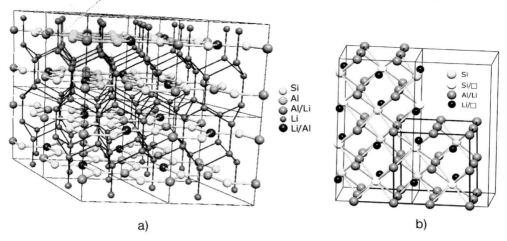

Si
Al
Al/Li
Li
Li/Al

Si
Si/□
Al/Li
Li/□

a)   b)

Fig. 9.4 Atomic packing in hexagonal Li_5AlSi_2 (a) and tetragonal Li_9AlSi_3 (b).

[58, 59] using an ordered supercell ($Li_{60}Al_{12}Si_{24}$) is characteristic of a narrow-gap semiconductor. The calculated Si–Li overlap population of 0.356 within the anionic layer, instead of 0.146 between Si and Li of the interlayer space, indicates that some covalent interactions actually occur between $AlSi_3^{9-}$ units and lithium within the layer, justifying the preferred formulation $[Li_3Al_3Si_6]^{12-}$ for the anionic network.

Li_9AlSi_3 crystallizes in the tetragonal system [52], space group $I4_1/amd$ (a = 6.179, c = 12.199 Å, Z = 2). This compound represents another example of atomic disorder and distribution of vacancies within a tetrahedral network. Its tetragonal unit cell derives from that of $Li_7Al_3Si_4$ by doubling the basic cubic unit in one direction (Fig. 9.4). Lithium atoms mix with Al at site 16 f and half fill the site 8e, Si atoms occupy two sites: 4a (full) and 4b (half-filled). Even with a high VEC (6.0), Li_9AlSi_3 is a nice illustration of how energetic stabilization can be achieved within a tetrahedral network by means of vacancies and atomic substitutions. In the structure of $Li_7Al_3Si_4$, the greater substitution of Li for Al has reduced the VEC to 4.57. Despite frequent occurrences of substitutional disorder and vacancies, the statistical stoichiometries found for these ternary LiAlSi phases always agree with formal oxidation states +1 (Li), +3 (Al) and −4 for the electronegative element Si [60]. For these reasons one probably should be free to refer to these materials as either compounds or phases. The electronic structure of Li_9AlSi_3 simulated in a reordered supercell using the DFT code CASTEP [61, 62] displays a partially filled valence band with a high density of states at Fermi level featuring good metallic properties.

9.4.4
The Tetragonal Compound Li₉AlSi₃, a Good Anodic Material

While the compounds LiAlSi, $Li_7Al_3Si_4$ and Li_5AlSi_2 behaved as poor negative electrodes for lithium secondary batteries, the tetragonal compound Li_9AlSi_3 displays interesting electrochemical properties. This material can reversibly accept large amounts of lithium giving a capacity of about 1040 mA.h.g^{-1} in standard conditions (Fig. 9.5), setting it among the best potential candidates to replace carbon anodes in lithium-ion batteries. Such excellent behavior is related to its good metallic conductivity, to the easy insertion of extra lithium in vacant sites and good ion dynamics owing to Li/Al substitutions.

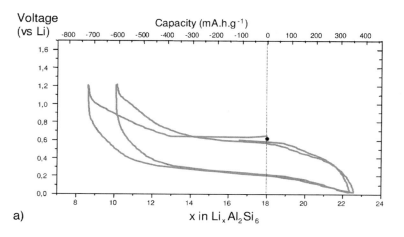

a)

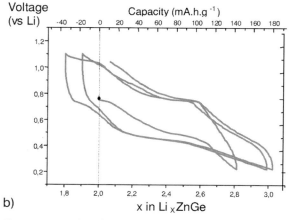

b)

Fig. 9.5 First cycles of the electrochemical lithium exchanges (insertion/removal) in compounds Li_9AlSi_3 (a) and Li_2ZnGe (b).

9.5
The Intermetallic Li-Zn-Ge System, from Electron-poor to Electron-rich Phases

The phase diagram of the pseudo-binary Li-(Zn,Ge) was published more than twenty years ago [63]. It reports the existence of two intermetallic compounds: $Li_{1.08}Zn_{0.92}Ge$ that decomposes peritectically at 620 °C, and Li_2ZnGe that crystallizes in two polymorphic forms: cubic above 502 °C and trigonal below, and melts congruently at 785 °C. This paper also refers to the Li-rich Li_3ZnGe, so-called δ-phase, melting peritectically at 497 °C [56].

$Li_{1.08}Zn_{0.92}Ge$, the crystal structure of which had been described as disordered in the trigonal space group P3 [64], was later reformulated LiZnGe and its structure refined as ordered in the hexagonal space group P$\bar{6}$m2 [65]. Li-Zn-Ge alloys were recently tested as possible candidates for negative electrodes in lithium secondary batteries [66]. The behavior of Li_2ZnGe has been compared with a pure lithium anode in a reversible electrochemical device. Starting from the cubic form of Li_2ZnGe, the removal/insertion of Li was found to proceed via complex equilibria accompanied by some release of germanium and zinc (Fig. 9.5). The system lithium-zinc-germanium was investigated again to afford additional compositional and structural information useful to the exploitation and rationalization of the electrochemical results and to the design of new syntheses.

9.5.1
The Electron-poor Hexagonal Phase LiZnGe

The hexagonal structure which has been recently refined more accurately [67] is built of two distinct parts: a ZnGe hexagonal flat layer in which the Zn–Ge bond length is 2.470 Å and a Zn_2Ge_2 corrugated twinned layer (Zn–Ge = 2.552 Å) (Fig. 9.6). These 2-D anionic units are connected parallel to the c-axis by a three center-one electron, Zn–Ge–Zn linear bond (Zn–Ge = 2.761 Å).

Results of the EHT calculations including all atoms (note that calculations performed without Li atoms, considered as fully ionized, give nearly the same

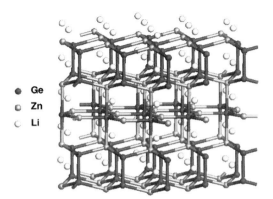

● Ge
◎ Zn
◖ Li

Fig. 9.6 The hexagonal structure (P$\bar{6}$m2) of LiZnGe.

results) are detailed elsewhere [68, 69]. As indicated by crystal orbital overlap populations computed for 51 to 54 electrons (three formula units per cell, Zn 3d electrons included), an electron count of 52 has been found optimal for the stabilization of the anionic network. In conclusion, bonding is assumed to be single (2c–2e) within the two ZnGe layer frameworks and strongly electron-deficient within cross-linking Zn–Ge–Zn bonds (3c–1e for 51 electrons).

The one-unit lowering from the ideal electron count represents a low energy penalty for bonding within the layers, but a slight gain in the interlayer bonding. Why the real material is 1/3 electron short per formula was attributed to some packing frustration effects, as there is not enough room to host an extra lithium in this structural arrangement. Let us imagine that by some means (for instance, electrochemical insertion of lithium) the anionic network could be reduced to 54 electrons (Li_2ZnGe). The interlayer interactions would become strongly anti-bonding with the effect of expanding the structure into well separated and then flattened ZnGe layers, making possible the accommodation of additional lithium. This is seemingly realized in the low-temperature polymorph of Li_2ZnGe ($P\bar{3}m1$, a = 4.326, c = 16.470 Å) [56].

9.5.2
The True Cubic Configuration of the Compound Li_2ZnGe

The high-temperature polymorph of Li_2ZnGe was reported to crystallize in cubic space group $Fm\bar{3}m$ (a = 6.142 Å), yet this structure was inferred from powder data only ($R = 11\%$, 10 reflections) [53, 54]. Surprisingly, after slow cooling from 780 °C (congruent melting point) to room temperature, we have found the cubic symmetry to hold well below the presumed cubic-trigonal transformation (502 °C) and no structural change was observed at −100 °C. This structure has been now determined accurately from single-crystal intensity measurements in the space group $F\bar{4}3m$ (a = 6.123 Å at 25 °C and 6.114 Å at −100 °C, $R1 = 1.7\%$, 60 unique reflections, 6 parameters varied). Rietveld analysis of the bulk powder ($Rp = 4.5\%$, $R_{Bragg} = 2.7\%$) is in good agreement with the single-crystal analysis [69].

CASTEP DFT geometry optimizations of the two Li_2ZnGe cubic configurations (Fig. 9.7), $Fm\bar{3}m$ and $F\bar{4}3m$, indicate that the latter is the more stable by 0.09 eV/atom. In this non-centrosymmetric configuration, Zn and Ge form a covalent tetrahedral network (Ge–Zn = 2.651 Å) providing narrow-gap and semiconducting properties (see band structure, Fig. 9.8). In the $Fm\bar{3}m$ centrosymmetric configuration where both the Ge-to-Zn separation (3.062 Å) and the overall coulombic repulsions between Li cations are increased, the electronic structure is metallic-like.

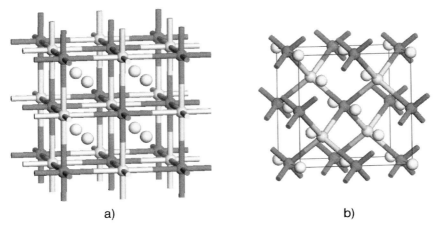

a)   b)

Fig. 9.7 Two cubic arrangements for Li₂ZnGe:
centrosymmetric Fm3̄m (a), non-centrosymmetric F4̄3m (b).

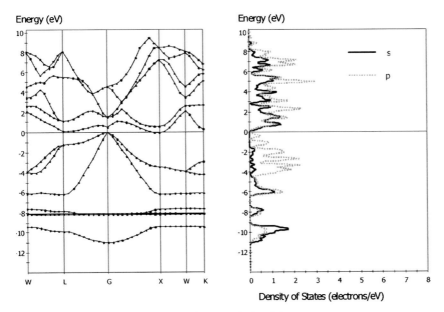

Fig. 9.8 Band structure and densities of states calculated for
the compound Li₂ZnGe in the non-centrosymmetric F4̄3m
cubic arrangement. Zn 3d inert orbitals (flat levels at –8 eV in
the band structure) are not represented in the DOS.

9.5.3
The Li-rich Compound Li$_8$Zn$_2$Ge$_3$ with an Open-layered Anionic Framework

New syntheses were carried out to isolate the Li-rich phase described by Schuster as the hexagonal (a=4.365, c=8.167 Å) δ-phase Li$_3$ZnGe with, as yet, no structure determined. We have instead obtained the new compound Li$_8$Zn$_2$Ge$_3$ that crystallizes in the trigonal space group R3c (a=7.555, c=24.449 Å). Its structure (Fig. 9.9) can be depicted as the stacking along the *c*-axis of six planar Zn$_2$Ge$_3$ layers separated by 4.05 Å, the interlayer space being stuffed with lithium cations. The anionic layer resembles a hetero-graphite ZnGe plane in which one-third of Zn is missing. Such an atom-deficient hexagonal layer is not unique and has already been observed in compounds Yb$_3$Si$_5$ [70] and Li$_{17}$Ag$_3$Sn$_6$ [71]. In this structure, each layer derives from the precedent by a c/6 translation and counter-clockwise six-fold rotation in such a manner that vacant sites are translated in order by 1/3, 2/3, 1/6. Interestingly, the Zn vacancy is compensated by two lithium atoms (Li dumbbell) placed 1.26 Å above and below the lacuna. The covalent Zn$_2$Ge$_3$ network contains two non-equivalent Zn–Ge bonds (2.537 and 2.547 Å) which, by comparison with Zn–Ge distances in

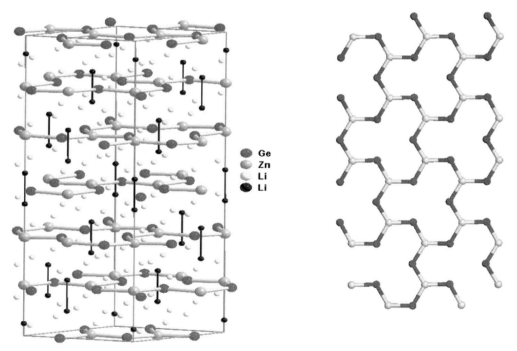

Ge
Zn
Li
Li

Fig. 9.9 Structure of the compound Li$_8$Zn$_2$Ge$_3$ (R3c hexagonal setting) with its lacunal Zn$_2$Ge$_3$ hexagon-based anionic layer.

Energy (eV)

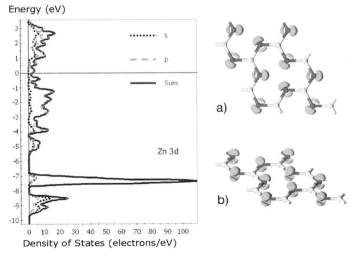

a)

b)

Zn 3d

Density of States (electrons/eV)

Fig. 9.10 CASTEP densities of states for $Li_8Zn_2Ge_3$. Non-bonding electron density distribution over selected bunches of crystal orbitals in the valence-band dispersion: (a) six orbitals between −2.4 and −0.7 eV, (b) four orbitals ranging from −3.3 to −1.9 eV.

Li_2ZnGe, LiZnGe and also with the sum of Pauling's covalent radii (2.472 Å), can be considered as single bonds.

The compound $Li_8Zn_2Ge_3$ can be expressed as $8\,Li^+$, $2\,Zn^{2+}$, $3\,Ge^{4-}$ in terms of oxidation numbers, or as $8\,Li^+$, $2\,Zn^-$, $3\,Ge^{2-}$ with regard to the Zintl-Klemm concept and valence considerations. It closely relates to Li_5AlSi_2 and the corresponding $[Li_3Al_3Si_6]^{12-}$ hexagonal layer in which Zn and Ge would play the role of Al and Si. With the Zn vacancy compensated by a lithium dumbbell, the $[Li_4Zn_4Ge_6]^{12-}$ anionic network is isoelectronic.

Periodic EHT and DFT calculations have been performed in order to analyze the electronic properties and bonding within the anionic sublattice. The DFT results indicate no gap but the density of states is very low at Fermi level; a situation somewhat different from the EHT calculation where the gap is 1.6 eV, but not incompatible, since gaps are usually overestimated in EHT and underestimated in DFT. At first sight, one should consider $Li_8Zn_2Ge_3$ as a narrow gap semiconductor in which Zn–Ge bonding is mainly σ through 4s, $4p_{x,y}$ (in-plane) orbital overlapping. The out-of-plane Zn $4p_z$ orbitals are not involved in bonding and are repelled well beyond the Fermi level into the conduction band. Electronic states lying just below the Fermi level are largely dominated by Ge 4p non-bonding electron pairs (Fig. 9.10).

9.6
Concluding Remarks

The few examples discussed above demonstrate how rich is the chemistry of the intermetallic phases of lithium. Tuning these structures by varying the stoichiometries and the electron contents via atomic substitutions still yields novel and often beautiful anionic frameworks. The main problem with these phases is to locate the lithium atoms precisely, especially in cases where they are mixed with other elements at the same crystallographic sites. Once the crystal structures are determined correctly, the understanding of chemical bonding therein, the determination of the electronic structures and properties may provide insight into the potential utilization of these materials in technological applications. Because they should be able to furnish high power capacity negative electrodes, products from the ternary combinations of lithium and light elements chosen among silicon, aluminum, boron, etc. are of great interest. Studies are to be pursued on how to incorporate them in not only better performing, but also still secure, lithium-ion batteries. This is a real challenge and the results obtained with the system lithium–aluminum–silicon are encouraging.

References

1 Zintl, E.; Brauer, G. *Z. Phys. Chem.* **1933**, *B20, 3/4*, 245.
2 Corbett, J. D. *Chem. Rev.* **1985**, *85*, 383 and references cited therein.
3 Corbett, J. D. *Angew. Chem., Int. Ed.* **2000**, *39*, 670.
4 Schäfer, H.; Eisenmann, B.; Müller W. *Angew. Chem., Int. Ed.* **1973**, *12–19*, 694.
5 Schäfer, H. *Ann. Rev. Mater. Sci.* **1985**, *15*, 1.
6 Von Schnering, H.G. *Angew. Chem., Int. Ed.* **1981**, *20*, 33.
7 Wade, K. *Adv. Inorg. Chem. Radiochem.* **1976**, *18*, 1.
8 Mingos, D. P.; Wales D. J. *Introduction to Cluster Chemistry* **1990**, Prentice Hall, Englewood Cliffs, New Jersey.
9 Teo, B. K. *Inorg. Chem.* **1984**, *23*, 1251; **1985**, *24*, 4209.
10 Teo, B. K.; Zhang, H. *Inorg. Chem.* **1988**, *27*, 414.
11 MacLachlan, M. J.; Coombs, N.; Ozin, G. A. *Nature* **1999**, *397*, 681.
12 Korlann, S. D.; Riley, A. E.; Kirsch, B. L.; Mun, B. S.; Tolbert, S. H. *J. Am. Chem. Soc.* **2005**, *127(36)*, 12516.

13 Mitzi, D. B.; Kosbar, L. L.; Murray, C. E.; Copel, L.; Afzali, A. *Nature* **2004**, *428*, 299.
14 Trikalitis, P. N.; Bakas, T.; Kanatzidis, M. G. *J. Am. Chem. Soc.* **2005**, *127(11)*, 3910.
15 Winter, M.; Besenhard, J.O.; Spahr, M.E.; Novak, P. *Adv. Materials* **1998**, *10/10*, 725.
16 Huggins, R. *J. Power Sources* **1999**, *81/82*, 13.
17 Nishi, Y. *J. Power Sources* **2001**, *100*, 101.
18 Tirado, J. L. *Mater. Sci. and Eng.* **2003**, *R40*, 103.
19 Goward, R.; Taylor, N.; Souza, D. C. S.; Nazar, L. F. *J. Alloys Compd.* **2001**, *329 (1–2)*, 82.
20 Schömaker, D. P.; Marsh, R. E.; Ewing, F. J.; Pauling, L. *Acta Cryst.* **1952**, *5*, 637.
21 Tillard-Charbonnel, M.; Belin, C. *C. R. Acad. Sci. Paris* **1998**, *306*, 1161.
22 Belin, C. H. E.; Tillard-Charbonnel, M. *Prog. Solid St. Chem.* **1993**, *22(2)*, 59.
23 Tillard-Charbonnel, M.; Belin, C.; Soubeyroux, J. L. *Eur. J. Solid State Inorg. Chem.* **1990**, *27*, 759.
24 King, R. B. *Inorg. Chem.* **1989**, *28*, 2796.

25 Belin, C.; Tillard-Charbonnel, M. *Coord. Chem. Rev.* **1998**, *178–180*, 529.

26 Tillard-Charbonnel, M.; Belin, C. *J. Solid State Chem.* **1991**, *90(2)*, 270.

27 Nesper, R. *Angew. Chem., Int. Ed.* **1994**, *33(8)*, 843.

28 Guo, T.; Jin, C.; Smalley, R. E. *J. Phys. Chem.* **1991**, *95*, 4948.

29 Audier, M.; Pannetier, J.; Leblanc, M.; Janot, C.; Lang, J. M.; Dubost, B. *Physica B* **1988**, *153*, 136.

30 Tillard-Charbonnel, M.; Chahine, A.; Belin, C. *Mater. Res. Bull.* **1993**, *28(12)*, 1285.

31 Tillard-Charbonnel, M.; Chahine, A.; Belin, C. *Z. Krist.* **1994**, *209*, 280.

32 Tillard-Charbonnel, M.; Chouaibi, N.; Belin, C.; Lapasset, J. *Eur. J. Solid State Inorg. Chem.* **1992**, *29*, 347.

33 Tillard-Charbonnel, M.; Belin, C. *Mater. Res. Bull.* **1992**, *27*, 1277.

34 Tillard-Charbonnel, M.; Chahine, A.; Belin, C. *Z. Krist.* **1993**, *208*, 372.

35 Cordier, G.; Muller, V. *Z. Naturforsch.* **1993**, *B48*, 1035.

36 Sevov, S. C.; Corbett, J. D. *Science* **1993**, *262*, 880.

37 Monconduit, L.; Tillard-Charbonnel, M.; Belin, C. *J. Solid State Chem.* **2001**, *156*, 37.

38 Monconduit, L.; Gillot, F.; Doublet, M. L.; Lemoigno, F. *Ionics* **2003**, *9(1/2)*, 56.

39 Crosnier, O.; Mounsey, C.; Herle, P.; Taylor, N.; Nazar, L.F. *Chem. Mater.* **2003**, *15(26)*, 4890.

40 Pauly, H.; Weiss, A.; Witte, H. *Z. Metallk.* **1941**, *33*, 391.

41 Villars, P.; Calvert, L.D. **1991** *Pearson's Handbook of Crystallographic Data for Intermetallic Phases*, second edition.

42 Lacroix-Orio, L.; Tillard, M.; Belin, C. *Solid State Sci.* **2004**, *6*, 1429.

43 Pearson, R. G. *Inorg. Chem.* **1988**, *27*, 734.

44 Nowotny, H.; Holub, F. *Monatsh. Chem.* **1960**, *91*, 877.

45 Schuster, H. U.; Hinterkeuser, H. W.; Schafer, W.; Will, G. *Z. Naturforsch.* **1976**, *B31*, 1540.

46 Pavlyuk, V. V.; Bodak, O. I. *Izv. Akad. Nauk. SSSR. Neorg. Mat.* **1992**, *28*, 988.

47 Kevorkov, D.; Gröbner, J.; Schmid-Fetzer, R. *J. Solid State Chem.* **2001**, *156*, 500.

48 Gröbner, J.; Kevorkov, D.; Schmid-Fetzer, R. *J. Solid State Chem.* **2001**, *156*, 506

49 Blessing, J. *Ph.D Thesis*, **1978**, University of Cologne.

50 Zürcher, S. *Dissertation* **2001**, Zürich, Switzerland.

51 Spina, L.; Tillard, M.; Belin, C. *Acta Cryst.* **2003**, *C59*, i9.

52 Spina, L.; Jia, Y. Z.; Ducourant, B.; Tillard, M.; Belin, C. *Z. Krist.* **2003**, *218*, 740.

53 Schuster, H. U. *Naturwissenschaften* **1966**, *53*, 361.

54 Schuster, H. U. *Z. Anorg. Allg. Chem.* **1969**, *370*, 149.

55 Schönemann, H.; Schuster, H. U. *Rev. Chim. Min.* **1976**, *13*, 32.

56 Cullmann, H. O.; Hinterkeuser, H. W.; Schuster, H. U. *Z. Naturforsch.* **1981**, *B36*, 917.

57 Tillard, M.; Belin, C.; Spina, L.; Jia, Y. Z. *Solid State Sci.* **2005**, *7*, 1125.

58 Whangbo, M. H.; Hoffmann, R. *J. Am. Chem. Soc.* **1978**, *100*, 6093.

59 CAESAR 1.0 package for Windows, *Crystal and Electronic Structure Analyser*, Ren, J.; Liang, W.; Whangbo, M.H. **1998**, North Carolina State University.

60 Electronegativities are 4.77 for Si, 3.23 for Al and 3.01 eV for Li from Pearson, R. G. *Inorg. Chem.* **1988**, *27*, 734.

61 Payne, M. C.; Teter, M. P.; Allan, D. C.; Arias, T. A.; Joannopoulos, J. D. *Rev. Mod. Phys.* **1992**, *64*, 1045.

62 CASTEP is a DFT program code distributed inside the Materials Studio package version 2.2.1 **2002** Accerys Inc., San Diego.

63 Cullmann, H. O.; Schuster, H. U. *Z. Anorg. Allg. Chem.* **1983**, *506*, 133.

64 Schönemann, H.; Schuster, H. U. *Z. Anorg. Allg. Chem.* **1977**, *432*, 87.

65 Belin, C.; Sportouch, S.; Tillard-Charbonnel, M. *C.R. Acad. Sci. Paris* **1993**, *317*, 769.

66 Alcantara, R.; Tillard-Charbonnel, M.; Spina, L.; Belin, C.; Tirado, J. L. *Electrochim. Acta* **2002**, *47*, 1115.

67 LiZnGe single-crystal structure was refined using 253 observed reflections to $R1 = 1.86\%$ from Xcalibur CCD X-ray

new data set, space group $P\bar{6}m2$, a = 4.2775(3), c = 9.3653(8) Å, Flack parameter 0.14(14). Ge atoms are found at 1a and 2i (z = 0.36328(7)) positions, Li atoms at 1d and 2h (z = 0.160(1)) and Zn atoms at 1e and 2g (z = 0.29483(10)).

68 Sportouch, S.; Thumim, J.; Tillard-Charbonnel, M.; Belin, C.; Rovira, M.C.; Canadell, E. *New J. Chem.* **1995**, *19*, 133.

69 Lacroix-Orio, L.; Tillard, M.; Belin, C. *Solid State Sci.* **2006**, *8*, 208.

70 Pöttgen, R.; Hoffmann, R.D.; Kussmann, D. *Z. Anorg. Allg. Chem.* **1998**, *624*, 945.

71 Lupu, C.; Downie, C.; Guloy, A.M.; Albright, T.A.; Mao, J.G. *J. Am. Chem. Soc.* **2004**, *126(13)*, 4386.

10
Polar Intermetallics and Zintl Phases along the Zintl Border

Arnold M. Guloy

10.1
"First comes the synthesis ..." – J. D. Corbett

The synthesis of intermetallic compounds provides a fertile ground for exploratory work. Not only does it offer a wide area for exploring the many elements,
but also provides a valuable testing ground for any new intuitive and innovative
concepts of chemical bonding. Inter-relationships between structure, composition and properties, rationalized by chemical bonding, have a central role in the
understanding of chemical systems. However, inorganic solid state chemistry,
unlike organic chemistry, suffers from the inability of current chemical principles and concepts to completely understand and predict the composition, structure and reactivity of solid-state materials. Hence the synthesis of novel solids is
as much an art as a science, and relies very much on empirical rules developed
from past serendipitous and intuitive discoveries. Thus, developing a rational
approach in the syntheses of new intermetallic phases offers challenges and opportunities, involving many inter-related and often complex factors. A particular
problem for chemists is the seemingly unpredictable behavior of intermetallic
compounds based on current chemical ideas derived from molecular chemistry.

10.2
What are Intermetallics?

When two metals A and B are melted together and the liquid mixture is then
slowly cooled, different equilibrium phases appear as a function of composition
and temperature. These equilibrium phases are summarized in a condensed
phase diagram. The solid region of a binary phase diagram usually contains
one or more intermediate phases, in addition to terminal solid solutions. In solid solutions, the solute atoms may occupy random substitution positions in the
host lattice, preserving the crystal structure of the host. Interstitial solid solutions also exist wherein the significantly smaller atoms occupy interstitial sites

Inorganic Chemistry in Focus III.
Edited by G. Meyer, D. Naumann, L. Wesemann
Copyright © 2006 WILEY-VCH Verlag GmbH & Co. KGaA, Weinheim
ISBN: 3-527-31510-1

in the larger host-metal lattice. The solid metal solutions, nominally called "alloys", normally exhibit wide compositional ranges. On the other hand, an intermediate phase has a narrow compositional range and a crystal structure unlike those of the component metals. These "narrow" intermediate phases are commonly known as "intermetallic compounds" or simply "intermetallics" [1, 2].

An intermetallic compound, CuZn, was first reported by Karl Karsten in 1839 after observing that CuZn reacted differently with acids from the other solutions of Cu-Zn of different proportions [3]. Subsequently, several systematic studies on intermetallic compounds ensued. The important compositional (phase diagram) studies by Tamman [4] and the physico-chemical studies of Kurnakov [5] and Desch [6] established the scientific studies on intermetallics. During the early stages, unaware of the atomic ordering that occurred in intermetallics, researchers were deeply puzzled by the physical anomalies exhibited by these phases. The introduction and the application of x-ray diffraction in structural chemistry proved to be the most valuable development in the study of these solid materials. This led to the recognition of long-range ordering in the structure of intermetallics. This enabled direct structural studies of solid-state compounds, and the appreciation of structure-bonding-property relationships was a natural result. In 1926 Hume-Rothery [7] and Westgren and Phragmen [8] made important contributions to the understanding of intermetallics by proposing an electron counting scheme for compounds between noble and s-p metals. Hume-Rothery correlated the crystal structures of these compounds, now collectively known as Hume-Rothery phases, with the average number of valence electrons per atom. This resulted in specific valence electron concentrations (VEC) for different types of crystal structures: 1.38 – (fcc) α-phase; 1.48 – (bcc) β-phase; 1.62 – γ-phase; 1.75 – (hcp) ε-phase. The Hume-Rothery scheme was rationalized by the "free-electron" model of Mott and Jones [9] and extended by Pauling [10] to other related transition-metal intermetallics. The free-electron model was later improved by pseudopotential calculations by Heine [11]. However, recent critiques by Pettifor [12], Massalski [13], and Mizutani [14] contend that an accurate explanation of the Hume-Rothery rules is still incomplete and that d-orbital participation, correct ionic effects and correlation should be included. Surprisingly, Lee showed that the Hume-Rothery rules can be rationalized by simple molecular orbital theory [15].

Another important concept in the formation and stability of intermetallics is based on the sizes of the constituent atoms. The idea was first introduced by Biltz [16] and Laves [17] and formalized by Frank and Kasper [18]. It states that complex structures found in intermetallics can be explained in terms of geometrical requirements of sphere close-packing that are associated with the atomic radius ratio of the component atoms. Restated, the coordination polyhedra exhibit the following features: (a) triangular faces due to effective tetrahedral close-packing; (b) convex polygon; and (c) five or six edges meet at every corner. Laves [19] further added that the geometrical factor is guided by three principles, namely: (1) the tendency to fill space as efficiently as possible; (2) the tendency to form arrangements of the highest possible symmetry; and (3) the tendency to form the closest connections between like atoms. These principles were best exemplified by the MgX_2 type compounds and later found to apply to other binary MX_2 and ternary derivatives. These phases, collectively known as Laves or Friauf-Laves phases [20], form the largest group of intermetallics [21].

The correlation between structural preference and radius ratios in the MX_2 Laves phases was surprisingly accurate, and theoretical explanations of the exhibited behavior have been proposed and have continued to evolve [21, 22]. Pseudopotential calculations have shown that, although geometrical factors are dominant, electrostatic effects can lead to distorted and defect structures [23]. These results have been supported by the impressive experimental work of Komura and coworkers [24]. Covalency was also proposed to play a significant role in the stability of Laves phases [25, 26], concomitant with the existence of non-classical polyhedral bonding. Recent thermodynamic studies on the prototypical Laves phases $MgZn_2$ and $MgCu_2$ strongly suggest the simultaneous existence of metallic, covalent and ionic bonding [27]. Hence the principle of geometric close-packing control in the formation of Laves phases can be interpreted in terms of the subtle interplay of electrostatic and electronic factors – generalized as "soft interatomic potentials" [28].

The synthesis of salt-like intermetallics like KBi_2, KPb_2 and $CsBi_2$ that violate the close-packing principle (more open structures), and the existence of MX phases, such as NaTl and LiCd, that presented violations to the Hume-Rothery rules led to ionic descriptions based on "charge-transfer" between constituent atoms. The concept of electronegativity differences was recognized as an important factor in the structural preference of intermetallics [29]. This is best illustrated by valence compounds [30], in which formation is associated with fulfilling the octet rule. Charge transfer is directed by electronegativity differences, i.e., electrons are transferred from the electropositive to the more electronegative component. A similar class of intermetallics is composed of those having polyanionic covalent moieties dictated by the 8-N valence rule – i.e., satisfying a "filled-octet" through the formation of homo- and heteroatomic covalent bonds. The idea of simultaneously applying ionic and covalent bonds in the description of intermetallic structures was first applied by Eduard Zintl in the bonding of NaTl and related MX phases [31]. Hence the term Zintl phase and the salt-like description of these phases were introduced [32].

Several attempts to classify intermetallics based on structure and bonding have led to popular generalizations. Hume-Rothery [33] developed a set of rules that included: (1) electronegativity differences between constituent atoms; (2) the tendency of main group and transition metals to fill their s-p and d shells, respectively; (3) size-factors; (4) electron concentration; and (5) "orbital restrictions" – related to the symmetry conditions for orbital hybridization. These factors were postulated to control the stability and formation of all intermetallics. Pearson [30] classified intermetallics into groups based on the most important factor that governed their crystal structures: (a) electron compound – governed by the valence electron concentration as in the Hume-Rothery phases; (b) geometric compounds – governed by the geometric requirements (radius ratios) of the crystal structures as in the Laves and Frank-Kasper phases; and (c) valence compounds – governed by the valence (8-N) rules as in the Zintl phases. However, Parthe and Girgis later showed that these groups do not have clear boundaries and that in most intermetallics all three principles are simultaneously involved, albeit in varying degrees [34].

The creative use of structure maps in predicting structures of binary and pseudo-binary intermetallics has been reported [35]. The semi-empirical approach involves the correlation or "threading onto a string" of different "chemical scales" of the constituent atoms (e.g., size, valence electrons, electronegativities, ionization potentials, electron counts, etc.) and establishing "regions of stability" for different crystal structures. Intermetallics with similar "map coordinates" have similar crystal structures. The use of structure maps has the advantage of simplicity; nevertheless, it suffers from the inability to offer a physical argument to explain the successes and failures of the approach. Furthermore, structure maps do not foresee the discovery of new structures. It must then be understood that a useful structure map should be treated as a good illustration of trends [36].

10.3
The Zintl-Klemm Concept

The Zintl-Klemm concept evolved from the seminal ideas of E. Zintl that explained the structural behavior of main-group (s-p) binary intermetallics in terms of the presence of both ionic and covalent parts in their bonding description [31, 37]. Instead of using Hume-Rothery's idea of a valence electron concentration, Zintl proposed an electron transfer from the electropositive to the electronegative partner (ionic part) and related the anionic substructure to known isoelectronic elemental structures (covalent part), e.g., Tl^- in NaTl is isoelectronic with C, Si and Ge, and consequently a diamond substructure is formed. Zintl hypothesized that the structures of this class of intermetallics would be salt-like [16b, 31f, 37e].

Klemm and Busmann [38] formalized Zintl's ideas and applied them to other binary and ternary phases. Following the Zintl picture, Klemm [38a] developed a general description of Zintl phases associating structure and bonding descriptions with physical properties. The charge-transfer as hypothesized by Zintl is manifested by a large volume contraction and large negative enthalpy of formation [38a, 39]. The covalent pseudo-element or anion framework exhibits a structure closely related to an isoelectronic element. Zintl phases are generally brittle, semiconducting and do not show temperature-independent paramagnetism [40]. These conclusions were based on many physico-chemical studies by Klemm and were rationalized by electron counting rules related to the filled-octet or 8-N rule.

The successful use of the Zintl-Klemm concept in the synthesis of complex intermetallics is represented by the work of Schaefer and co-workers in Darmstadt [41]. The extensive investigations on Zintl phases mainly by von Schnering and Nesper in Stuttgart and Zurich [42, 43] and Corbett in Ames [44] have led to the refinement and evolution of the original Zintl-Klemm ideas to include cation metal–metal bonding, multi-center anion cluster bonding and non-stoichiometry. These ideas and the polar description of Zintl phases have been validated by numerous theoretical and physical studies [45]. Modern band-structure calculations on NaTl-type binary and ternary Zintl phases confirmed conclusions drawn from a bonding scheme that is simultaneously ionic and covalent [45 a]. Altogether, quantum mechanical studies support the polar/covalent assumptions of Zintl, Klemm and Brauer and underscore the simplicity and brilliance of the Zintl-Klemm concept.

The effective charge transfer and bonding character between atoms largely depend on differences in electronegativities. Consequently, a criticism of the Zintl concept is the unreasonable assignments of ionic charges [46]. However, the description of Na^+Tl^- and Li^+In^- is not contrary to the accurate theoretical and experimental picture if the assignment of "charges" were only in a *formal oxidation* sense. What is actually represented by the Zintl formalism is the number of occupied Wannier-like electronic states [47]. In Zintl phases these states are mostly bonding or non-bonding, and are largely derived from the metalloid atomic states. The unoccupied anti-bonding states are mainly of the electropositive metal partner. It may then be assumed that, if the Zintl electronic scheme could be extended to other intermetallic phases regardless of "type", then it could be used to probe into their chemical bonding and understand their electronic properties. It is anticipated that metals (main group and transition metals and metalloids) with sufficient electronegativity differences will form what we term as *polar intermetallics* wherein strong homo- and heterometallic bonding occurs within the covalent partial structure, and the valence generalities about Zintl phases still apply.

In the spectrum from classical intermetallics to valence compounds to insulators, a smooth transition in their chemical bonding (metallic to ionic) is observed. At the border between Zintl phases and metallic phases, the typical properties of Zintl phases diminish and metallic conductivity appears. However, it is inaccurate to impose and define a sharp boundary between classical Zintl phases and the metallic phases (e.g., Laves and Hume-Rothery phases), and it is in the overlapping regimes where much chemistry still remains to be discovered and understood.

10.4
"Electron-poor" Polar Intermetallics

The validity (or lack thereof) of the classical Zintl formalism as applied to less polar intermetallics, involving metals along the Zintl border, is nicely probed by "electron-poor" trelides. Seminal work by Corbett [44] and Belin [48] recognized the proclivity of trelides (Ga, In, Tl) to form cluster-based anion structures. The apparent "electron deficiency" in the chemical bonding of these cluster compounds is offset by the formation of three-center-two electron bonds, following the classic Wade-Mingos rules for boranes and extended through isolobal analogies. However, cluster-based intermetallic phases that violate even these rules have also appeared. These violations arose with extremely "electron-poor" trelides that featured closed deltahedral frameworks, but did not exhibit their canonical polyhedral form [44, 50, 51]. Representative examples of these include $K_{10}Tl_7$ [50b], K_8In_{11} [50c], Sr_3Sn_5 [50d], and $Na_8K_{23}Cd_{12}In_{48}$ [50e]. The unusual class of "electron poor" polar intermetallics are collectively known as hypoelectronic (yet electron-precise) systems. Molecular hypoelectronic analogs, such as metalloboranes of early transition metals, do exist and these exhibit accompanying geometrical irregularities [51]. In many of these cluster-based intermetallic compounds the cluster anions fulfill the electronic requirements for bonding, albeit in the presence of excess electrons delocalized in a metal-based conduction band. The term "metallic Zintl phase" has been proposed to characterize the metallic hypoelectronic phases and is increasingly being recognized in describing many polar intermetallics that lie along the Zintl border [44].

A more unusual class of "electron-poor" Zintl phases are collectively known as *real* hypoelectronic systems with open valence bands, represented by compounds such as La_3In_4Ge, $SrIn_4$ and Sr_3In_5 [52]. These constitute recent examples of truly hypoelectronic intermetallic systems with open and incompletely filled valence bands, having one- or two-electron deficiencies per formula unit. The unusual existence of an open valence band is responsible for the characteristic metallic behavior of "real" hypoelectronic systems. The remarkable stability of these hypoelectronic systems has been attributed to the significant role of geometric effects, as well as Madelung contributions within the polar intermetallic structure. The metallic hypoelectronic phases, including "metallic" Zintl phases, represent polar intermetallics that behave closest to the "nearly free-electron" intermetallic compounds.

10.5
Intermetallic π-Systems

As in molecular chemistry, an alternative path to compensate for electron deficiency is the formation of multiple bonds, through π-interactions, as in unsaturated and aromatic molecular systems. Our work in Houston focuses on probing the efficacy of the Zintl concept in rationalizing stoichiometries, crystal structures and chemical bonding of complex "electron-poor" Zintl phases that exhibit novel π-systems. Their chemical bonding is reflected by their unusual crystal structures related to unsaturated hydrocarbons [53].

A first example is the diamagnetic semiconductor $SrCa_2In_2Ge$, synthesized from the elements at high temperatures, that features novel anionic zigzag chains, $[In=In–Ge]^{6-}$ [54]. Its crystal structure is derived from that of CaGe (CrB-type) through a tripling of the chain axis revealed by the ordering of the Ge and In atoms within the chains, as shown in Fig. 10.1. In the spirit of the

Zintl concept, corresponding to $[In_2Ge]^{6-}$, the formal electron count does not satisfy the (8-N) valence rule if In and Ge were singly bonded. Electronic band structure calculations and MO orbital analyses, as shown in Fig. 10.2, reveal the anionic chains of the "electron-deficient" Zintl phase, with short In–In bonding distances (2.772(2) Å), were characterized as being analogous and isoelectronic with the allyl chain $[CH=CH–CH_2]_\infty$. Similarly, the $[Ga]^{2-}$ zigzag chains in CaGa have been likened to polyacetylene [53a].

Later, the metallic $Ca_5In_9Sn_6$ was discovered during our exploratory high-temperature synthesis and found to exhibit an unprecedented intergrowth structure of hexagonal $CaIn_3$ (Ni_3Sn-type) and cubic $Ca(In/Sn)_3$ ($AuCu_3$-type) slabs in a 1:4 ratio [55]. The crystal structure of $Ca_5In_9Sn_6$ is shown in Fig. 10.3. Unlike the normal cubic $Ca(In/Sn)_3$ slabs, the hexagonal slabs feature distorted closed-packed arrangements of indium that effectively result in isolated In_3 triangles. Band-structure calculations and orbital analyses, as illustrated in Fig. 10.4, reveal a surprising result in that the In_3 trimers in the hexagonal slabs actually have optimized In–In bonds and can be formally assigned as anionic $[In_3]^{5-}$ units that are isoelectronic with the aromatic cyclopropenium, $C_3H_3^+$. The $[In_3]^{5-}$ anions are also analogous to the $[Sn_3]^{2-}$ units in the superconducting $BaSn_3$ [56]. The metallic $Ca_5In_9Sn_6$ can be described a an intergrowth of Zintl and metallic layers, where the valence states of the Zintl layers lie deep below the Fermi level. The excellent metallic behavior of $Ca_5In_9Sn_6$ was attributed to the well-dispersed electronic states of the intermetallic layers that dominate the Fermi level.

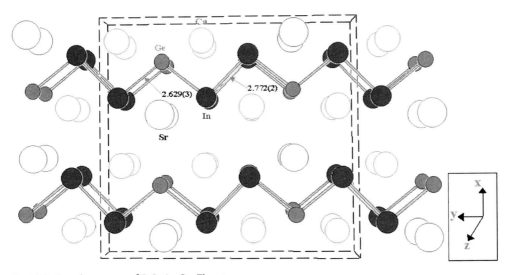

Fig. 10.1 Crystal structure of $SrCa_2In_2Ge$. The atoms are represented as follows: Sr – large open circles; Ca – small open circles; Ge – small shaded circles; In – large shaded circles.

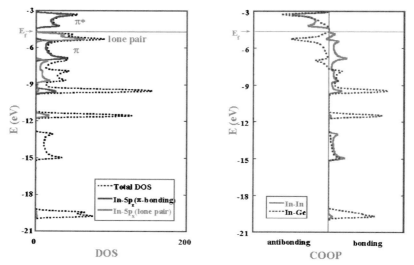

Fig. 10.2 Crystal Orbital Overlap Population (COOP) and Densities of States (DOS) plots for SrCa$_2$In$_2$Ge: (a) COOP plots of the In–In (solid) and In–Ge (dashed) interactions; (b) DOS plots of the total DOS (dotted), In-5py lone pair (dashed), and In-5px p-states (solid).

Similar exploratory work by von Schnering and Nesper on ternary tetrelides (Si and Ge) resulted in the synthesis of arene-like π-systems, represented by Ba$_4$Li$_2$Si$_6$ and Li$_8$MgSi$_6$ [57]. The novel lithium-based ternary Ba$_4$Li$_2$Si$_6$ features [Si$_6$]$^{10-}$ units related to aromatic hydrocarbon rings like benzene [57a]. While quasi-aromatic Si five-membered rings, [Si$_5$]$^{6-}$, are featured in Li$_8$MgSi$_6$ [57b]. Moreover, condensed systems consisting of hexagonal Si$_6$ rings similar to arenes and conjugated polyaromatics have also been isolated and characterized [58].

The analogy of intermetallic to molecular π-systems was also found to extend into organometallic systems. Our work on alkaline earth metal nickel silicides has led to the discovery of Ba$_2$NiSi$_3$ – a novel polar transition-metal intermetallic that features unusual metallocene-like [NiSi$_3$]$^{4-}$ chains [59]. Obtained from high-temperature reactions Ba$_2$NiSi$_3$ exhibits a crystal structure reminiscent of Laves phases. However, it exhibits a narrow homogeneity range and a unique crystal that features chains of Ni-centered Si$_{6/2}$ trigonal prisms, as shown in Fig. 10.5. The NiSi$_3$ chains, effectively isolated by the large Ba atoms, are similar to the novel charge-density-wave NbSe$_3$ chains [60]. It is interesting to note that the electronegativities of Ni and Si are essentially identical and one would then expect strong covalent bonds between them, thus low-dimensional structures are favored [61]. Using extended Hückel theory, the chemical bonding of the novel metallic chains were nicely rationalized as a one-dimensional solid-state analog of an extended metallocene, as shown in Fig. 10.6. This leads to the analogy with infinitely stacked face-to-face metallocene polymers [62].

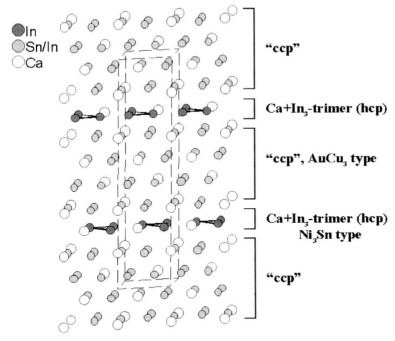

○ In
○ Sn/In
○ Ca

"ccp"

Ca+In₃-trimer (hcp)

"ccp", AuCu₃ type

Ca+In₃-trimer (hcp)
Ni₃Sn type

"ccp"

Fig. 10.3 Crystal structure of Ca₅In₉Sn₆ shown as stacking of slabs along the *c*-axis. The atoms are represented as follows: Ca, open circles; In/Sn, grey shaded circles; In, dark shaded circles.

Our discussion on novel intermetallic π-systems culminates in the very novel Zintl phase Li₁₇Ag₃Sn₆ [63]. The complex polar intermetallic exhibits a unique structure (Fig. 10.7) that features an anionic substructure derived from the intergrowth of defect graphite-like and Kagome nets. The AgSn₃ Kagome nets, sandwiched between Li layers, Li₁₀[AgSn₃], effectively distort to form isolated trigonal planar AgSn₃ units. The units are analogous to the carbonate ion and thus formally assigned as [AgSn₃]¹¹⁻. The Kagome slabs are alternately stacked with Li₇[Ag₂Sn₃] slabs that feature defect honeycombed Ag₂Sn₃ nets sandwiched by Li atoms. The [Ag₂Sn₃]⁶⁻ nets of 3-bonded Ag and 2-bonded Sn atoms appear to be electron deficient with respect to the classical Zintl definition. Extended Hückel and DFT calculations indicate that the seeming electron deficiency in Li₁₇Ag₃Sn₆ is accommodated by a unique metallic bonding scheme associated with aromatic and unsaturated hydrocarbons and that the chemical bonding is optimized. The calculations confirm that the isolated [AgSn₃]¹¹⁻ assignment is isoelectronic and analogous with CO₃²⁻. More interestingly, the [Ag₂Sn₃]⁶⁻ nets are found to exhibit significant π-bonding as in BN [64], as well as its bonding being further optimized by the formation of a "trefoil" (σ-aromaticity) bonding

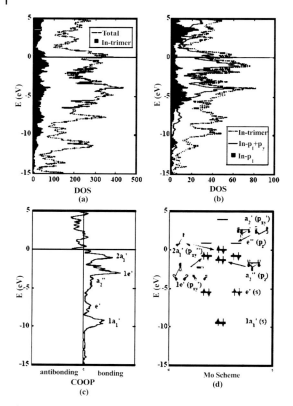

Fig. 10.4 COOP and DOS plots for $Ca_5In_9Sn_6$. (a) DOS plots. Total DOS – dotted line; In_3 trimer states – solid area; (b) projected DOS of In_3-derived states. Total PDOS – dotted line; P(x,y) σ contributions – solid line; P(z) π contributions.

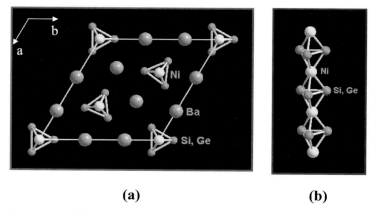

Fig. 10.5 (a) [001] View of the crystal structure of Ba_2NiSi_3. (b) A $[NiSi_3]^{4-}$ chain in Ba_2NiSi_3.

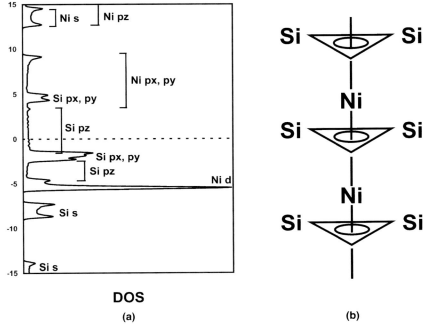

Fig. 10.6 (a) Density of states of [NiSi$_3$]$^{4-}$ in Ba$_2$NiSi$_3$; the bands are labeled accordingly. The Fermi level is set to 0 eV. (b) Schematic representation of the metallocene-like bonding of [NiSi$_3$]$^{4-}$ in Ba$_2$NiSi$_3$.

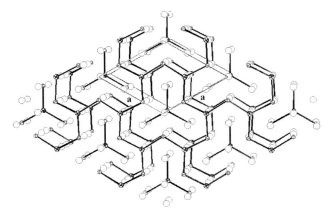

Fig. 10.7 Crystal structure of Li$_{17}$Ag$_3$Sn$_6$ as viewed along the c-axis. Sn (crossed ellipsoids); Ag (octant-shaded); Li (isotropic). The a and b axes are shown.

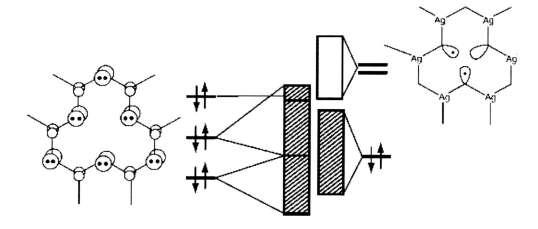

π-bands

σ–trefoil bands

Fig. 10.8 Schematic diagram of the bands in $[Ag_2Sn_3]^{6-}$. On the left are the three filled π-orbitals, two of which show significant dispersion. On the right are the three in-plane hybrid orbitals, which make up the

trefoil state. The lowest combination (bonding) is filled while the doubly degenerate *e* set (anti-bonding) is empty.

state, as shown in Fig. 10.8 [65]. The "trefoil" bonding state, previously proposed and predicted to exist in aromatic hydrocarbon annulene molecules, is finally encountered, albeit in an extended intermetallic network [65]. This also highlights the important role of interactions between incompletely filled lone pairs in the stabilization of low-dimensional anion structures.

Our work described in this section clearly illustrates the importance of the nature of the cations (size, charges, electronegativities), electronegativity differences, electronic factors, and matrix effects in the structural preferences of polar intermetallics. Interplay of these crucial factors lead to important structural adaptations and deformations. We anticipate exploratory synthesis studies along the Zintl border will further result in the discovery of novel crystal structures and unique chemical bonding descriptions.

10.6
Some Final Remarks

The unprecedented Zintl phases reported herein confirm our original ideas on the richness of intermetallic chemical systems and the possibilities of unique inorganic π-systems among "electron-deficient" Zintl phases. The discovery of these extraordinary compounds strongly suggests the possibility of finding other unusual inorganic "electron-poor" Zintl phases with novel bonding. Overall, our studies emphasize the importance of correct synthesis, and structural, physical and theoretical characterizations in understanding the chemical bonding in Zintl phases, which lie along the border between semiconductors and metals. Moreover, the possibilities of novel electronic properties are emphasized by the variety of electronic behavior exhibited by polar intermetallic π-systems. These range from narrow-gap semiconductors to excellent metals, and even superconductors.

References

1 a) Grube, G. *Angew. Chem.* **1935**, *48*, 714; b) Hägg, G. *Angew. Chem.* **1935**, *48*, 720.

2 *Intermetallic Compounds*; Westbrook, J.H., Ed.; Wiley, New York, 1967.

3 Karsten, K. *Pogg. Ann.* **1839**, *2*, 160.

4 a) Tamman, G. *Z. Anorg. Chem.* **1905**, *45*, 24; b) Tamman, G. *Z. Anorg. Chem.* **1906**, *47*, 296; c) Tamman, G. *Z. Anorg. Allg. Chem.* **1903**, *37*, 303; d) Tamman, G. *Z. Anorg. Chem.* **1919**, *107*, 1.

5 a) Kurnakov, N.S. *Z. Anorg. Chem.* **1914**, *88*, 109; b) Smirnov, V.I.; Kurnakov, N.S. *Z. Anorg. Allg. Chem.* **1911**, *72*, 31.

6 a) Desch, C.H. *Intermetallic Compounds*; Longmans Green, London, 1914; b) Desch, C.H. *Chemistry of Solids*; Cornell University Press, Ithaca, NY, 1934, pp. 153–160.

7 Hume-Rothery, W. *J. Inst. Metals* **1926**, *35*, 307.

8 Westgren, A.; Phragmen, G. *Arkiv. Mat. Astron. Fysik.* **1926**, *19B*, 1.

9 Mott, N.F.; Jones, H. *Theory of the Properties of Metals and Alloys*; Clarendon, Oxford, 1936.

10 Pauling, L.; Ewing, F.J. *Rev. Modern. Phys.* **1948**, *20*, 112.

11 Heine, V. In *Phase Stability in Metals and Alloys*; Rudman, P.S.; Stringer, J.S.; Jafee, R.I., Eds.; McGraw-Hill, New York, 1966, pp. 135–142.

12 Pettifor, D.G. In *Physical Metallurgy*, Cahn, R.W.; Haasen, P., Eds.; Elsevier, Amsterdam 1983, Chapter 3.

13 Massalski, T.B.; Mizutani, V. *Prog. Mater. Sci.* **1972**, *22*, 151.

14 Mizutani, U.; Takeuchi, T.; Sato, H. *Prog. Mater. Sci.* **2004**, *49*, 227.

15 Hoistad, M.; Lee, S. *J. Am. Chem. Soc.* **1991**, *113*, 8216.

16 Biltz, W. *Angew. Chem.* **1935**, *48*, 729.

17 a) Laves, F.; Witte, H. *Metallwirtschaft* **1935**, *14*, 645; b) Laves, F. *Naturwissenschaften* **1939**, *27*, 65; c) Laves, F. in *Intermetallic Compounds*; Westbrook, J.H., Ed.; Wiley, New York, 1967, pp. 129–143.

18 a) Frank, F.C.; Kasper, J.S. *Acta Crystallogr.* **1958**, *11*, 184; b) Frank, F.C.; Kasper, J.S. *Acta Crystallogr.* **1959**, *12*, 483.

19 Laves, F. In *Intermetallic Compounds*; Westbrook, J.H., Ed.; Wiley, New York, 1967, pp. 85–99.

20 Berry, R.L.; Raynor, G.V. *Acta Crystallogr.* **1953**, *6*, 178.

21 a) Stein, F.; Palm, M.; Sauthoff, G. *Intermetallics* **2004**, *12*, 713; b) Stein, F.; Palm, M.; Sauthoff, G. *Intermetallics*, **2005**, *13*, 1056.

22 a) Sinha, A.K. *Prog. Mater. Sci.* **1972**, *15*, 79; b) Simon, A. *Angew. Chem.* **1983**, *95*, 94.

23 a) Johannes, R.L.; Haydock, R.; Heine, V. *Phys. Rev. Lett.* **1976**, *36*, 372; b) Machlin, E.S.; Loh, B.; *Phys. Rev. Lett.* **1981**, *47*, 1087.

24 a) Ohba, T.; Kitano, Y.; Komura, Y. *Acta Crystallogr.* **1984**, *C40*, 1; b) Komura, Y.; Kitano, Y. *Acta Crystallogr.* **1977**, *B33*, 2496; c) Komura, Y.; Mitarai, M.; Nakatani, I.; Iba, H.; Shimizu, T. *Acta Crystallogr.* **1970**, *B26*, 666.

25 Pauling, L. *Phys. Rev. Lett.* **1981**, *47*, 277.

26 a) Nesper, R.; Miller, G.J. *J. Alloys Compds.* **1993**, *197*, 109; b) Häussermann U.; Worle, M.; Nesper, R. *J. Am. Chem. Soc.* **1996**, *118*, 11789.

27 Zhu, J.H.; Liu, C.T.; Pike, L.M.; Liaw, P.K. *Intermetallics* **2002**, *10*, 579.

28 a) Hafner, J. *From Hamiltonians to Phase Diagrams*; Solid State Sciences, Vol. 70; Springer, Berlin, 1987, Chapter 8 and references therein; b) Amerioun, S.; Yokoshawa, T.; Lidin, S.; Häussermann, U. *Inorg. Chem.* **2004**, *43*, 4751.

29 a) Pauling, L. *The Nature of the Chemical Bond*; Cornell Univ. Press, Ithaca, NY, 1960, Chapter 3; b) Watson, R. E.; Bennet, L. H.; *Phys. Rev. B* **1978**, *18*, 6439.

30 Pearson, W. B. *The Crystal Chemistry and Physics of Metals and Alloys*; Wiley-Interscience, New York, 1972, Chapter 5.

31 a) Zintl, E.; Dullenkopf, W. *Z. Phys. Chem.* **1932**, *B16*, 195; b) Zintl, E.; Neumayr, S. *Z. Phys. Chem.* **1933**, *B20*, 272; c) Zintl, E.; Schneider, A. *Z. Electrochem.* **1934**, *40*, 588; d) Zintl, E.; Schneider, A. *Z. Electrochem.* **1934**, *40*, 107; e) Zintl, E.; Harder, A. *Z. Phys. Chem.* **1936**, *B34*, 238; f) Zintl, E. *Angew. Chem.* **1939**, *52*, 1.

32 Laves, F. *Naturwissenschaften* **1941**, *29*, 244.

33 Hume-Rothery, W. In *Phase Stability in Metals and Alloys*; Rudman, P. S.; Stringer, J.; Jaffee, R. I., Eds.; McGraw-Hill, New York, 1967, pp. 3–23.

34 a) Parthe, E. *Z. Krist.* **1961**, *115*, 52; b) Girgis, K. In *Physical Metallurgy*, Cahn, R. W.; Haasen, P., Eds.; Elsevier: Amsterdam, 1983, Vol. 1, pp. 220–269.

35 a) Pettifor, D. G. *Phys. Rev. Lett.* **1984**, *53*, 1080; b) Pettifor, D. G. *J. Phys. C* **1986**, *19*, 285; c) Guenee, L.; Yvon, K. *J. Alloys Compds* **2003**, *356/357*, 114; d) Harada, Y.; Morinaga, M.; Saito, J.; Takagi, Y. *J. Phys. C.* **1997**, *9*, 8011; e) Clark, P. M.; Lee, S.; Fredrickson, D. C. *J. Solid St. Chem.* **2005**, *178*, 1269.

36 Pettifor, D. G. *J. Phys. Conden. Matter,* **2003**, *15*, V13.

37 a) Zintl, E.; Brauer, G. *Z. Phys. Chem.* **1933**, *B20*, 245; b) Zintl, E.; Kaiser, H. *Z. Anorg. Chem.* **1933**, *211*, 113; c) Zintl, E.; Harder, A. *Z. Phys. Chem.* **1932**, *B16*, 206; d) Brauer, G.; Zintl, E. *Z. Phys. Chem.* **1937**, *B37*, 323; f) Zintl, E.; Harder, A.; Dauth, B. *Z. Electrochem.* **1934**, *40*, 588.

38 a) Klemm, W. *Proc. Chem. Soc. London* **1958**, 329; b) Busmann, E. *Z. Anorg. Allg. Chem.* **1961**, *313*, 90; c) Klemm, W.; Busmann, E. *Z. Anorg. Allg. Chem.* **1963**, *319*, 297.

39 Kubaschewski, O.; Villa, H. *Z. Electrochem.* **1949**, *53*, 32.

40 Klemm, W. *Z. Electrochem.* **1945**, *51*, 14.

41 a) Schäfer, H.; Eisenmann, B.; Müller, W. *Angew. Chem. Int. Ed. Engl.* **1973**, *12*, 694; b) Schäfer, H.; Eisenmann, B. *Rev. Inorg. Chem.* **1981**, *3*, 29; Schäfer, H. *Ann. Rev. Mater. Sci.* **1985**, *15*, 1.

42 a) von Schnering, H. G. *Angew. Chem. Int. Ed. Engl.* **1981**, *20*, 33; b) von Schnering, H. G.; Hönle, W. *Chem. Rev.* **1988**, *88*, 243.

43 a) Nesper, R. *Prog. Sol. St. Chem.* **1990**, *20*, 1; b) Nesper, R. *Angew. Chem. Int. Ed. Engl.* **1991**, *30*, 789; c) Nesper, R.

44 a) Corbett, J. D. *Angew. Chem. Int. Ed. Engl.* **2000**, *39*, 670; b) Corbett, J. D. *Inorg. Chem.* **2000**, *39*, 5178; c) Corbett, J. D. *Struct. Bonding* **1997**, *87*, 157.

45 a) Schmidt, P. C. *Struct. Bonding* **1987**, *65*, 91; b) Genser, O.; Hafner, J. *J. Phys. Cond. Mat.* **2001**, *13*, 959; c) Seifert-Lorenz, K.; Hafner, J. *Phys. Rev. B* **1999**, *59*, 829; d) Pawlowska, Z.; Christensen, N. E.; Satpathy, S.; Jepsen, O. *Phys. Rev. B* **1986**, *34*, 7080.

46 a) McNeil, M. B.; Pearson, W. B.; Bennet, L. H.; Watson, R. E. *J. Phys. C* **1973**, *6*, 1; b) Watson, R. E.; Bennet, L. H. In *Charge Transfer/Electronic Structure of Alloys*; Bennet, L. H.; Willens R. H., Eds.; Metall. Soc. AIME, New York, 1974, pp. 1–25.

47 a) Anderson, P. W. *Proc. Robert A. Welch Found. Conf. Chem. Res.* **1988**, *32*, 1; b) van Schilfgaarde, M.; Harrison, W. A. *Phys. Rev. B* **1986**, *33*, 2653; c) Thygesen, K. S.; Hansen, L. B.; Jacobsen, K. W. *Phys. Rev. Lett.* **2005**, *94*, 026405.

48 Belin, C. H. E.; Tillard-Charbonnel, M. *Prog. Sol. St. Chem.* **1993**, *22*, 59.

49 Wade, K. *Adv. Inorg. Chem. Radiochem.* **1976**, *18*, 1.

50 a) Dong, Z. C.; Corbett, J. D. *J. Am. Chem. Soc.* **1993**, *115*, 11299; b) Kaskel, S.; Corbett, J. D. *Inorg. Chem.* **2000**, *39*, 778; c) Sevov, S.; Corbett, J. D. *Inorg. Chem.* **1991**, *30*, 4875; d) Klemm, M. T.; Vaughey, J. T.; Harp, J. G.; Corbett, J. D. *Inorg. Chem.* **2001**, *40*, 7020; e) Flot, D. M.; Tillard-Charbonnel, M.; Belin,

C. H. E. *J. Am. Chem. Soc.* **1996**, *118*, 5229.

51 a) Ghosh, S.; Beatty, A. M.; Fehlner, T. P. *J. Am. Chem. Soc.* **2001**, *123*, 9188; b) King, R. B.; *Inorg. Chem.* **2002**, *41*, 4722; c) Wadepohl, H. *Angew. Chem. Int. Ed. Engl.* **2002**, *41*, 4220.

52 a) Guloy, A. M.; Corbett, J. D. *Inorg. Chem.* **1996**, *35*, 2616; b) Seo, D. K.; Corbett, J. D. *J. Am. Chem. Soc.* **2000**, *122*, 9621; c) Seo, D. K.; Corbett, J. D. *J. Am. Chem. Soc.* **2001**, *123*, 415; d) Mao, J. G.; Guloy, A. M. *J. Alloys Compds* **2004**, *363*, 143.

53 a) Miller, G. J. in *Chemistry, Structure and Bonding of Zintl Phases and Ions*; Kauzlarich, S. M., Ed.; VCH Publishers, New York, 1996; b) Guloy, A. M.; Xu, Z.; Goodey, J. *ACS Symp. Ser.* **1999**, *727*, 2.

54 Xu, Z.; Guloy, A. M. *J. Am. Chem. Soc.* **1997**, *119*, 10541.

55 Xu, Z.; Guloy, A. M. *J. Am. Chem. Soc.* **1998**, *120*, 7349.

56 Fässler, T. F.; Kronseder, C. *Angew. Chem. Int. Ed. Engl.* **1997**, *36*, 2683.

57 a) von Schnering, H. G.; Bolle, U.; Curda, J.; Peters, K.; Carillo-Cabrera, W.; Somer, M.; Schultheiss, W.; Wedig, U. *Angew. Chem. Int. Ed. Engl.* **1996**, *35*, 984; b) Nesper, R.; Curda, J.; von Schnering, H. G. *J. Solid State Chem.* **1986**, *62*, 199.

58 a) Müller, W.; Schäfer, H.; Weiss, A. *Z. Naturforsch. B* **1970**, *25*, 1371; b) Nesper, R.; Currao, A.; Wengert, S. *Chem. Eur. J.* **1998**, *4*, 2251; c) Currao, A.; Wengert, S.; Nesper, R.; Curda, J.; Hillebrecht, H. *Z. Anorg. Allg. Chem.* **1996**, *622*, 501; d) Currao, A.; Nesper, R. *Angew. Chem. Int. Ed.* **1998**, *37*, 841; e) Wengert, S.; Nesper, R. *Inorg. Chem.* **2000**, *39*, 2861; f) Wörle, M.; Nesper, R. *Angew. Chem. Int. Ed. Engl.* **2000**, *39*, 2349.

59 Goodey, J.; Mao, J.; Guloy, A. M. *J. Am. Chem. Soc.* **2000**, *122*, 10478–10479.

60 Meerchaut, A.; Rouxel, J. in *Crystal Chemistry and Properties of Materials with Quasi-one Dimensional Structures*; Rouxel, J., Ed.; Reidel, Dordrecht, 1986, p. 205.

61 Goodey, J. *PhD Dissertation*, University of Houston, **2001**.

62 Lichtenberger, D. L.; Hoppe, M. L.; Subramanian, L.; Kober, E. M.; Hughes, R. P.; Hubbard, J. L.; Tucker, D. S. *Organometallics* **1993**, *12*, 2025.

63 Lupu, C.; Downie, C.; Guloy, A. M.; Albright, T. A.; Mao, J. G. *J. Am. Chem. Soc.* **2004**, *126*, 4386.

64 Burdett, J. K. *Chemical Bonding in Solids*; Oxford University Press, New York, 1995, 69.

65 Fukunaga, T.; Simmons, H. E.; Wendoloski, J. J.; Gordon, M. D. *J. Am. Chem. Soc.* **1983**, *105*, 2729.

11
Rare-earth Zintl Phases:
Novel Magnetic and Electronic Properties

Susan M. Kauzlarich and Jiong Jiang

11.1
Introduction

As a postdoctoral associate in Professor John Corbett's group in the late 1980s, I experienced the excitement of synthesizing new compounds whose structures could not have been predicted and were totally unknown. The idea of being able to produce new compounds has inspired my research pursuits ever since. While much of my interest is focused on the properties of compounds, this research must always start with synthesis. I have been fortunate to find a formula that provides new compounds, sometimes with unexpected structures, and most certainly with unexpected properties. I have exploited the Zintl concept towards this end by investigating rare-earth and transition-metal-containing Zintl phases [1–4]. The classical Zintl concept requires completed charge transfer from an alkali or alkaline earth metal to a post-transition element from group 13–15 to form a valence-precise intermetallic compound. Applying this concept more broadly has provided a mechanism for the discovery of more complicated intermetallic compounds with novel physical properties such as thermoelectricity [5, 6] and colossal magnetoresistance [7–10].

One family of compounds that has received significant attention are the rare-earth-containing transition-metal analogs of $Ca_{14}AlSb_{11}$ structure type [11], $A_{14}MnPn_{11}$ [2, 4, 12–15]. The structure can be rationalized as an ionic compound composed of $14\,Ca^{2+}$ ions, $4\,Sb^{3-}$ anions, a Sb_3^{7-} triatomic unit, and a $AlSb_4^{9-}$ tetrahedron. The bonding can be considered within the extended description of a Zintl compound [16]. Compounds of this structure type with Mn replacing Al show long-range magnetic ordering, either ferro- or anti-ferromagnetic, when the band gap is small enough. We have suggested that the long-range magnetic ordering involves spin exchange mediated by conduction electrons, such as a Rudderman-Kittel-Kasuya-Yosida (RKKY) type of interaction [17]. Substitution of the Ca sites by rare-earth metal adds more magnetic ions into the structure and makes the magnetic interaction more complicated. The magnetic interaction between a rare-earth and a transition metal is an indirect

Inorganic Chemistry in Focus III.
Edited by G. Meyer, D. Naumann, L. Wesemann
Copyright © 2006 WILEY-VCH Verlag GmbH & Co. KGaA, Weinheim
ISBN: 3-527-31510-1

interaction which involves an intra-atomic f-d interaction within the rare-earth metal and an inter-atomic d-d interaction between a rare-earth and a transition metal [18]. At the same time, RKKY interaction remains the major interaction between transition-metal sites.

Of the various phases with the $Ca_{14}AlSb_{11}$ structure type, the $Eu_{14}MnPn_{11}$ shows the largest magnetoresistance (116–300% at 5 T) [10]. Eu Mössbauer studies suggested that Eu may also be present in this compound as both Eu^{2+} and Eu^{3+} [13]. In order to investigate this further, we decided to try various flux methods [19, 20] to synthesize large crystals. To date, we have not been successful in preparing large crystals of $Eu_{14}MnP_{11}$. However, we were successful in synthesizing large crystals of a number of compounds [21–26].

In particular, we have found unusual magnetic properties within a series of Eu-In-P compounds that we have recently synthesized: Eu_3InP_3, $Eu_3In_2P_4$, and $EuIn_2P_2$ [24–26]. The first two can be described as classical Zintl phases and the third shows semi-metallic properties. $EuIn_2P_2$ may also be a Zintl phase with the semi-metallic properties attributed to adventitious crossing of the valence band with the conduction band.

Synthesis The Eu-In-P phases are prepared from an indium flux. In a typical synthesis, the elements are placed in a 2 ml cylindrical alumina crucible with the In on both the top and bottom of the crucible. A second crucible is filled with quartz wool, inverted and put on top of the first crucible and the entire system is sealed in a fused silica ampoule under 1/5 atm of argon. The sealed reaction vessel was then heated, typically to 1100 °C for a period of 1 hour or more in order that all components are in the melt, then cooled slowly over a particular temperature range, with the hope that crystals of the particular phase that we are interested in will precipitate from the flux and grow. Finally, the reaction vessel is removed from the furnace at an elevated temperature, inverted, and centrifuged. This removes the metal flux from the surface of the crystals. When first exploring a phase diagram, crystals of 1–2 mm^3 can be obtained, which are sufficient for X-ray crystallography. After optimization, large crystals, 200 mg and heavier, can be grown for further physical characterization. Eu_3InP_3, and $Eu_3In_2P_4$ are black needle-shaped crystals that are air- and moisture-sensitive. $EuIn_2P_2$ grows as black plates that are air- and moisture-stable.

11.2
Structure

The structures of Eu_3InP_3, $Eu_3In_2P_4$, and $EuIn_2P_2$ are shown in Fig. 11.1. Eu_3InP_3 is isostructural to the orthorhombic structures of Ca_3InP_3 and Sr_3InP_3 [27, 28]. The structure consists of InP_4 tetrahedra that are corner-shared to form linear chains, isolated by Eu cations, as shown in Fig. 11.1. The In–P bond distances range from 2.585(2) to 2.617(2) Å and are similar to those found in the Ca and Sr analogs. The P–In–P bond angles are also similar to those of the alkaline

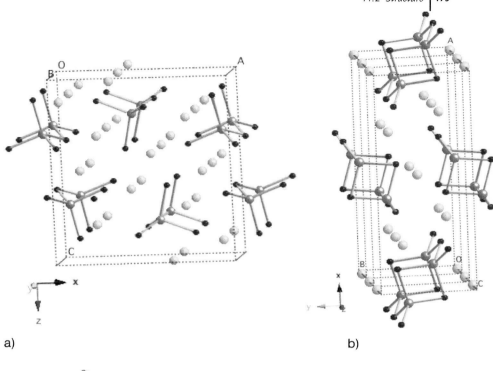

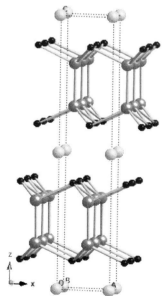

Fig. 11.1 Crystal structures of (a) Eu_3InP_3, (b) $Eu_3In_2P_4$, and
(c) $EuIn_2P_2$. The large, medium, and small balls represent Eu,
In, and P, respectively.

earth analogs. There are three crystallographically unique Eu ions. Eu(1) is coordinated by five phosphorus ions in a distorted square pyramidal coordination environment. Eu(2) is coordinated by six phosphorus ions in a distorted octahedral environment with a set of five between 2.989(2) and 3.166(2) Å with the sixth at 3.423(2) Å. Eu(3) is similar to Eu(2), being in a distorted octahedral environment. However, it is much more distorted with five distances between 3.001(2) and 3.155(2) Å and a sixth at 3.644(2) Å. There is also a short Eu–Eu distance between Eu(1) and Eu(2) of 3.5954(7) Å, significantly shorter than the 3.967 Å distance observed in europium metal. The Eu ions form triangles with edge distances of 3.5954(7), 3.8331(8), and 4.0851(7) Å. These triangles can be further connected into a 3D network containing tunnels that are outlined as edge-shared bi-pentagons in which the $[InP_3]^{2-}$ chains reside, shown in Fig. 11.2. In addition, there is a zig-zag chain of Eu atoms along the b axis with a distance of 3.9052(6) Å.

$Eu_3In_2P_4$ is isostructural with $Sr_3In_2P_4$ [29]. The compound is composed of edge-shared tetrahedral that are corner-shared and stacked to form a chain, isolated by Eu ions, as shown in Fig. 11.1. The In–P bond distances range from 2.5618(16) to 2.6369(10) Å, almost identical to those of $Sr_3In_2P_4$. The P–In–P angles are also almost identical to those of the Sr analog. There are two crystallographically unique Eu sites in this structure.

The structures of both Eu_3InP_3 and $Eu_3In_2P_4$ can be understood within the Zintl concept, as has been described for the alkaline earth analogs [27–29]. In

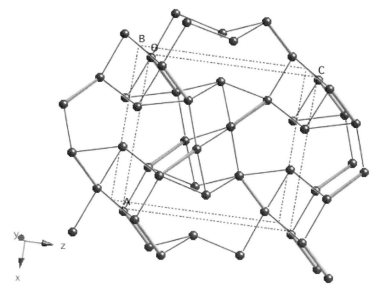

Fig. 11.2 A view of Eu_3InP_3 showing only Eu atoms. The thick lines indicate the short Eu–Eu bond distance and the thin lines indicate distances between 3.6 and 4.2 Å.

both cases, Eu is considered to be Eu^{2+}, confirmed by both magnetic and Mössbauer data and electron counting confirms the closed-shell valence precise nature of the anionic chains.

$EuIn_2P_2$ is a new structure type and is quite different from either of the two most common AB_2X_2 structure types, $ThCr_2Si_2$ and $CaAl_2Si_2$ [30]. $EuIn_2P_2$ contains alternating Eu and In_2P_2 layers. The oxidation state of Eu was determined to be +2, based on magnetic susceptibility measurements (described below). Therefore, each In_2P_2 unit has a charge of −2. The In–In bond distance is 2.7608(10) Å, a fairly typical In–In bond length in intermetallics such as SrInGe [31] and slightly longer than in coordination compounds such as $[(C_6H_5)_4P]_2[In_2Cl_6]$, which are considered to be In^{+1} [32]. The In–P bond distance is 2.6161(8) Å, in good agreement with the In–P bond distances described above. This structure has just one crystallographic Eu site. The Eu–Eu distance in this compound is 4.0829(6) Å, much longer than the shortest Eu–Eu distances in Eu_3InP_3 (3.5954(7) Å) and $Eu_3In_2P_4$ (3.7401(6) Å).

11.3
Resistivity

Temperature-dependent resistivity data ($\ln \rho$ vs $1/T$) for both Eu_3InP_3 and $Eu_3In_2P_4$ are shown in Fig. 11.3 and indicate that they are semiconductors. The room-temperature resistivities are on the order of 1–100 Ω cm. Band gaps were determined by fitting the data from about 130–300 K to the relationship, $\ln \rho = E_g/2k_BT + f$, providing a band gap, E_g, of approximately 0.5 eV for both samples. Since these two compounds can be rationalized as electron-precise Zintl phases, semiconducting behavior is expected.

$EuIn_2P_2$ shows a rather low resistance at room temperature of 10^{-3} Ω cm. The temperature dependence, shown in Fig. 11.4, is suggestive of a semi-metal. The resistivity decreases as a function of decreasing temperature until about 100 K, at which temperature it increases dramatically to a maximum coincident

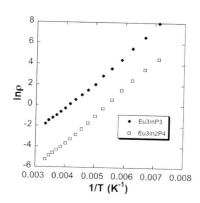

Fig. 11.3 $\ln \rho$ versus T^{-1} for Eu_3InP_3 and $Eu_3In_2P_4$.

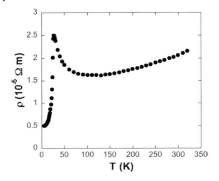

Fig. 11.4 Resistivity (ρ) versus temperature (T) for $EuIn_2P_2$.

with the magnetic ordering (see below) and then sharply decreases. There are some similarities in the temperature dependence of these data to the semiconductor–semimetal transition observed in EuB_6 [33] and in $Eu_{14}MnPn_{11}$ (Pn = Sb, As, P) [10, 34]. The transition in the resistivity at the magnetic ordering temperature suggests a strong correlation between resistivity and magnetism.

11.4
Magnetic Properties

Figure 11.5 shows the low-temperature magnetic behavior of Eu_3InP_3, $Eu_3In_2P_4$, and $EuIn_2P_2$. In the case of Eu_3InP_3, there are clearly two magnetic ordering transitions observed at about 14 K and 10.4 K. These transitions have been confirmed as long-range magnetic ordering by both temperature-dependent Eu Mössbauer spectroscopy and heat capacity [24]. Eu Mössbauer spectroscopy could be fitted with three components which are assigned to the three crystallographically inequivalent Eu sites. A model for the magnetic behavior has been suggested with the two Eu sites that have a close Eu–Eu distance ordering first. The third site may experience some magnetic frustration resulting from the anti-ferromagnetic coupling of the initially ordered sites and a triangular arrangement of this third Eu site with respect to the ordered sites. This may give rise to the slight differences observed in the field-cooled (FC) and zero-field-cooled (ZFC) data presented in Fig. 11.5.

Eu$_3$In$_2$P$_4$ shows long-range magnetic ordering at 14.5 K. The data for the three directions of the single crystal are consistent with anti-ferromagnetic behavior. However, magnetic susceptibility as a function of field shows that the magnetic moment can be saturated at approximately 1 Tesla for all crystal orientations. In addition, the Weiss constant is positive, indicating ferromagnetic interactions at high temperatures. A model where the Eu moments are in a canted anti-ferromagnetic alignment at low fields that can easily be saturated in a ferromagnetic fashion, has been proposed to account for the magnetic data.

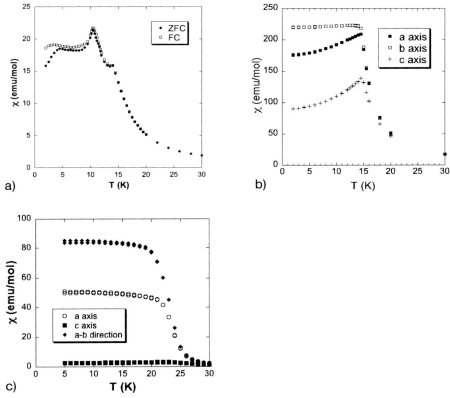

Fig. 11.5 Magnetization versus temperature for (a) Eu_3InP_3 powder, (b) $Eu_3In_2P_4$ single crystal, and (c) $EuIn_2P_2$ single crystal.

$EuIn_2P_2$ magnetic susceptibility shows a clear magnetic transition at about 24 K. In Fig. 11.5 c, magnetic susceptibility as a function of the a, $a–b$, and c direction were obtained on a plate crystal. The a axis exhibits the largest susceptibility and is the easy magnetization direction. The c axis is the hard magnetization direction.

In all cases, the high-temperature magnetic susceptibility data are consistent with Eu being all Eu^{2+}.

11.5
Magnetoresistance

Figure 11.6 provides the magnetoresistance data for $Eu_3In_2P_4$ and $EuIn_2P_2$. Both compounds show negative magnetoresistance. In the case of $Eu_3In_2P_4$, the effect occurs at the magnetic ordering temperature. Since the compound is a semiconductor, the magnetoresistance is attributed to the reduction of spin disorder due to

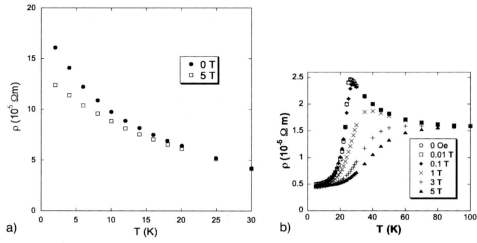

a)   b)

Fig. 11.6 Resistivity (ρ) versus temperature (T) and applied field (0–5 Tesla) for (a) $Eu_3In_2P_4$ and (b) $EuIn_2P_2$.

the presence of the ferromagnetic state of the compound at 3 Tesla. The magnetic model described above would be consistent with the magnetoresistance observed. At 5 Tesla, the maximum magnetoresistance is −30% (magnetoresistance = $[(\rho(H)-\rho(0))/\rho(H)]\times100\%$] where ρ = resistivity, H = applied field; $\rho(0)$ = resistivity at zero applied field) at the lowest temperature measured, 2 K.

$EuIn_2P_2$ shows colossal magnetoresistance with a negative magnetoresistance onset at about 90 K. The magnetoresistance increases to reach a maximum of −398% at approximately 25 K with an applied field of 5 Tesla. In this case, similar to colossal magnetoresistance observed in $Eu_{14}MnPn_{11}$ (Pn = Sb, As, P) and EuB_6, there must be correlation between the spin of the electrons in the conduction band and the spin on the Eu^{2+} cations. This may be due to an RKKY type interaction, as described for $Eu_{14}MnPn_{11}$ compounds [7, 10], or band-splitting, such as proposed for EuB_6 [33].

11.6
Summary

We have shown that even relatively simple Zintl phases can have complex magnetic and electronic properties. Eu_3InP_3 and $Eu_3In_2P_4$ can be regarded as classical Zintl phases and because they are semiconductors, long-range magnetic ordering was unexpected. However, both show unusual behavior. Eu_3InP_3 has three Eu^{2+} sites that can independently order due to the different local environment and energy landscape of each site. $Eu_3In_2P_4$ shows one magnetic ordering. The combination of magnetic and transport data suggests that this system is a canted anti-ferromagnet which ferromagnetically orders at fields above 1

Tesla, giving rise to negative magnetoresistive effects. $EuIn_2P_2$ is a unique example among the three in that it does not conform to Zintl-type formulism and shows semi-metallic properties and colossal magnetoresistance. More detailed studies of the $EuIn_2P_2$ system to find analogs and correlate structure and properties may provide new insights into structural and electronic requirements necessary for colossal magnetoresistance.

While our original goal of this study, to produce large crystals of $Eu_{14}MnP_{11}$ for detailed characterization, has been unsuccessful to date, we have discovered a number of new compounds with unique magnetic and electronic properties, illustrating the complexity of synthesis and the richness of solid-state structures. This work illustrates the point that new and unexpected magnetic and electronic behavior can be found where it is least expected, providing further justification for synthetic exploration.

Acknowledgments

We thank Marilyn Olmstead, Han-Oh Lee, Peter Klavins, and Zachary Fisk for useful discussions. We acknowledge the funding of the NSF (DMR- 0120990). JJ was supported in part by a Tyco Fellowship.

References

1 Kauzlarich, S. M. *Comments Inorg. Chem.* **1990**, *10*, 75.

2 Kauzlarich, S. M. In Chemistry, Structure, and Bonding of Zintl Phases and Ions; Kauzlarich, S. M., Ed.; VCH Publishers, Inc., New York, 1996, p 245–274.

3 Kauzlarich, S. M.; Chan, J. Y.; Taylor, B. R. In Inorganic Materials Synthesis; Winter, C. H., Hoffman, D. M., Eds.; American Chemical Society, Washington D.C., 1999; Vol. ACS Symposium Series Vol. 727, p 15–27.

4 Kauzlarich, S. M.; Payne, A. C.; Webb, D. J. In Magnetism: Molecules to Materials III; Miller, J. S., Drillon, M., Eds.; Wiley-VCH, Weinheim, 2002, p 37–62.

5 Mahan, G.; Sales, B.; Sharp, J. *Physics Today* **1997**, *50*, 42–47.

6 Kim, S.-J.; Hu, S.; Uher, C.; Kanatzidis, M. G. *Chem. Mater.* **1999**, *11*, 3154–3159.

7 Chan, J. Y.; Kauzlarich, S. M.; Klavins, P.; Shelton, R. N.; Webb, D. J. *Chem. Mater.* **1997**, *9*, 3132–3135.

8 Chan, J. Y.; Kauzlarich, S. M.; Klavins, P.; Shelton, R. N.; Webb, D. J. *Phys. Rev. B* **1998**, *57*, 8103–8106.

9 Chan, J. Y.; Kauzlarich, S. M.; Klavins, P.; Liu, J.-Z.; Shelton, R. N.; Webb, D. J. *Phys. Rev. B* **2000**, *61*, 459–463.

10 Payne, A. C.; Olmstead, M. M.; Kauzlarich, S. M.; Webb, D. J. *Chem. Mater.* **2001**, *13*, 1398–1406.

11 Cordier, G.; Schäfer, H.; Stelter, M. *Z. Anorg. Allg. Chem.* **1984**, *519*, 183–188.

12 Sánchez-Portal, D.; Martin, R. M.; Kauzlarich, S. M.; Pickett, W. E. *Phys. Rev. B* **2002**, *65*, 144414.

13 Hermann, R. P.; Grandjean, F.; Kauzlarich, S. M.; Jiang, J.; Brown, S.; Long, G. J. *Inorg. Chem.* **2004**, *43*, 7005–7013.

14 Holm, A. P.; Ozawa, T. C.; Kauzlarich, S. M.; Morton, S. A.; Waddill, G. D.; Tobin, J. G. *J. Solid State Chem.* **2005**, *178*, 262–269.

15 Kim, H.; Kauzlarich, S. M. *J. Solid State Chem.* **2005**, *178*, 1935–1939.

16 Chemistry, Structure, and Bonding of Zintl Phases and Ions; Kauzlarich, S. M.,

Ed.; VCH Publishers, Inc., New York, 1996.

17 Rehr, A.; Kuromoto, T. Y.; Kauzlarich, S. M.; Del Castillo, J.; Webb, D. J. *Chem. Mater.* **1994**, *6*, 93–99.

18 de Boer, F. R.; Zhao, Z. G. *Physica B* **1995**, *211*, 81–86.

19 Fisk, Z.; Canfield, P. C. In Handbook on the Physics and Chemistry of Rare Earths; Gschneidner, J., K. A., Eyring, L., Eds.; Elsevier, Amsterdam, 1989, Vol. 12.

20 Canfield, P. C.; Fisk, Z. *Philosophical Magazine B* **1992**, *65*, 1117–1123.

21 Payne, A. C.; Sprauve, A. E.; Holm, A. P.; Olmstead, M. M.; Kauzlarich, S. M.; Klavins, P. *J. Alloys Compd.* **2002**, *338*, 229–234.

22 Payne, A. C.; Sprauve, A. E.; Olmstead, M. M.; Kauzlarich, S. M.; Chan, J. Y.; Reisner, B. A.; Lynn, J. W. *J. Solid State Chem.* **2002**, *163*, 498–505.

23 Holm, A. P.; Park, S. M.; Condron, C. L.; Olmstead, M. M.; Kim, H.; Klavins, P.; Grandjean, F.; Hermann, R. P.; Long, G. J.; Kanatzidis, M. G.; Kauzlarich, S. M.; Kim, S. J. *Inorg. Chem.* **2003**, *42*, 4660–4557.

24 Jiang, J.; Payne, A. C.; Olmstead, M. M.; Lee, H. O.; Klavins, P.; Fisk, Z.; Kauz-

larich, S. M.; Hermann, R. P.; Grandjean, F.; Long, G. J. *Inorg. Chem.* **2005**, *44*, 2189–2197.

25 Jiang, J.; Olmstead, M. M.; Kauzlarich, S. M.; Lee, H. O.; Klavins, P.; Fisk, Z. *Inorg. Chem.* **2005**, *44*, 5322–5327.

26 Jiang, J.; Kauzlarich, S. M. *Chem. Mater.* in press.

27 Cordier, G.; Schäfer, H.; Stelter, M. *Z. Naturforsch.* **1985**, *40b*, 1100–1104.

28 Cordier, G.; Schäfer, H.; Stelter, M. *Z. Naturforsch.* **1987**, *42b*, 1268–1272.

29 Cordier, G.; Schäfer, H.; Stelter, M. *Z. Naturforsch.* **1986**, *41b*, 1416–1419.

30 Zheng, C.; Hoffmann, R. *J. Solid State Chem.* **1988**, *72*, 58–71.

31 Mao, J.-G.; Goodey, J.; Guloy, A. M. *Inorg. Chem.* **2002**, *41*, 931–937.

32 Bubenheim, W.; Frenzen, G.; Mueller, U. *Acta Crystallogr., Sect. C: Cryst. Struct. Commun.* **1995**, *C51*, 1120–1124.

33 Guy, C. N.; von Molnar, S.; Etourneau, J.; Fisk, Z. *Solid State Commun.* **1980**, *33*, 1055–1058.

34 Chan, J. Y.; Wang, M. E.; Rehr, A.; Kauzlarich, S. M.; Webb, D. J. *Chem. Mater.* **1997**, *9*, 2131–2138.

12

Understanding Structure-forming Factors and Theory-guided Exploration of Structure–Property Relationships in Intermetallics

Dong-Kyun Seo, Li-Ming Wu and Sang-Hwan Kim

12.1
Introduction

Intermetallic compounds are fascinating because they provide great challenges in the understanding of their chemical structures, bonding and properties, and hence in establishing their structure–property relationships. There are over 2500 structure types that have been identified [1], and yet the majority of the structure types have not been studied closely from the electronic point of view. The question of which structures are interesting can be answered differently, of course, depending on the physical, electronic or chemical properties of interest, and the focus of the studies may vary substantially. In any event, however, systematic exploration of materials properties can be helped by a certain degree of predictability in structure formations upon compositional changes and by well-established structure–property relationships. Since many important physical and chemical properties are governed by the electronic structure in a narrow energy region near the Fermi level, their studies require us to recognize the factors that are important in structure formation and how they can change the electronic structures near the Fermi level. Given the fact that the numerical solutions of quantum-mechanical equations have their limitation in structure predictions, one practical approach is the exploration of the structure-formation principles, by compositional modifications of known compounds, which is buttressed by theoretical analysis of structures and properties. With two representative examples, this work illustrates our recent effort in examining structure-forming factors in intermetallics that contain d- and/or f-elements. We start by examining the two important structure-forming factors of inorganic solid-state compounds: covalent bonding and close packing.

While sharing of electrons, i.e., covalent bonding, is the major component of the cohesive force in intermetallics, rationalization of their structure formation based on such chemical bonding is not trivial, because of the failure of the common electron counting rules that chemists have developed over the years from the studies of covalent compounds. The origin of the problem is the well-delo-

Inorganic Chemistry in Focus III.
Edited by G. Meyer, D. Naumann, L. Wesemann
Copyright © 2006 WILEY-VCH Verlag GmbH & Co. KGaA, Weinheim
ISBN: 3-527-31510-1

calized multi-center bonding between metal atoms in intermetallics in contrast to the localized two-center two-electron bonding that is prevalent in highly covalent structures [2]. Therefore it is not, in general, possible to construct or understand electronic structures by using a simple description of local structures within the principle of pairwise bond formations. In addition, the multi-center bonding in intermetallics is relatively less directional, so that one specific structure may not be dominant in the landscape of covalent energy. The "nearly-free-electron" model describes the strong delocalization of electrons throughout the structure and justifies the reciprocal-space approach to the electron counting rule on main-group intermetallic compounds, particularly for Hume-Rothery electronic compounds [3]. Closely related to this is the moments method, which correlates structure with electron count for elemental metals and main-group intermetallics within the tight-binding approximation [4].

Close packing of atoms is broadly defined here as optimization of atomic arrangements by non-directional pairwise attractions and repulsions, and it has been utilized in the traditional way of describing and understanding crystal structures of intermetallics in the simplest form of hard-sphere packing among the nearest neighbors. The concept of pairwise interactions can be utilized beyond the hard-sphere approximation. For example, for main-group intermetallics formed from the elements with similar electronegativities and sizes, effective pair potentials obtained from pseudopotential theory may provide structural predictions by including the free-electron response to the pseudopotential in the form of damped oscillations [5]. In polar intermetallics, where electronegativities and sizes are much different between active metals and p-metals, close packing of cations and polyanions has been implicitly assumed by considering cations as passive space-fillers. However, recent advances in the area have signified the important role of active cations beyond close packing in structure formations [6]. Cation–anion interactions can also occur through orbital interactions, especially when the cations are alkaline-earth or rare-earth metals, and such interactions can influence the physical properties of the compounds by modifying electronic structure significantly near the Fermi level. The nature of close packing of cations amongst themselves may be different from that found in ionic compounds such as oxides or nitrides. Namely, the effective charges on cations in polar intermetallics can be significantly smaller than their formal charges and hence provide much weaker cation–cation repulsion.

12.2
$Mn_{14}Al_{56+x}Ge_{3-x}$ (x = 0.00, 0.32, 0.61)

The effect of chemical bonding on magnetism is particularly striking among the p-metal-rich transition-metal intermetallics in which the formation of spin polarization faces a tough battle against the electron pairing driven by bonding optimization in the intermetallic structures. This is in contrast to other classes of transition-metal compounds such as oxides, nitrides and chalcogenides in which the presence of magnetic moments are taken granted whenever the transition metal atoms have an open-shell d-electron configuration. Not to mention the prediction of magnetic coupling patterns, the survival of magnetic moment (spin polarization) itself and the prediction of the degree of spin polarization, which are already complicated problems that require various theories at different levels of sophistication [7]. The essence of the relationship between chemical bonding and spin polarization can be understood by applying Stoner theory within a rigid-band approximation. According to the theory, ferromagnetism can be predicted for a metal when $DOS(E_F) \cdot I_S > 1$ where $DOS(E_F)$ and I_S are the density of states at the Fermi level calculated for a non-magnetic state and the Stoner parameter, respectively [8]. When the covalent bonding is optimized in the structure, it is likely that the d-orbitals of transition metal atoms are strongly mixed with the s- and p-orbitals to result in a (pseudo) bandgap at the Fermi level [9]. Spin polarization is prevented by the chemical bonding optimization due to depletion of the $DOS(E_F)$ values.

Discovery of the aluminum-rich Mn$_{14}$Al$_{56+x}$Ge$_{3-x}$ alloys occurred from our intention to replace the silicon atoms in a quasicrystal approximant a-Mn-Al-Si (a close composition: Mn$_{17.4}$Al$_{72.5}$Si$_{10.1}$) by germanium [10]. The approximant and its related icosahedral quasicrystal i-Mn-Al-Si (a close composition: Mn$_{20}$Al$_{74}$Si$_6$) are known to exhibit a pseudo bandgap with low $DOS(E_F)$, which is consistent with their Pauli-paramagnetic behavior within the Stoner theory [11]. Disregarding the detailed variation of local structures over a long periodic or quasiperiodic undulation, however, it is also easy to recognize that the atoms are, in principle, close-packed tetrahedrally as in Frank-Kasper phases [12]. Compared to silicon, which is close to aluminum in size, a substantially large germanium atom can influence the packing of the constituting atoms without changing the valence electron concentrations, given the same composition. It was anticipated that any new structures in the aluminum-rich Mn-Al-Ge system might exhibit ill-optimized covalent bonding due to the influence of close packing and show a relatively large $DOS(E_F)$ resulting in a spin polarization.

A [001] projection view of the structure of Mn$_{14}$Al$_{56+x}$Ge$_{3-x}$ (space group: P$\bar{3}$) is shown in Fig. 12.1. A convenient way of describing the structure is first to identify structural building units that are reminiscent of the high-symmetry polyhedra found in the approximants and presumably quasicrystals. Such polyhedral structures exist around only one atomic site in the Mn$_{14}$Al$_{56+x}$Ge$_{3-x}$ structure, which is one of three inequivalent Mn positions. As shown in Fig. 12.2 a, the Mn position is surrounded by ten-membered Al$_9$Ge polyhedron as its first shell to form a [Mn@Al$_9$Ge] cluster (d(Mn–Al) = 2.43–2.54 Å; d(Mn–Ge) = 2.85 Å). The cluster can be considered as an incomplete variant of an icosahedron; i.e., it is derived by replacing one triangular face at the bottom of the icosahedron with one Ge atom.

The most remarkable feature in the Mn$_{14}$Al$_{56+x}$Ge$_{3-x}$ structure is noted in the second shell structure that surrounds the [Mn@Al$_9$Ge] clusters (Fig. 12.2 b). The top half of the shell manifests a dome structure made of all triangular faces, while the ill-shaped bottom half is formed with randomly fused triangular and

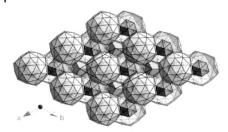

Fig. 12.1 Polyhedral view of the trigonal $Mn_{14}Al_{56+x}Ge_{3-x}$ structure along [001].

quadrangular faces. The apparently crushed shape of the bottom part is caused by the close-contact requirement for the Ge atom in the [Mn@Al$_9$Ge], with the surrounding Al atoms. The triangular faces of the dome structure share their edges and corners to form pentagons and hexagons, and indeed the structure corresponds to one-half part of a Mackay icosahedron (Fig. 12.2c) in the α-MnAl(Si) structure [13]. Fig. 12.1 shows the spatial arrangement of the half-broken Mackay icosahedra as the dome structures.

To examine the electronic structure of the $Mn_{14}Al_{56+x}Ge_{3-x}$, tight-binding linear muffin-tin orbital atomic sphere approximation (TB-LMTO-ASA) calculations were performed on the stoichiometric $Mn_{14}Al_{56}Ge_3$ as a representative model for the title phase. As shown in Fig. 12.3, the calculated total DOS reveals a pseudogap at the Fermi energy with a width of ~ 0.3 eV, indicative of a weakly conducting behavior. The pseudogap is much narrower than the ones

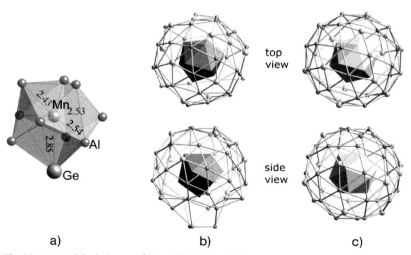

a)   b)   c)

Fig. 12.2 (a) Polyhedral view of [Mn@Al$_9$Ge]. (b) Ball-and-stick view of the shell structure around the [Mn@Al$_9$Ge] polyhedral cluster. The top and side views are along $\langle 001 \rangle$ and $\langle 110 \rangle$, respectively. In (c) a Mackay icosahedron in α-MnAl(Si) is oriented along the corresponding directions.

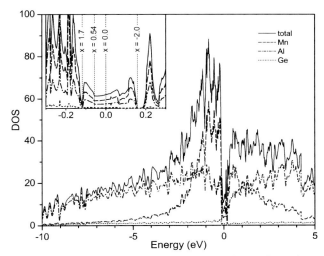

Fig. 12.3 DOSs (states/eV cell) for the $Mn_{14}Al_{56}Ge_3$ compound from TB-LMTO-ASA calculations. The Fermi energy (E_F) = 0 eV. The pseudogap region near the E_F is enlarged in the inset. The vertical dashed lines indicate the Fermi levels for the corresponding x values in the formula $Mn_{14}Al_{56+x}Ge_{3-x}$, within a rigid-band approximation.

found in other aluminides whose complicated structures do not contain high-symmetry polyhedra [11]. The calculated DOS(E_F) is ~0.15 states/eV per atom. The DOS(E_F) value is too small to provide a spontaneous spin polarization, and this is consistent with the physical properties observed in the alloys. The calculated DOS in Fig. 12.3 further provides insight into the question of the non-stoichiometries and partial substitution observed in our $Mn_{14}Al_{56+x}Ge_{3-x}$ phase. Within a rigid-band approximation, the VEC, and hence the E_F decreases as x increases from zero, i.e., as the Ge sites become partially substituted by Al atoms. As is clearly seen in the inset of Fig. 12.3, this results in a gradual decrease in the total DOS to a minimum at an energy of –0.054 eV which corresponds to x=0.54 (indicated by a vertical dashed line). This compares well with x=0.61(2) (the corresponding E_F=–0.059 eV), the highest partial occupancy observed crystallographically in our experiments. A similar explanation has already been suggested regarding the non-stoichiometry of α-Mn-Al(Si). The partial substitution of Al sites by Si atoms increases the Fermi energy to the point where the total DOS diminishes, and hence achieves structural stabilization with an optimum VEC [14, 15]. Such energy points, with zero DOS values, exist also in the total DOS curve of the $Mn_{14}Al_{56}Ge_3$ above and below the E_F, as shown in the inset of Fig. 12.3. With corresponding x values of –2.0 and 1.7 (67% increase and 57% decrease in Ge content, respectively). However, we suspect that such a degree of atomic substitutions may not maintain the $Mn_{14}Al_{56+x}Ge_{3-x}$ structure because of the more drastic size difference between Ge and Al atoms than between Si and Al.

In summary, it was found that the small size perturbation by Ge atoms in the Al-rich Mn–Al system could be easily accommodated in bonding optimization and that the presence of Mackay icosahedra was not important for the existence of the pseudo bandgap. Further studies will elucidate the extent of the bonding optimization capability of the Al-rich Mn–Al system upon introduction of various structure-perturbing factors such as differences in size, VEC and electronegativities. Interestingly, the $Mn_{14}Al_{56.6}Ge_{2.4}$ shows a gradual metal-to-semi-conductor-to-metal transition, which could be from the enhanced scattering of the electrons by the partial occupancy in the structure.

12.3
$La_{5-x}Ca_xGe_4$ (x = 3.37, 3.66, 3.82) and $Ce_{5-x}Ca_xGe_4$ (x = 3.00, 3.20, 3.26)

Metal-rich R_5Tt_4 compounds (R = rare earth metal; Tt = Ge, Si), especially Gd_5 $Si_{4-x}Ge_x$, are fascinating because of their closely related structural variations with common robust building units as well as their structural transformations accompanied by dramatic changes in their physical properties [16, 17]. R_5Tt_4 metal-rich Zintl compounds [18] exhibit both cation–cation and anion–anion bonding in their closely related structural variations, a-, β- and γ-types [17]. The alkaline-earth analogues are not known in the literature. All the structures contain two different kinds of Tt atoms based on their bonding behavior: two Tt1 and two Tt2 atoms (Fig. 12.4 a) in a formula unit. While all Tt2 atoms are dimerized in the structures (d (Tt–Tt) = 2.5–2.7 Å), the portions of dimerized Tt1 atoms are different: all Tt1 atoms in the a-type (Gd_5Si_4-types); only a half in the β-type (the monoclinic modification of $Gd_5Si_2Ge_2$); none in the γ-type (Sm_5Ge_4-type). Following the Zintl–Klemm concept [19], the a-, β- and γ-type structures are all *electron-rich* with chemical formulas $(R^{3+})_5(Tt_2^{6-})_2(3e^-)$, $(R^{3+})_5(Tt_2^{6-})_{1.5}(Tt^{4-})(2e^-)$, and $(R^{3+})_5(Tt_2^{6-})(Tt^{4-})_2(e^-)$, respectively. The bottom of the conduction bands consists of R–R and R–Tt bonding and Tt–Tt anti-bonding states [17 d]. The suggested electron-counting scheme of the R_5Tt_4 structures implies that the structure formations may also be affected by changes in VEC. Indeed, it has been demonstrated that the bond formation between the Tt1 atoms can be induced when a γ-type compound is deprived of some of the excess electrons by partial substitutions of Ge atoms with size-equivalent triel Ga in $Gd_5Ga_xGe_{4-x}$ ($x = 0$–2.2; $31 \geq VEC \geq 28.8$) [17 d] and in $La_5Ga_xGe_{4-x}$ ($x = 1$) [20]. When $x \geq 1$ ($30 \geq VEC$), the alloys formed in an a-type structure, and the highest Ga content ($x = 2.2$; VEC = 28.8) was still electron-rich. We were interested in examining the structural and electronic robustness of the alloys to the lowest VEC in experiments.

In our design, divalent Ca was chosen to partially substitute the trivalent atoms, and La and Ce were selected for a trivalent element because their ionic size ($r_{La3+} = 1.50$ Å; $r_{Ce3+} = 1.48$ Å) was close to that of Ca^{2+} ($r_{Ca2+} = 1.48$ Å) [21]. Like La, the Ce element also generally shows a formal +3 oxidation state in intermetallics. From the reactions of the elements, we have identified as major phases the electron-precise/deficient alloys, $Ln_{5-x}Ca_xGe_4$ (Ln = La, Ce; $x = 3.37$,

3.66, 3.82 for La; $x = 3.00$, 3.20, 3.26 for Ce) ($28 \geq \text{VEC} \geq 27.2$) which exhibit the lowest VECs observed in the R_5Tt_4 system [22]. With the observed VEC values, all the alloys are found to exhibit the a-type structure, and this is in line with the behavior of the $Gd_5Ga_xGe_{4-x}$.

In Fig. 12.4b, the dimer bond distances are shown with different VECs. Overall, the bond distances are shorter in electron-precise or deficient alloys than the electron-rich ones ($Gd_5Ga_xGe_{4-x}$). The dramatic decrease in Tt1–Tt1 in $Gd_5Ga_xGe_{4-x}$ is caused by the dimer bond formation, while the gradual decrease in Tt2–Tt2 can be understood from the previous theoretical results that the bottom of the conduction bands contain the σ_p^* anti-bonding character of the Ge dimers [17d]. From VEC = 29 to 28, bond shortening is likely due to the complete removal of the conduction electrons. Interestingly, further gradual decreases in the dimer bonds are apparent in the $Ln_{5-x}Ca_xGe_4$. This is understandable when we consider the filled π^* states present in the electron-precise Ge_2^{6-} dimers ($\sigma_s^2 \sigma_s^{*2} \sigma_p^2 \pi^4 \pi^{*4} \sigma_p^{*0}$) in a simplistic bonding scheme. The electron deficiency in the new alloys should occur in the π^* states that will be located largely in the top region of the valence bands. Therefore, the alloys can be described as $(Ln^{3+})_{5-x}(Ca^{2+})_x(Ge_2^{6-})_2(\gamma h^+)$, where h^+ represents a hole in the valence bands and $\gamma = x - 3$ ($0 \leq \gamma \leq 0.8$).

The TB-LMTO-ASA calculations were carried out on an electron-precise $La_2Ca_3Ge_4$ whose structure was modeled from the $La_{1.63}Ca_{3.37}Ge_4$ by having La and Ca atoms ordered according to the observed site preference [22]. In Fig. 12.5, the band structure shows an energy gap at the Fermi level for the electron-precise model alloy, predicting a semiconducting behavior. The Ge

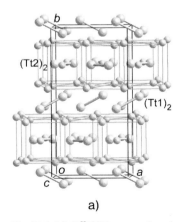

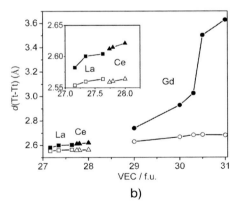

a)

b)

Fig. 12.4 (a) Off-[001] perspective view of the a-type orthorhombic structure of R_5Tt_4. Small and large spheres represent R and Tt, respectively. The β- and γ-type structures also have similar atomic arrangements but with different Tt–Tt bonding behavior.

(b) The dimer bond distances in $Ln_{5-x}Ca_xGe_4$ and $Gd_5Ga_xGe_{4-x}$ [17d] alloys plotted as a function of VEC. The solid and open circles correspond to Tt1–Tt1 and Tt2–Tt2 distances in the text.

atoms contribute much less compared to the La/Ca in the conduction bands, and the COHP curves indicate that the Ge–Ge anti-bonding character is well spread out over the conduction bands as well as at the top of the valence bands. This is consistent with the electron configuration of the Ge_2 dimers adopted in our previous discussions for the explanation of the trend in the bond distance changes.

The existence of the bandgap should be taken cautiously because the calculations were carried out on a model structure. Although not shown here, however, we consistently observed the energy gap in the band structures calculated for all the model structures with $x \geq 3$ even when the La and Ca atom positions were exchanged. This bandgap formation might be due to the smaller electronegativity of Ca and the shortened Ge–Ge distances. In any event, this is an interesting contrast to the behavior of the "same" Ge_2^{6-} in the likewise "precise" Ca_5Ge_3 in a different structure in which strong cation interactions evidently lead to the loss of the π^* shell, and Ge states appear above the Fermi level [6g].

In summary, the new $Ln_{5-x}Ca_xGe_4$ alloys show that the formation of the metal-rich Zintl compounds R_5Tt_4 can be extended to an electron-deficient region

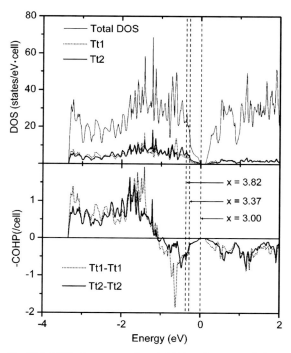

Fig. 12.5 DOSs (upper panel) and COHPs (lower panel) for a model compound, $La_2Ca_3Ge_4$. The vertical lines are the Fermi levels for the corresponding *x* values within a rigid-band approximation.

with *varied* VEC, and precise satisfaction of the Zintl-Klemm concept is found to be inessential in the structure formation. It is suspected that the existence of cation–cation bonding is not required for the stabilization of the R_5Tt_4-type structures either. The robust nature of the structures is likely due to the strong cation–anion bonding of the rare-earth ions both covalently and electrostatically. It is important that this provides at least some clue to the recently raised questions: how "extensively" can the active cations play their role in determining structure formations [6e, f]? The existence of the electron-deficient $Ln_{5-x}Ca_xGe_4$ alloys demonstrates that cation–anion bonding is the prevalent structure-forming factor at least in the R_5Tt_4 system and that closed-shell electron configuration is not essential. In addition, the curious formation of an energy gap between the valence and conduction bands, which does not appear originally in the R_5Tt_4, indicates that the electronic and physical properties can be modified dramatically in such a Zintl system even without changing the structure. We anticipate more of this type of example when they are *purposely* sought through specially designed synthesis, and it may provide a viable route in modifying the physical properties of the compounds.

12.4
Concluding Remarks

In studies of structure–property relationships, the designed synthesis of intermetallic materials requires good control over the structure formations on compositional changes as well as in-depth electronic analysis of the resulting properties. Compared to other areas, the chemistry of intermetallics is still in its early stages in the sense that the identified structure-forming factors have not been fully examined in the different structures to provide general answers to those structures which are useful and can be modified as desired. While examples are not unprecedented, the presented works emphasize careful studies on the structure formations and property changes by combining experiment and theory in order to explore and exploit the unique physicochemical properties of intermetallics. As once stated, opportunities clearly abound [23]!

Acknowledgments

D.-K. S. is grateful for financial support from the National Science Foundation in USA through his CAREER Award (DMR – Contract No. 0239837). He also greatly acknowledges helpful discussions with Profs. John D. Corbett and Ulrich Häussermann.

References

1 Villars, P. in *Crystal Structures of Intermetallic Compounds*, Westbrook, J. H.; Fleischer, R. L. Eds., John Wiley & Sons, New York, 2000.

2 (a) Zheng, C.; Hoffmann, R. *Z. Naturforsch.* **1986**, *41b*, 292. (b) Miller, G. J. in *Chemistry, Structure, and Bonding of Zintl Phases and Ions*, Kauzlarich, S. M. Ed., VCH, New York, 1996.

3 Hume-Rothery, W. *Electrons, Atoms, Metals, and Alloys*, 3 ed., Dover, New York, **1963**.

4 (a) Cyrot-Lackman, F. *Adv. Phys.* **1967**, *16*, 393. (b) Burdett, J. K.; Lee, S. *J. Am. Chem. Soc.* **1985**, *107*, 3063. (c) Lee, S. *Acc. Chem. Res.* **1991**, *24*, 249.

5 (a) Hafner, J.; Heine, V. *J. Phys. F* **1983**, *13*, 2479. (b) Hafner, J. *From Hamiltonians to Phase Diagrams*; Springer, Berlin, 1987.

6 (a) Corbett, J. D. *Angew. Chem. Int. Ed. Eng.* **2000**, *39*, 670. (b) Seo, D.-K.; Corbett, J. D. *J. Am. Chem. Soc.* **2000**, *122*, 9621. (c) Seo, D.-K.; Corbett, J. D. *J. Am. Chem. Soc.* **2001**, *123*, 4512. (d) Seo, D.-K.; Corbett, J. D. *Science* **2001**, *291*, 841. (e) Seo, D.-K.; Corbett, J. D. *J. Am. Chem. Soc.* **2002**, *124*, 415. (f) Häussermann, U.; Amerioun, S.; Eriksson, L.; Lee, C.-S.; Miller, G. *J. Am. Chem. Soc.* **2002**, *124*, 4371. (g) Mudring, A.-V.; Corbett, J. D. *J. Am. Chem. Soc.* **2004**, *126*, 5277.

7 (a) Kübler, J. *Theory of Itinerant Electron magnetism*; Clarendon Press, Oxford, 2000. (b) Mohn, P. *Magnetism in the Solid State*; Springer, Berlin, 2003.

8 (a) Stoner, E. C. *Proc. Roy. Soc. London A* **1938**, *165*, 372. (b) Stoner, E. C. *Proc. Roy. Soc. London A* **1939**, *169*, 339. (c) Vosko, S. H.; Perdew, J. P. *Can. J. Phys.* **1975**, *53*, 1386. (d) Gunnarsson, O. *J. Phys. F* **1976**, *6*, 587. (e) Janak, J. F. *Phys. Rev. B* **1977**, *16*, 255. (f) Andersen, O. K.; Madsen, J.; Poulsen, U. K.; Jepsen, O.; Kollar, J. *Physica B* **1977**, *86–88*, 249. (g) Kulatov, E.; Mazin, I. I. *J. Phys.: Condens. Matter* **1990**, *2*, 343.

9 (a) Fujiwara, T. in *Physical Properties of Quasicrystals*, Stadnik, Z. M. Ed.; Springer, Berlin, 1999. (b) Fujiwara, T. *Curr. Opin. Solid State Mater. Sci.* **1999**, *4*, 295.

10 Wu, L.; Seo, D.-K. *J. Am. Chem. Soc.* **2004**, *126*, 4398.

11 De Laissardiere, G. T.; Manh, D. N.; Magaud, L.; Julien, J. P.; Cyrot-Lackmann, F.; Mayou, D. *Phys. Rev. B* **1995**, *52*, 7920.

12 (a) Frank, F. C.; Kasper, J. S. *Acta Crystallogr.* **1959**, *12*, 483. (b) Shoemaker, D. P.; Shoemaker, C. B. *Mater. Sci. Forum* **1987**, *22–24*, 67.

13 (a) Robinson, K. *Acta Crystallogr.* **1952**, *5*, 397. (b) Cooper, M.; Robinson, K. *Acta Cryst.* **1966**, *20*, 614. (c) Sugiyama, K.; Kaji, N.; Hiraga, K. *Acta Cryst. C* **1998**, *54*, 445.

14 Poon, S. J. *Adv. Phys.* **1992**, *41*, 303 and references therein.

15 Stadnik, Z. M., Ed. *Physical Properties of Quasicrystals*; Springer, New York, 1999.

16 (a) Pecharsky, V. K.; Gschneidner, K. A., Jr. *Phys. Rev. Lett.* **1997**, *78*, 4494. (b) Choe, W.; Pecharsky, V. K.; Pecharsky, A. O.; Gschneidner, K. A., Jr.; Young, V. G. *Phys. Rev. Lett.* **2000**, *84*, 4617. (c) Pecharsky, V. K.; Samolyuk, G. D.; Antropov, V. P.; Pecharsky, A. O.; Gschneidner, K. A., Jr. *J. Solid State Chem.* **2003**, *171*, 57.

17 (a) Le Roy, J.; Moreau, J.; Paccard, D.; Parthe, E. *Acta Crystallogr.* **1978**, *B34*, 3315. (b) Choe, W.; Pecharsky, A. O.; Worle, M.; Miller, G. J. *Inorg. Chem.* **2003**, *42*, 8223. (c) Choe, W.; Miller, G. J.; Meyers, J.; Chumbley, S.; Pecharsky, A. O. *Chem. Mater.* **2003**, *15*, 1413. (d) Mozharivskyj, Y.; Choe, W.; Pecharsky, A. O.; Miller, G. J. *J. Am. Chem. Soc.* **2003**, *125*, 15183. (e) Mozharivskyj, Y.; Choe, W.; Pecharsky, A. O.; Miller, G. J. *J. Am. Chem. Soc.* **2005**, *127*, 317.

18 Regardless of their possible metallic properties, "metal-rich" Zintl system or phases are defined here as cation-rich compounds exhibiting anionic moieties of metal or metalloid elements whose structures can be generally understood by applying the classical or modern electron counting rules for molecules.

19 Chemistry, Structure, and Bonding of Zintl Phases and Ions, Kauzlarich, S. M., Ed.

20 Smith, G. S.; Tharp, A. G.; Johnson, Q. *Acta Crystallogr.* **1967**, *22*, 940.

21 Shannon, R. D. *Acta Crystallogr.* **1976**, *324*, 751.

22 Wu, L.-M.; Kim, S.-W.; Seo, D.-K. *J. Am. Chem. Soc.* **2005**, submitted.

23 Corbett, J. D. *Angew. Chem. Int. Ed.* **2000**, *39*, 670.

13

Ternary and Quaternary Niobium Arsenide Zintl Phases

Franck Gascoin and Slavi C. Sevov

13.1
Introduction

Since Zintl's pioneering work on polar main-group intermetallic compounds that now carry his name, Zintl phases [1], the number of such compounds has grown almost exponentially throughout the years [2]. Perhaps the greatest evolution of Zintl's concept took place when compounds containing transition metals were included in the family. Compared to the large number of main-group Zintl compounds, there were only a few such compounds that contained transition metals [3]. This is not surprising since the traditional definition of Zintl phases automatically excludes transition metals. However, many such compounds with transition metals can qualify for Zintl phases when a broader definition is used. Initially "admitted" were transition-metal phases that were isostructural with known main-group Zintl phases [3]. These included, for example, Na_8TiAs_4 and Na_5HfAs_3 isostructural with A_8SnSb_4 (A = Na, K) and Na_5SnAs_3, respectively [4]. Next came all electronically-balanced compounds that contain transition-metal ions with d^0 or d^{10} closed-shell configurations. These involve elements of the Ti-, V- and Cr-groups as fully oxidized d^0 ions [5], and of the Ni-, Cu- and Zn-groups as d^{10} ions [6]. These electronic configurations mimic main-group elements, and therefore the compounds, when electronically balanced, can reasonably be considered Zintl phases [3]. All but one of the pnictides (Pn) of transition metals in the d^0 state contain isolated tetrahedra MPn_4^{n-} or edge-sharing dimers of tetrahedra $M_2Pn_6^{m-}$ (M = Ti, Hf, Nb, Ta, W) [4, 5]. The exception is $Rb_5TaTl_2As_4$ where two opposite edges of the tetrahedral $TaAs_4$ are bridged by thallium atoms, $(\mu\text{-Tl})[As_2TaAs_2](\mu\text{-Tl})^{5-}$ [7]. Finally, the definition of Zintl phases was broadened even further by adding compounds containing transition-metal ions with partially-filled d-orbitals as long as the compounds were electronically balanced. This latter group has recently grown to include five members, all of them based on manganese. One of the structure types, that of $A_{14}MnPn_{11}$ (A = Ca, Sr, Ba, Eu and Pn = As, Sb, Bi) [3], was initially thought to contain Mn(III), but more recent studies with X-ray magnetic circular dichroism

Inorganic Chemistry in Focus III.
Edited by G. Meyer, D. Naumann, L. Wesemann
Copyright © 2006 WILEY-VCH Verlag GmbH & Co. KGaA, Weinheim
ISBN: 3-527-31510-1

suggested that it might be Mn(II) [8]. The oxidation state of manganese in the remaining four compounds of this group, $Sr_{21}Mn_4Sb_{18}$ [9], $Sr_{21}Mn_4Sb_{18}$ [10], $Eu_{10}Mn_6Sb_{13}$ [11], and AMnPn (A = K, Rb, Cs and Pn = P, As, Sb, Bi) [12], is clearly 2+, i.e., Mn(II). Thus, despite the large variety of transition metals and numerous feasible combinations with main-group elements, the studies have dealt only with manganese-based phases, and the field has remained somewhat underdeveloped and unexplored.

We have been interested in the solid-state chemistry of the alkali metal–niobium–arsenic systems and have undertaken extensive and systematic studies of the quaternary systems A–Nb–As–E where A = alkali metal and E = post-transition element. The results indicate that, structurally, these systems are very promising, and many new compounds with eventually novel properties are to be expected. Similar compounds may be accessible by utilizing the large number of transition metals and the possible combinations with pnictogens and a variety of s-elements. Furthermore, many of these compounds may be soluble in polar solvents such as ethylenediamine, liquid ammonia, pyridine, THF, etc., that are stable to highly reduced species. The possibility of extracting mixed transition-metal/pnictogen anions in solution is of great interest. For example, $Cs_4Na_3NbAs_4$ which contains $NbAs_4$ tetrahedra in the solid state dissolves readily in ethylenediamine, and Nb-centered crowns of cyclic As_8, i.e., $[Nb@As_8]^{8-}$, can be crystallized with the help of 2,2,2-crypt (4, 7, 13, 16, 21, 24-hexaoxa-1,10-diazabicyclo-[8.8.8]-hexacosane) in the compound $(Na\text{-}crypt)Cs_5[Nb@As_8]$. This is a clear indication that similar compounds containing transition metals would be good possibilities for the formation of novel anions.

In the past 10–15 years there have been very extensive studies, carried out almost exclusively by Kauzlarich's group, on the magnetic and transport properties of the large class of isostructural compounds with formula $A_{14}MnPn_{11}$ (A = alkaline- or rare-earth element or a combination of the two) [3, 13]. Manganese in these compounds is tetrahedrally coordinated by pnictogen, $MnPn_4$, and the structure contains also linear trimers and isolated pnictogen atoms, Pn_3^{7-} and Pn^{3-}, respectively. The formula could be written as $(A^{2+})_{14}(Mn^{III}Pn_4^{9-})(Pn_3^{7-})(Pn^{3-})_4$. This is consistent with an electronically balanced compound and suggests semiconducting properties. However, recent studies using X-ray magnetic circular dichroism revealed that manganese is very likely Mn(II) and the overall magnetic moment produced by the five unpaired d-electrons is reduced by an anti-aligned moment on the Pn_4 tetrahedron. Furthermore, it has been shown that, although the As-compounds are semiconducting, the Sb- and Bi-compounds are conducting at low temperatures with electron concentrations presumably similar to those of d-band metals. This conductivity has been explained as the result of overlap of the conduction and valence bands in these compounds. At low temperatures, the electrons localized at the manganese centers couple ferromagnetically in the antimonides but antiferromagnetically in the bismutides. This coupling occurs despite the very long distances between the manganese centers, in the order of 10 Å, and has been explained by the Ruderman-Kittel-Kasuya-Yosida (RKKY) theory as the coupling of magnetic centers via the available conduction electrons [3]. Furthermore, the compounds with rare-earth elements are even more complex magnetically and often show more than one magnetic transition [3, 13]. More recently, colossal negative magnetoresistance was discovered for the europium and strontium analogs (Eu or Sr)$_{14}$Mn(Sb or Bi)$_{11}$ and the mixed-cation Ca$_{14-x}$Eu$_x$MnSb$_{11}$ with magnetoresistance ratios of up to −70 % [13]. Apparently, the mechanism is different from the double-exchange mechanism between Mn(III) and Mn(IV) proposed for the perovskites but is similar to the superexchange arising from s-d scattering of conduction electrons (low density) and unpaired spins at the Mn-centers of the manganese pyrochlores [13]. An important requirement for magnetoresistance in such systems is, apparently, the existence of localized spins and low overall conductivity so that the disorder of the spins can cause localization of the charge carriers and lowering of the overall conductivity. Appropriate magnetic fields applied at appropriate temperatures cause a decrease in the disorder of the spins and, as a result of this, the conductivity increases. Such materials have some potential for application as magnetoresistive devices in the data-storage industry and for other technological advances. Clearly, transition-metal Zintl phases have the potential to introduce diverse structures and properties, and their further study is of great importance.

13.2
New Main-group Arsenides

While complex systems based on alkali-metal chalcogenides mixed with elements from various main- and/or transition-metal groups have been studied very extensively [14], similar systems based on alkali-metal pnictides have not received nearly as much attention. The former have shown great structural richness, mainly due to the complexity of the mixtures that have been studied, usually quaternary, quintenary, and phases of even higher order. For the arsenides, we initially studied main-group systems of alkali metal, arsenic, and a third p-element or elements that can modify the electronic requirements of the anionic part of the structure. The original goal of these studies was to explore for heteroatomic deltahedral clusters made of group 13 (Tr) and group 15 (Pn), [Tr$_4$Pn$_5$]$^{3-}$, that would be isoelectronic and isostructural with the known E$_9^{4-}$ (E = Si, Ge, Sn, Pb) in A$_4$E$_9$ and A$_{12}$E$_{17}$ (A = alkali metal) [15]. The results, however, showed that such clusters do not exist in the corresponding A-Tr-Pn compounds, although they form upon dissolution of such precursors in ethylenediamine and can be crystallized with the aid of 2,2,2-crypt [16]. Nonetheless, the studies of the A-In-As solid-state systems and, more specifically, the observed structural features, clearly indicated that, similar to the chalcogenides (the selenides in particular), the arsenides provide equally impressive structural richness, even for relatively simpler systems. One such compound with novel structural features is Cs$_5$In$_3$As$_4$ [17]. The anionic part of its structure consists of both chains and layers made of indium and arsenic, and, more importantly, these chains and layers are of exactly the same composition and charge, $[In_3As_4]^{5-}$ (Fig. 13.1). The layers, $^2_\infty[In_3As_4]^{5-}$ (Fig. 13.1a), are parallel to the chains, $^1_\infty[In_3As_4]^{5-}$ (Fig. 13.1b), which, in turn, are parallel to each other, and the regions of layers and chains alternate and are separated by the cations. Therefore, the structure can be viewed as if two polymorphic forms of the compound have co-crystallized together. Also, a closer look at the chains and the layers reveals that they are built of one and the same building block, which is then connected differently in the two fragments. This building block is a dimer of semicubanes (a cubane with one missing vertex) made of three indium and four arsenic atoms each, In_3As_4, where the semicubanes are connected to each other by two parallel In–As bonds (related by an inversion center). The dimers in the chains are connected to each other by the same pairs of In–As bonds as those forming the dimer itself and, therefore, the chains can also be described simply as repeating monomers of semicubanes. The dimers in the layers, on the other hand, are connected to each other by four In–In bonds at the four indium atoms that are not involved in intra-dimer bonding.

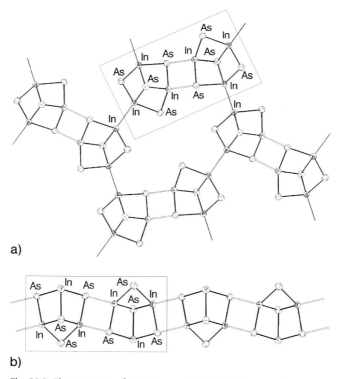

a)

b)

Fig. 13.1 The structure of $Cs_5In_3As_4$ contains both layers and chains of $[In_3As_4]^{5-}$ shown in (a) and (b), respectively. Shown in boxes is the same building unit of the chains and the layers, i.e., a dimer of semicubanes of In_3As_4.

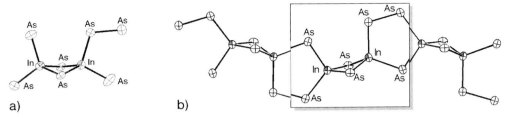

Fig. 13.2 Views of (a) the monomer of $[In_2As_7]^{13-}$ and (b) the polymeric chain of $_\infty[In_2As_5]^{13-}$ in KBa_2InAs_3. Highlighted in (b) is a motif of the chain that clearly resembles the monomer in (a).

More complex quaternary systems were also studied. They were achieved by mixing alkali and alkaline-earth cations and therefore the systems could, to some extent, be viewed as pseudo-ternary. It should be pointed out that this approach has proved to be very effective in exploratory synthesis and has resulted in many novel compounds that could not be achieved otherwise [2]. Mixing K and Ba as counter-cations in the In–As system provided the new compound KBa_2InAs_3 [18]. The anionic part of its structure is made of isolated units of $[In_2As_7]$ and polymeric chains made by polymerization of the same units (Fig. 13.2). The isolated species (Fig. 13.2 a) can be described as analogous to 1-ethyl-1,3,3-trimethyl-cyclobutane with indium at the positions of the quaternary carbon atoms and diarsenic playing the role of the ethyl substituent. Taking into account that As^-, As^{2-}, and In^- are isoelectronic with CH_2, CH_3, and C, respectively, we can easily determine the charge of the species as 13-, i.e., $[In_2As_7]^{13-}$. The "inorganic" description is to view the formation as made of two edge-sharing tetrahedra of arsenic that are centered by indium, and where an arsenic atom is exo-bonded to a corner of one of the tetrahedra, $[(As-As)(As)]In(\mu-As)_2$ $In(As)_2$. The chains in the structure are made of alternating five- and four-membered rings of $[In_2As_3]$ and $[In_2As_2]$, respectively (Fig. 13.2 b), where the indium corners are shared between the rings. One can easily recognize the monomer described above as the repeating unit in the polymer (highlighted in Fig. 13.2 b). Thus, the all-organic analog of the chain would be poly-1-ethyl-1,3,3-trimethyl-cyclobutane where the "polymerization" occurs at all terminal arsenic atoms. Due to this sharing of the four arsenic atoms, the formula of the repeating unit is two arsenic atoms short from the formula of the monomer, $[In_2As_5]^{7-}$.

13.3
Compounds Based on Isolated [NbAs₄] Tetrahedral Centers

The first alkali metal-niobium-arsenic compounds were synthesized by accident while attempting the synthesis of alkali-metal main-group arsenides at relatively high temperature. It turns out that niobium and tantalum containers react readily with arsenic in the presence of alkali metals at temperatures above 700 °C. The advantages of this were quickly recognized, and more focused exploration was carried out in niobium-containing arsenide Zintl phases. These studies were later extended into more and more complex systems. Thus, series of compounds based on niobium centers coordinated tetrahedrally by arsenic, [NbAs$_4$], were synthesized. These are the d^0-compounds A$_6$NbAs$_5$ (A=K, Rb, Cs), K$_6$TlNbAs$_4$, and K$_8$PbNbAs$_5$ (Fig. 13.3) [19]. In all of them one of the edges of the NbAs$_4$-tetrahedron is bridged by another atom: As in [NbAs$_5$]$^{6-}$ (Fig. 13.3a), Tl in [TlNbAs$_4$]$^{6-}$ (Fig. 13.3b), and Pb in [PbNbAs$_5$]$^{8-}$ (Fig. 13.3c). The lead atom in the latter is additionally bonded to one more arsenic atom. The charges of the anions can be ra-

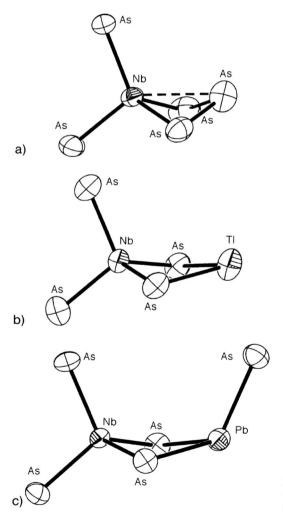

Fig. 13.3 Views of the anions in (a) Cs$_6$NbAs$_5$, (b) K$_6$TlNbAs$_4$, and (c) K$_8$PbNbAs$_5$.

tionalized by considering niobium as Nb(V), lead as Pb(II), thallium as Tl(I), and arsenic as isolated As^{3-} anions and As_3^{5-} trimers. The anions can be then written as $(Nb^{5+})(As^{3-})_2(As_3^{5-})$, $(Nb^{5+})(Tl^+)(As^{3-})_4$, and $(Nb^{5+})(Pb^{2+})(As^{3-})_5$. The arsenic-bridged anion, $[NbAs_5]^{6-}$, is quite interesting from the theoretical point of view. It is bent in such a way that the bridging arsenic becomes quite close to the niobium at a distance of 2.814 Å, and the As_3-ligand acts almost as tridentate (Fig. 13.3a). Molecular orbital analysis indicates that this is possible due to interactions across the ring that involve overlap between the empty d_{xz} and to some extent $d_{x^2-y^2}$ of niobium with the filled p_z orbital of the arsenic [19]. The latter is a part of the filled π-system of the trimer made of the p_z orbitals. The $NbAs_2$ fragment can be easily derived from the MO diagram of a square-planar ML_4 by removing two of the ligands (Fig. 13.4). The resulting metal-based frontier orbitals have predominantly non- and anti-bonding d- and p-character and are empty for Nb(V). The frontier orbitals of the bent As_3 trimer are also easily envisioned as the system of the isostructural allyl where the three combinations of bonding, non-bonding, and anti-bonding interactions are all filled. All three π-orbitals of the trimer have appropriate matches with the ML_2 fragment and form three bonding orbitals with $d_{x^2-y^2}$ (mixed with p_x), d_{xy} (mixed with p_y), and d_{xz}. However, only two of these interactions, those with d_{xz} and to some extent with $d_{x^2-y^2}$, involve the middle atom of

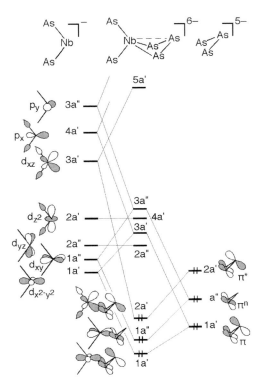

Fig. 13.4 An orbital interaction diagram for $NbAs_5^{6-}$ made of the fragments As_2Nb^- and bent $[As-As-As]^{5-}$.

the trimer. Thus, based purely on this MO diagram, i.e., not considering the underlying σ- and π-bonding Nb–As molecular orbitals, the Nb–As bond order for the middle arsenic atoms can be calculated as 0.667. The latter correlates very well with the Nb–As distance of 2.814(2) Å to the middle arsenic atom when compared with the single-bond distance of 2.720 Å and corresponds to Pauling's bond order of 0.697.

The three new anions $[NbAs_5]^{6-}$, $[TlNbAs_4]^{6-}$, and $[PbNbAs_5]^{8-}$ can be viewed also as tetrahedral niobium complexes with two monodentate (isolated As^{3-} anions) and one bidentate (As_3^{5-} or $TlAs_2^{5-}$ or $PbAs_3^{7-}$) ligands each. As discussed in the introduction, the simple case of a tetrahedral complex with four monodentate arsenic ligands, $[NbAs_4]^{7-}$, has been observed in a few compounds [5]. The two remaining possible combinations are a monodentate with tridentate ligands and a tetradentate ligand. Obviously the latter would require a very exotic ligand and should be considered practically impossible. It turns out that the former combination of one monodentate and one tridentate ligand is possible and is realized in the new compound $Cs_7NbIn_3As_5$ [20]. It contains an isolated anion in the shape of a cubane with a handle (Fig. 13.5). One way to view the species is as a cubane made of four arsenic, three indium and one niobium corners with an additional arsenic that is exo-bonded to the niobium corner, the "handle". The charge of the anion can be rationalized as $[(Nb^{5+})(In^+)_3(As^{3-})_5]^{7-}$. Another way to describe the anion is to consider it as a tetrahedral niobium complex with one monodentate As^{3-} ligand and one tridentate ligand of $[In_3As_4]^{9-}$ coordinated to Nb^{5+}. The tridentate ligand is exactly the same semicubane species $[In_3As_4]$ found as is interconnected into chains and layers in $Cs_5 In_3As_4$ (Fig. 13.1). The Nb–As exo-bond is extremely short, 2.390(2) Å, and according to molecular orbital analysis corresponds to a triple bond [21]. The new anion $[NbIn_3As_5]^{7-}$ resembles known molecules with multiple bonds such as $[N(CH_2CH_2(Me_3Si)N)_3]W\equiv As$ and $(tBu_3SiO)_3Ta=As\text{-}Ph$ [22]. The multiple bonding in these compounds and in $[NbIn_3As_5]^{7-}$ is based on the available π-interactions between the empty d_{xz} and d_{yz} on the transition metal and the filled p_π orbitals on the arsenic. Furthermore, the monodentate arsenic ligand

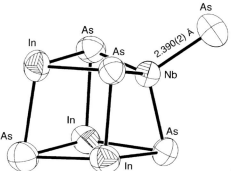

Fig. 13.5 $[As(InAs)_3Nb\equiv As]^{7-}$ – a cubane with a "handle".

in $[\text{NbIn}_3\text{As}_5]^{7-}$ is terminal and can be considered sp-hybridized where both p_x and p_y orbitals are available for π-interactions. At the same time, the arsenic atoms of the semicubane tridentate ligand are "forced" to be sp^3 hybridized and cannot engage any of the available niobium d orbitals for π-interactions.

The semicubane building block $[\text{In}_3\text{As}_4]$ found as a free ligand linked into chains and layers in $\text{Cs}_5\text{In}_3\text{As}_4$ and as a tridentate ligand coordinated to niobium in $\text{Cs}_7\text{NbIn}_3\text{As}_5$ is apparently a very stable and reoccurring unit. It was found in two more compounds where it is both coordinated to niobium and interconnected at the same time. The two compounds are $\text{Cs}_{13}\text{Nb}_2\text{In}_6\text{As}_{10}$ and $\text{Cs}_{24}\text{Nb}_2\text{In}_{12}\text{As}_{18}$ with isolated oligomers of $[\text{Nb}_2\text{In}_6\text{As}_{10}]^{13-}$ (Fig. 13.6a) and $[\text{Nb}_2\text{In}_{12}\text{As}_{18}]^{24-}$ (Fig. 13.6b), respectively [23]. The former is a dimer of cubanes with handles that are linked by an In–In bond. The two "prefabricated" semicubanes of $[\text{In}_3\text{As}_4]$ are clearly recognizable. They are linked into what can be viewed as a hexadentate ligand $[\text{As}_4\text{In}_3\text{–In}_3\text{As}_4]$ that coordinates to two niobium centers. The second anion is a linear tetramer of the same semicubanes where the end members are bonded to their neighbors via the same In–In bond as in the dimers in Fig. 13.6a, while the middle two members are bonded to each other via the same pair of In–As bonds as in the dimers of semicubanes of the chains and layers in $\text{Cs}_5\text{In}_3\text{As}_4$ (Fig. 13.1). The two end members are coordinated to two niobium centers as tridentate ligands. The Nb–As distances to the

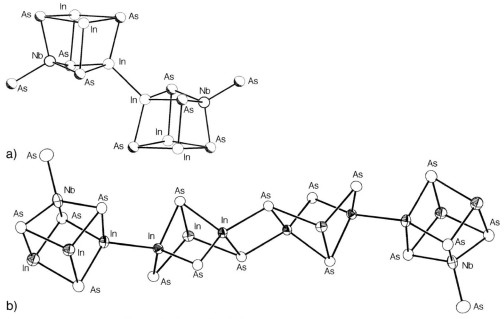

Fig. 13.6 Shown are (a) a dimer of cubanes (bonded via indium vertices) found in $\text{Cs}_{13}\text{Nb}_2\text{In}_6\text{As}_{10}$; (b) a tetramer of two cubanes and two semicubanes found in $\text{Cs}_{24}\text{Nb}_2\text{In}_{12}\text{As}_{18}$.

terminal arsenic in both anions are very short, 2.357(1) Å in $[Nb_2In_6As_{10}]^{13-}$ and 2.348(1) Å in $[Nb_2In_{12}As_{18}]^{24-}$, i.e., they are shorter than the distance of 2.390(2) Å in the single cubane of $Cs_7NbIn_3As_5$ (Fig. 13.5), and should qualify as triple bonds as well.

The anions have three types of indium atoms. Those that are bonded only within one semicubane can be considered as In^+ because of the remaining inert lone pair. The pairs of In–In are formally In_2^{4+}, and the indium atoms that are four-bonded to arsenic are In^{3+}. This leads to the following charge formulas of the anions, $(Nb^{5+})_2(In^+)_4(In_2^{4+})(As^{3-})_{10}$ and $(Nb^{5+})_2(In^+)_6(In_2^{4+})_2(In^{3+})_2(As^{3-})_{18}$. Notice, however, that while these charge distributions provide the correct charge of the latter anion, i.e., overall 24-, the charge of the former sums up to 12- instead of the observed 13-. The two possible explanations are that: a) the charge of the anion is 12- and the thirteen cesium cations provide one extra electron that is delocalized over the structure, and b) the charge of the anion is indeed 13- because one of the niobium atoms is Nb(IV) and not Nb(V). These two possibilities would give rise to different magnetic properties of the compound. Electron delocalization over the whole compound will result in a metallic compound and, therefore, temperature independent and positive magnetic susceptibility. The presence of Nb(IV) in the compound, on the other hand, will provide Curie-Weiss temperature dependence of the magnetic susceptibility due to the magnetic moment of the localized one d-electron. The measurements showed the latter case, i.e., a Curie-Weiss behavior with one unpaired spin per anion. Therefore, the formula of this mixed-valence anion should be rewritten as $[(Nb^{5+})(Nb^{4+})(In^+)_4(In_2^{4+})(As^{3-})_{10}]^{13-}$. The need of Nb(IV) for the formation of the compound is not clear. As in the other mixed-valence transition-metal Zintl phases described below, as well as in many metallic Zintl phases with one or more delocalized "extra" electrons, the most likely reason is packing requirements. Perhaps an extra cation is required for the crystallization of $Cs_{13}Nb_2In_6As_{10}$ with anions that are both well packed and well separated. Similarly, extra cations are required for metallic Zintl phases, but the extra electron in these compounds cannot localize on one atom since these phases are made of main-group elements only. The only choice for the extra electron or electrons is to delocalize over the whole structure and to provide the metallic properties of the corresponding compound.

13.4
Compounds Based on Edge-sharing Dimers of [NbAs₄] Tetrahedra

Three compounds fall into this category, $Cs_9Nb_2As_6$, $K_9Nb_2As_6$, and $K_{38}Nb_7As_{24}$ [24]. In addition to these, a forth compound, $K_{10}NbInAs_6$, with dimers of heteroatomic edge-sharing tetrahedra of [NbAs₄] and [InAs₄], can also be viewed as a member of this group [24]. The dimers can be represented as $[As_2Nb(\mu\text{-As})_2NbAs_2]$ and $[As_2Nb(\mu\text{-As})_2InAs_2]$. Assuming that niobium is Nb(V) in both types, their corresponding charges are calculated as 8- and 10-, respectively, i.e.,

$[(Nb^{5+})_2(As^{3-})_6]^{8-}$ and $[(Nb^{5+})(In^{3+})(As^{3-})_6]^{10-}$. The latter is exactly the case in $K_{10}NbInAs_6$ and the compound is measured as diamagnetic and therefore electronically balanced. $Cs_9Nb_2As_6$, $K_9Nb_2As_6$, and $K_{38}Nb_7As_{24}$, however, exhibit Curie-Weiss paramagnetism with magnetic moments that correspond to one unpaired electron per formula. Furthermore, there are clearly nine counter-cations per dimer in the first two compounds. The third compound, $K_{38}Nb_7As_{24}$, contains both dimers of edge-sharing tetrahedra as well as isolated tetrahedra, $K_{38}[Nb_2As_6]_2[NbAs_4]_3$. The dimers in the three compounds are of two different types, flat and bent (Fig. 13.7). Planar dimers involving transition metals have been characterized before, for example $[Hf_2As_6]^{10-}$ in Na_5HfAs_3 with d^0 hafnium(IV) [4]. $K_9Nb_2As_6$ has only flat dimers while $Cs_9Nb_2As_6$ contains only bent dimers. The dimers in $K_{38}Nb_7As_{24}$, on the other hand, are of both types in equal ratio. Apparently both types of dimers can exist with mixed-valent niobium and this is exemplified by $K_9Nb_2As_6$ and $Cs_9Nb_2As_6$. Their formula can be written as $[(Nb^{5+})(Nb^{4+})(As^{3-})_6]^{9-}$. One of the two types of dimers in $K_{38}Nb_7As_{24}$, however, should be of Nb(V) only in order to keep the charges balanced in the compound, i.e., $[(Nb^{5+})(Nb^{4+})(As^{3-})_6]^{9-}[(Nb^{5+})_2(As^{3-})_6]^{8-}\{[(Nb^{5+})(As^{3-})_4]^{7-}\}_3$. All three compounds have one mixed-valent dimer per formula and this is in agreement with the observed magnetic moment.

In addition to Curie-Weiss magnetic behavior, $Cs_9Nb_2As_6$ exhibits a broad maximum at ~36 K in its magnetic susceptibility. Initially, this was interpreted as the onset of long-range antiferromagnetic order. However, it was difficult to rationalize such a high ordering temperature in light of the fact that the compound is electronically balanced and insulating, and the magnetic centers are very "diluted" with an inter-dimer Nb–Nb distance of 7.2 Å. As discussed above, a similar effect observed for $(AE)_{14}MnPn_{11}$ was explained as the result of overlap between the valence and conduction bands where the delocalized electrons

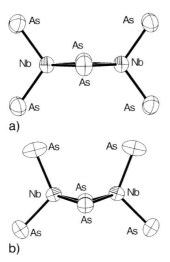

a)

b)

Fig. 13.7 Flat (a) and bent (b) dimers of [Nb₂As₆].

provide the means for "magnetic communication" between the localized spins. Similar interpretation is not applicable for the interactions in $Cs_9Nb_2As_6$ because of its non-metallic properties, i.e., black color and two-probe resistance of more than 30 kΩ measured for a crystal length of about 0.5 mm. Detailed measurements and further studies of this compound showed that the observed magnetic properties can be explained by statistical occupation of the equivalent Nb sites in the dimer by equal concentrations of Nb(IV) and Nb(V) [25]. From this analysis the broad maximum was determined to arise from intra-dimer antiferromagnetic exchange with an exchange constant of $J/k = -40$ K, and no long-range magnetic order except, possibly, below 5 K.

The niobium-arsenic compounds described here, with novel anionic structures, clearly demonstrate the many possibilities for structural motifs in this system. Similarly high potential for diverse structures and properties should be expected for the heavier pnictides when combined with other transition metals and the alkali-metal group.

References

1 E. Zintl, W. Dullenkopf, *Z. Phys. Chem.* **1932**, *B16*, 183; E. Zintl, G. Brouer, *Z. Phys. Chem.* **1933**, *B20*, 245; E. Zintl, *Angew. Chem.* **1933**, *52*, 1.

2 J.D. Corbett in *Chemistry, Structure and Bonding of Zintl Phases and Ions* (Ed.: S.M. Kauzlarich), VCH Publishers, New York, 1996; p. 139; J.D. Corbett, *Angew. Chem. Int. Ed.* **2000**, *39*, 670; B. Eisenmann, G. Cordier in *Chemistry, Structure and Bonding of Zintl Phases and Ions* (Ed.: S.M. Kauzlarich), VCH Publishers, New York, 1996; p. 61; C. Belin, M. Tillard-Charbonnel, *Prog. Solid St. Chem.* **1993**, *22*, 59.

3 S.M. Kauzlarich in *Chemistry and Bonding of Zintl Phases and Ions* (Ed.: S.M. Kauzlarich), VCH Publishers, New York, 1996; p. 245 and the references therein.

4 J. Stuhrmann, A. Adam, H.U. Schuster, *Z. Naturforsch. B* **1993**, *48*, 898; A. Adam, H.U. Schuster, *Z. Naturforsch. B* **1990**, *45*, 559.

5 K. Vidyasagar, W. Hönle, H.G. von Schnering, *J. Alloys Compd.* **1996**, *236*, 38; J. Nuss, W. Hönle, H.G. von Schnering, *Z. Anorg. Allg. Chem.* **1997**, *623*, 1763; J.H. Lin, W. Hönle, H.G. von Schnering, *J. Alloys Compd.* **1992**, *183*, 403; J. Nuss, W. Hönle, K. Peters, H.G. von Schnering, *Z. Anorg. Allg. Chem.*

1996, *622*, 1879; J. Nuss, R.H. Cardoso Gil, W. Hönle, K. Peters, H.G. Schnering, *Z. Anorg. Allg. Chem.* **1996**, *622*, 1854; K. Vidyasagar, W. Hönle, H.G. von Schnering, *Z. Anorg. Allg. Chem.* **1996**, *622*, 518; D. Huang, J.D. Corbett, *Inorg. Chem.* **1998**, *37*, 4006.

6 M. Somer, *Z. Naturforsch. B* **1994**, *49*, 1203; Z.C. Dong, R.W. Henning, J.D. Corbett, *Inorg. Chem.* **1997**, *36*, 3559; S.C. Sevov, J.D. Corbett, *J. Am. Chem. Soc.* **1993**, *115*, 9089; R.W. Henning, J.D. Corbett, *Inorg. Chem.* **1999**, *38*, 3883; S.S. Dhingra, R.C. Haushalter, *J. Am. Chem. Soc.* **1994**, *116*, 3651; U. Zachwieja, J. Muller, J. Wlodarski, *Z. Anorg. Allg. Chem.* **1998**, *624*, 853; D.P. Huang, Z.C. Dong, J.D. Corbett, *Inorg. Chem.* **1998**, *37*, 5881; E. Todorov, S.C. Sevov, *Angew. Chem. Int. Ed.* **1999**, *38*, 1775; S. Kaskel, J.D. Corbett, *Inorg. Chem.* **2000**, *39*, 3086; V. Queneau, S.C. Sevov, *J. Am. Chem. Soc.* **1997**, *119*, 8109.

7 D. Huang, J.D. Corbett, *Inorg. Chem.* **1998**, *37*, 4006.

8 A.P. Holm, S.M. Kauzlarich, S.A. Morton, G.D. Waddill, W.E. Pickett, J.G. Tobin, *J. Am. Chem. Soc.* **2002**, *124*, 9894.

9 H. Kim, C.L. Condron, A.P. Holm, S.M. Kauzlarich, *J. Am. Chem. Soc.* **2000**, *122*, 10720.

10 A.P. Holm, M.M. Olmstead, S.M. Kauzlarich, *Inorg. Chem.* **2003**, *42*, 1973.

11 A.P. Holm, S.M. Park, C.L. Condron, M.M. Olmstead, H. Kim, P. Klavins, F. Grandjean, R.P. Hermann, G.J. Long, M.G. Kanatzidis, S.M. Kauzlarich, S.J. Kim, *Inorg. Chem.* **2003**, *42*, 4660; D.E. Brown, C.E. Johnson, F. Grandjean, R.P. Hermann, S.M. Kauzlarich, A.P. Holm, G.J. Long, *Inorg. Chem.* **2004**, *43*, 1229.

12 W. Bronger, P. Müller, R. Höppner, H.U. Schuster, *Z. Anorg. Allg. Chem.* **1986**, *539*, 175; W. Bronger, *Ber. Bunsenges. Phys. Chem.* **1992**, *96*, 1572; F. Schucht, A. Dascoulidou, R. Müller, W. Jung, H.U. Schuster, W. Bringer, P. Müller, *Z. Anorg. Allg. Chem.* **1999**, *625*, 31.

13 J.Y. Chan, M.E. Wang, A. Rehr, S.M. Kauzlarich, *Chem. Mater.* **1997**, *9*, 2131; J.Y. Chan, S.M. Kauzlarich, *Chem. Mater.* **1997**, *9*, 3132; J.Y. Chan, M.M. Olmstead, S.M. Kauzlarich, *Chem. Mater.* **1998**, *10*, 3538; A.C. Payne, M.M. Olmstead, S.M. Kauzlarich, *Chem. Mater.* **2001**, *13*, 1398; D.J. Webb, R. Cohen, P. Klavins, R.N. Shelton, J.Y. Chan, S.M. Kauzlarich, *J. Appl. Phys.* **1998**, *83*, 7192; J.Y. Chan, S.M. Kauzlarich, P. Klavins, R.N. Shelton, D.J. Webb, *Phys. Rev. B* **1998**, *57*, R8103; I.R. Fisher, T.A. Wiener, S.L. Bud'ko, P.C. Canfield, J.Y. Chan, S.M. Kauzlarich, *Phys. Rev. B* **1999**, *59*, 13829; J.Y. Chan, S.M. Kauzlarich, P. Klavins, J.Z. Liu, R.N. Shelton, D.J. Webb, *Phys. Rev. B* **2001**, *61*, 459; H. Kim, J.Y. Chan, M.M. Olmstead, P. Klavins, D.J. Webb, S.M. Kauzlarich, *Chem. Mater.* **2002**, *14*, 206; D. Sanchez-Portal, R.M. Martin, S.M. Kauzlarich, W.E. Pickett, *Phys. Rev. B* **2002**, *65*, 1444414; H. Kim, M.M. Olmstead, P. Klavins, D.J. Webb, S.M. Kauzlarich, *Chem. Mater.* **2002**, *14*, 3382; H. Kim, Q. Huang, J.W. Lynn, S.M. Kauzlarich, *J. Solid. State Chem.* **2002**, *168*, 162.

14 M.G. Kanatzidis, S.P. Huang, *Coordin. Chem. Rev.* **1994**, *130*, 509.

15 V. Quéneau, S.C. Sevov, *Angew. Chem. Int. Ed. Engl.* **1997**, *36*, 1754; V. Quéneau, E. Todorov, S.C. Sevov, *J. Am. Chem. Soc.* **1998**, *120*, 3263; V. Quéneau, S.C. Sevov, *Inorg. Chem.* **1998**, *37*, 1358; E. Todorov, S.C. Sevov, *Inorg. Chem.* **1998**, *37*, 3889; H.G. von Schnering, M. Baitinger, U. Bolle, W. Carrillo-Cabrera, J. Curda, Y. Grin, F. Heinemann, L. Llanos, K. Peters, A. Schmeding, M. Somer, *Z. Anorg. Allg. Chem.* **1997**, *623*, 1037; H.G. von Schnering, M. Somer, M. Kaupp, W. Carrillo-Cabrera, M. Baitinger, A. Schmeding, Y. Grin, *Angew. Chem., Int. Ed.* **1998**, *37*, 2359; C. Hoch, M. Wendorff, C. Rohr, *Acta. Cryst.* **2002**, *C58*, 45; C. Hoch, M. Wendorff, C. Rohr, *J. Alloys Compd.* **2003**, *361*, 206.

16 L. Xu, S.C. Sevov, *Inorg. Chem.* **2000**, *39*, 5383.

17 F. Gascoin, S.C. Sevov, *Inorg. Chem.* **2001**, *40*, 6254.

18 F. Gascoin, S.C. Sevov, *Inorg. Chem.* **2002**, *41*, 2292.

19 F. Gascoin, S.C. Sevov, *Inorg. Chem.* **2002**, *41*, 2820.

20 F. Gascoin, S.C. Sevov, *Angew. Chem. Int. Ed.* **2002**, *41*, 1232.

21 Z. Lin, M.B. Hall, *Coord. Chem. Rev.* **1993**, *123*, 149.

22 M. Scheer, J. Müller, M. Häser, *Angew. Chem. Int. Ed.* **1996**, *35*, 2492; J.B. Bonanno, P.T. Wolczanski, E.B. Lobkovsky, *J. Am. Chem. Soc.* **1994**, *116*, 11159.

23 F. Gascoin, S.C. Sevov, *Inorg. Chem.* **2003**, *42*, 8567.

24 F. Gascoin, S.C. Sevov, *Inorg. Chem.* **2002**, *41*, 5920; F. Gascoin, S.C. Sevov, *Inorg. Chem.* **2003**, *42*, 904.

25 A.S. Sefat, J.E. Greedan, *Inorg. Chem.* **2003**, *43*, 142.

14

The Building-block Approach to Understanding Main-group-metal Complex Structures – More than just "Attempting to Hew Blocks with a Razor" *

Peter K. Dorhout

For John D. Corbett, March 23, 2006

Felix qui potuit rerum cognoscere causas. Fortunatus et ille deos qui novit agrestis.
"Lucky is he who has been able to understand the causes of things. Fortunate, too, is the man who has come to know the gods of the countryside".

Virgil – *Georgics*, no. 2.

What we are today comes from our thoughts of yesterday, and our present thoughts build our life of tomorrow: our life is the creation of our mind.

Pali Tipitaka – *Dhammapada*, v. 1.

14.1
Introduction

Over the course of the last 14 years, we have been able to prepare new rare-earth-metal materials using main-group metal ion building blocks. During my tenure in the Corbett group (1989–1991), the concept of *die Bausteine* or the building blocks approach to understanding structures was emerging, as the group employed computational methods to understand the structures of new reduced rare-earth-metal compounds and their physical properties. The major aims of our research programs since then have been to apply this critical thinking process towards developing hypotheses that enable us to examine the structural chemistry of more oxidized rare-earth-metal thio- and selenophosphates, $(P_nQ_m)^{x-}$ [1–6], thio- and selenoantimonites, $(Sb_nQ_m)^{x-}$ [7–9] and thio- and selenotetrelates, $(Td_nQ_m)^{x-}$ [8, 10] where Td = Si, Ge. Our unique approach to studying the composition phase space of these quaternary systems has enabled us to map where, under certain chemical and thermodynamic conditions, particular structural building blocks may be predicted to form; thus, we are poised on the

---

* Alexander Pope, *Miscellanies* (1727), vol. 2.

Inorganic Chemistry in Focus III.
Edited by G. Meyer, D. Naumann, L. Wesemann
Copyright © 2006 WILEY-VCH Verlag GmbH & Co. KGaA, Weinheim
ISBN: 3-527-31510-1

brink of a moderate level of structural predictability. Without the critical thinking foundations cast in the Corbett group, we and many others would not be in a position to approach solid state chemistry in this manner, and we certainly would still be courting the goddess *Fortuna* for our inspiration.

14.2
The Building-block Approach

14.2.1
Quaternary Rare-earth Metal Chalcophosphates

We were one of the first groups to report a ternary selenophosphate of a rare-earth metal [11–13]. Since that time, we have uncovered a host of rare-earth metal chalcophosphates [1, 13, 14] that complement the transition-metal compounds found by the Kanatzidis group [15–31]. Despite the host of publications in the area of metal chalcophosphate chemistry, there have only been our systematic studies of the quaternary phase space of the rare-earth metal chalcophosphates [1, 13, 14].

The chemistry and the structures in this report have been organized by the main-group/chalcogen building block units (Fig. 14.1). At the start of 2000, we were beginning to understand aspects of the chemistry of chalcophosphates both as selenophosphates and thiophosphates. The thesis of Dr. Carl Evenson [14] and his subsequent papers discuss the composition phase diagrams of Y,

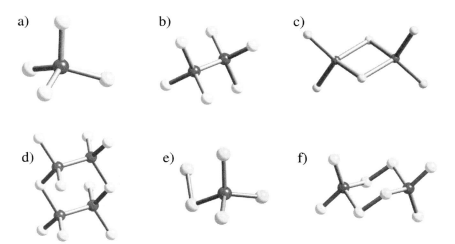

Fig. 14.1 Representations of main group chalcogen polyhedra. In each figure, the main group element (Si, Ge) is darker, and the chalcogen (S, Se, Te) is shown lighter. (a) $M^{IV}Q_4^{4-}$, (b) $M_2^{III}Q_6^{6-}$ (c) $M_2^{IV}Q_6^{4-}$ (d) $M_4^{III}Q_{10}^{8-}$ (e) $M^{IV}Q_5^{4-}$ (f) $M_2^{IV}Q_8^{4-}$.

La, and Eu chalcophosphates [1, 3–5]. These composition phase diagrams have allowed us to understand how the composition of a given molten flux reaction using the flux K_2Q_x (Q = S, Se; x = 2, 3) yields phases that possess predictable building blocks – either PS_4^{3-} or $P_2S_6^{4-}$, for example. An example of such phase diagrams is shown in Fig. 14.2 of Ref. [3] where the reactions are indicated with numbers along the relative oxidation states of phosphorous observed in the phases. We clearly distinguished how phases relied not only on the oxidizing power of a given potassium sulfide or selenide flux, but also on the identity of the rare-earth element – note how the phase regions for P^{5+} and P^{4+} differ for La and Eu in Fig. 14.2.

In each of the composition diagrams in Fig. 14.2, the numbers represent a series of reactions run at a defined composition and temperature. These are isometric sulfur "slices" through three-dimensional K/P/RE/S quaternary phase diagrams. As just one example of what we have studied, Table 14.1 identifies the compositions at each point and the resulting phase(s). We have rigorously studied how phase formation is dependent upon the compositions of reactions for the rare-earth elements Y, Eu, and La and we have also discovered key structural relationships between the rare-earth elements, indicating a significant dependence on rare-earth and alkali-metal size for sulfides and selenides.

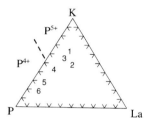

Fig. 14.2 Composition phase diagrams for reactions containing K/P/Eu and K/P/La at constant sulfur concentrations and at a constant temperature of 725 °C. Numbers indicate given reactions tested; the dashed line indicates the separation of P^{5+} and P^{4+} phases. Table 14.1 provides composition data for K-Eu-P-S as a representative example.

Table 14.1 Composition and phases for points in Fig. 14.2 for the K/P/Eu diagram.

Phase diagram point	K/K+P+Eu	P/K+P+Eu	Eu/K+P+Eu	Reaction product
1	0.6667	0.1667	0.1667	$K_4Eu(PS_4)_2$
2	0.5714	0.2857	0.1429	
3	0.4	0.4	0.2	$KEuPS_4$
4	0.3	0.6	0.1	$Eu_2P_2S_6$
5	0.2	0.7	0.1	
6	0.1	0.8	0.1	

The phase relationships between the rare-earth element chalcophosphates are easily mapped on a two-dimensional grid. For example, in the series of layered thiophosphate phases indicated by point 5 on the K/P/La phase diagram in Fig. 14.2, for different rare-earth elements and alkali metals (K, Cs; La, Y as a representative example), four interrelated phases can be easily identified belonging to a nearly isopointal family of structures. These phases, while they possess identical building blocks (PS_4^{3-} tetrahedra and $P_2S_6^{4-}$ ethane-like blocks), do have distinct attributes that can be identified in the images in Fig. 14.3. For example, $K_2Y(P_2S_6)_{1/2}(PS_4)$ is shown in the upper left-hand corner of the figure with dimeric edge-sharing YS_8 polyhedra (striped) connected in chains via PS_4 tetrahedra (black) and then connected into sheets through the P_2S_6 units (P atoms are black spheres). The related $K_2La(P_2S_6)_{1/2}(PS_4)$ phase is shown in the upper right-hand corner. Distinct structural differences due to the coordination preferences for the rare-earth elements were noted. Changing the cation from K to Cs changed the structure of the yttrium compound significantly but not so for the lanthanum phase.

We initially prepared two unique quaternary rare-earth metal selenophosphates in a study of rare-earth metal chalcogen compounds: $KLnP_2Se_6$, where one phase comprised the ceria rare earths and the other phase comprised the yttria rare earths [11, 12, 32]. The family of rare-earth chalcophosphates has been expanded recently by our group and the Kanatzidis group to include phases of $K_4Eu(PSe_4)_2$, $K_9Ce(PSe_4)_4$, $Rb_2LnP_2Se_9$, and $Rb_2LnP_2Se_7$ [1, 3, 5, 6, 14, 29, 33–37].

A detailed study of two rare-earth metals under one set of reaction conditions, for example, yielded the two composition phase diagrams, shown in Fig. 14.2, for the Eu and La thiophosphate systems [3]. To prepare these phase diagrams, we varied the alkali metal, the rare-earth metal, and the phosphorous concentration to kept the sulfur concentration constant. We prepared similar studies in

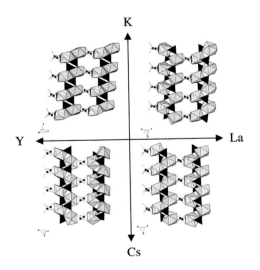

Fig. 14.3 Polyhedral packing plots for the two-dimensional layers of $[RE(P_2S_6)_{1/2}(PS_4)]^{2-}$ in the series of solids $A_2RE(P_2S_6)_{1/2}(PS_4)$, where A = K, Cs; RE = Y, La. Rare-earth polyhedra are striped; PS_4 polyhedra are black; phosphorous atoms in P_2S_6 are shown as black circles. Alkali atoms are not shown for clarity. Although these phases have distinctly different structures based on space group symmetry and atomic positions, the compounds are clearly related upon close inspection of the building blocks.

Table 14.2 Predicted formulae and reported chalcophosphate structures from Eq. (1).

l	m	n	p	Predicted formula	Reported structure
0	1	0	0	A_3PSe_4	K_3PSe_4 [38]
1	0	0	0	$A_4P_2Se_6$	$K_4P_2Se_6$ [39]
0	0	1	0	$La_4(P_2Se_6)_3$	unknown
0	0	0	1	$LaPSe_4$	unknown
1	0	1	0	$A_4La_4(P_2Se_6)_4$	$KLaP_2Se_6$ [11]
1	0	0	2	$A_4La_2(P_2Se_6)(PSe_4)_2$	$K_2La(P_2Se_6)_{1/2}(PSe_4)$ [1]
0	1	0	1	$A_3La(PSe_4)_2$	$K_3La(PSe_4)_2$ [1]
0	2	0	1	$A_6La(PSe_4)_3$	$K_4La_{0.67}(PSe_4)_2$ [1]
0	3	0	1	$A_9La(PSe_4)_4$	$K_9La(PSe_4)_4$ [1]

the equivalent selenide systems [1, 14]. We were able to find four new compounds in the lanthanum set based on careful analysis of the phase diagrams and the development of a revised set of chemical family formulae, Table 14.2 of Ref. [1] based on the linear combination of structural building blocks, where **A** is an alkali-metal atom. A similar set of formulae can be constructed for the thiophosphates [3].

$$(A_4P_2Se_6)_l(A_3PSe_4)_m[La_4(P_2Se_6)_3]_n(LaPSe_4)_p \tag{1}$$

In our analysis of these phases, we observed key differences between these structures and those for a set of phosphate minerals. For example, $K_3La(PQ_4)_2$ is uniquely different from the *glaserite*-type phases of $K_3Ho(PO_4)_2$ [40, 41]. In addition, phases like *vitusite* ($Na_3Ce(PO_4)_2$) [42] and *eulytite* ($Sr_3La(PO_4)_3$) [43] are not found in our studies. Indeed, building blocks which we have seen in other chemical systems such as $(P_2S_7)^{4-}$ and $(P_3Se_{10})^{5-}$ are not observed with the rare-earth elements we have studied to date and there is no clear reason why they should not be found in continuing composition phase-space searches [44, 45]. Nevertheless, *there are clearly unique structures that can be prepared from the chalcophosphate building blocks.*

14.2.2
Quaternary Rare-earth Metal Chalcoarsenites and Antimonites

Our initial work demonstrated that there is a wide-open field for discovery in this part of the periodic table. We have begun to develop ternary and quaternary phase diagrams using a series of peritectically melting materials as reagents in synthesis [7–9]. There are examples recently of complex materials prepared from arsenite or antimonite building blocks [46–68] and fewer with rare-earth metals [69–74]. We have begun to extend the known chemistry of rare-earth metal chalcoarsenites and antimonites through a systematic study of their quaternary phase diagrams.

We have been building on the concept that we can use mineralogical structures as targets by recognizing that building blocks in the families of arsenites and antimonites are less well-defined. Despite this, they may yield interesting structures related to minerals. For example, it has been shown that mixing phases to create complex structures – even incommensurate or "modulated structures" [75, 76] – can be facilitated by combining the building-block approach with a larger structural model. Most of these analyses have focused on understanding the complex structures of minerals but *none has been used to predict the formation of new materials*, in particular, rare-earth-metal substituted systems. Our scheme, below, has been used to predict possible structures which thus enables us to then target and prepare new materials in the laboratory.

The list of compounds in Table 14.3 illustrates known and target compounds in a selected family of structures related to the *bismuthite–aikinite* solids. This approach to synthesis is different from those used by other groups to target chalcoarsenites and chalcoantimonites [65–68, 77]. We have focused on the *stibnite* and *aikinite* family substitutions. Our primary building blocks comprise *bismuthite*, Bi_2S_3, *aikinite*, $CuPbBiS_3$, and *krupkite*, $CuPbBi_3S_6$. A mathematical model has been proposed in order to understand these mineral systems that we have modified and adapted *to predict possible structures and their space group symmetries* [78]. The systematic description begins with Eq. (2) where $Pn = As$, Sb; $Q = S$, Se:

$$(Pn_4Q_6)_m((I)_2(II)_2Pn_2Q_6)_n((I)(II)Pn_3Q_6)_p \text{ where } (I) = Cu, Ag; (II) = Eu, Yb, Sr \quad (2)$$

For this scheme, m, n, and p have been defined for minerals as being even numbers. Notice also that the formula for *bismuthite* and *aikinite* have been doubled in keeping with the reported model *where only* (I) = Cu *and* (II) = Pb *are known* [78]. Also defined are the terms A ($A = m/4$) and B ($B = n/2$). It has been observed that for most mineralogical systems, when $A + B$ is *even*, the corresponding space group was the non-centrosymmetric, $Pmc2_1$, and when $A + B$ is

Table 14.3 Target chalcopnictite compounds – $Pn = As^{III}$, Sb^{III}.

mnp	Compound (empirical)	A : B ratio	Space group	Substitution monovalent	Substitution divalent	Substitution trivalent
200	Pn_2Q_3	1:0	Pnma	N/A	N/A	N/A
020	$(I)(II)PnQ_3$	0:1	Pnma	Ag, Cu	Eu, Yb, Sr	Gd, Tb
002	$(I)(II)Pn_3Q_6$	1:1	$Pmc2_1$	Ag, Cu	Eu, Yb, Sr	Gd, Tb
022	$(I)_3(II)_3Pn_5Q_{12}$	1:2	$Pmc2_1$	Ag, Cu	Eu, Yb, Sr	Gd, Tb
202	$(I)(II)Pn_7Q_{12}$	3:1	$Pmc2_1$	Ag, Cu	Eu, Yb, Sr	Gd, Tb
024	$(I)_2(II)_2Pn_4Q_9$	1:2	Pmna	Ag, Cu	Eu, Yb, Sr	Gd, Tb
042	$(I)_5(II)_5Pn_7Q_{18}$	1:5	$Pmc2_1$	Ag, Cu	Eu, Yb, Sr	Gd, Tb
204	$(I)(II)Pn_5Q_9$	2:1	Pmna	Ag, Cu	Eu, Yb, Sr	Gd, Tb
402	$(I)(II)Pn_{11}Q_{18}$	5:1	$Pmc2_1$	Ag, Cu	Eu, Yb, Sr	Gd, Tb
026	$(I)_5(II)_5Pn_{11}Q_{24}$	3:5	$Pmc2_1$	Ag, Cu	Eu, Yb, Sr	Gd, Tb
062	$(I)_7(II)_7Pn_9Q_{24}$	1:7	$Pmc2_1$	Ag, Cu	Eu, Yb, Sr	Gd, Tb
206	$(I)_3(II)_3Pn_{13}Q_{24}$	5:3	$Pmc2_1$	Ag, Cu	Eu, Yb, Sr	Gd, Tb
602	$(I)(II)Pn_{15}Q_{24}$	7:1	$Pmc2_1$	Ag, Cu	Eu, Yb, Sr	Gd, Tb

odd, the reported space group was the centrosymmetric, Pnma. Using these ideas, the following table could be constructed that illustrates how we targeted specific compounds and predicted their space group symmetry. (Note that even when m = 0, one can deduce a value for A by dividing the compound into its Pn_2Q_3 and $(I)(II)PnQ_3$ components.)

However, it may seem that we have ignored the m, n, and p *odd* combinations that are also possible. For example, in *jamesonite*, $Fe^{II}Pb_4Sb_6S_{14}$ [80], the mnp values are 034 (or 068), respectively (assuming one $Fe(II) = 2\,Cu(I)$). We have successfully substituted europium into the structure to create a congruently melting, semiconducting $FePb_2Eu_2Sb_6S_{14}$ [79]. This has a value of $A:B =$ 4:10, indicating that this compound, as we found, has a space group of $Pmc2_1$. It is clear that targeting only the combinations in this table, with no substitution of a trivalent lanthanide for the pnictide, Pn, there are 72 different reactions to monitor, for each sulfur or selenium and arsenic or antimony system, at varying temperatures (our compounds were prepared between 750 and 1200 °C).

14.2.3
Quaternary Rare-earth Metal Chalcotrielates and Tetrelates

We have developed a systematic study of quaternary chalco-gallates, indates (trielates) and chalco-silicates, and germinates (tetrelates) of the rare-earth metals. We have demonstrated that a series of new compounds could be formed in these families of materials [1, 3, 9, 10, 14, 81, 82]. There have been reports in the literature of a few examples of these types of materials, but there is sufficient evidence to support the fact that this area is wide open for exploring the likelihood for new rare-earth metal-based materials [83–90]. Our studies have focused on the gallium, indium, germanium and silicon sulfides, selenides, and tellurides.

$KLaGeSe_4$, synthesized in 1993, was the first member in a new class of layered quaternary materials, which can be generally formulated $AM^{III}M^{IV}Q_4$ ($A = K$; $M^{III} = Ln = La$, Nd, Gd, Y; $M^{IV} = Si$, Ge; $Q = S$, Se) [87]. This phase was engineered to make use of the unique coordination preferences adopted by the alkali, lanthanide, and main group elements to form a structure with no disorder. A second phase, $CsSmGeS_4$, synthesized in 1994, is a closely related, but structurally distinct phase [91]. Due to differences in the reaction conditions used to generate these two phases, as well as the variation in the alkali metal, lanthanide, and chalcogen, it is difficult to draw any conclusions as to why the two structures differ. A systematic study was undertaken by our group to determine the effect of the alkali metal identity in $ASmGeSe_4$ ($A = Na$, K, Rb, Cs) under identical synthetic conditions, where the typical building block comprises $GeSe_4^{4-}$ tetrahedral [82]. All of the reactions were performed at 725 °C through the combination of elemental Sm, Ge, and Se, with the reactive flux A_2Se_n ($n =$ 2–3). This study confirmed the identity of four new layered quaternary structures (Fig. 14.4). $KSmGeSe_4$ is isostructural to $KLaGeSe_4$ in the non-centrosym-

a)   b)   c)

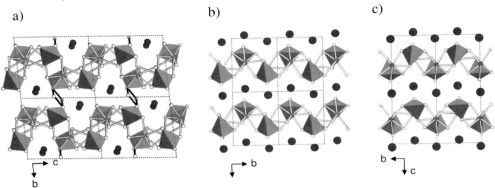

Fig. 14.4 ASmGeSe$_4$ compounds viewed parallel to the SmGeSe$_4$ layers. Dashed lines indicate unit cell boundaries in each case. GeSe$_4$ tetrahedra are drawn as tetrahedral solids, and the SmSe$_n$ environments are drawn using a ball-stick model. In all structures, Sm is dark grey, and Se is light grey, and the alkali is black.
(a) Na$_{1-2x}$SmGeSe$_{4+x}$ viewed along a, inter-layer Se–Se bonds are shaded black.
(b) KSmGeSe$_4$ viewed along a.
(c) RbSmGeSe$_4$ viewed along a.

metric space group P2$_1$. RbSmGeSe$_4$ and CsSmGeSe$_4$ are isostructural to CsSmGeS$_4$ in the non-centrosymmetric space group P2$_1$2$_1$2$_1$. The sodium ana-logue, Na$_{1-2x}$SmGeSe$_{4+x}$ (x=0.10), is unique in that it contains a disordered Se$_3^{-2}$ bridge closely related to the GeSe$_5^{4-}$ unit found in K$_2$EuGeSe$_5$ [4]. The differing size of the alkali metal in these structures determines the organization of the layers of [SmGeSe$_4$]$^{1-}$.

While investigating the effects of lanthanide substitution using Yb, Ho, and Gd, we noticed that, under reaction conditions identical to those used to form KSmGeSe$_4$, the quaternary phases KLnGeSe$_4$ (Ln=Yb, Ho, Gd) did not form, but rather the reactants phase separated into KLnSe$_2$, and assorted ternary potas-sium selenogermanates. This has led to an ongoing investigation of lanthanide size effects on the quaternary phase stability. It should be noted that we did pre-pare the sulfur analog, KYbGeS$_4$, so it seems that the larger cavity formed from the selenium anions is responsible for the observed phase behavior.

Europium forms two different structures with a similar formula, NaEuGeQ$_4$ (Q=S, Se). NaEuGeSe$_4$ crystallizes in a structure that appears to be more closely related to Na$_{1.5}$Pb$_{0.75}$PSe$_4$ than KLaGeSe$_4$ [92]. NaEuGeS$_4$ is a high symmetry structure, crystallizing in the R3c space group [93]. This structure contains one-dimensional tubes of EuGeS$_4$, with sodium occupying positions within the channel and between the tubes (Fig. 14.5). At this time, these structures appear to be unique to the europium system.

Our group has observed a class of ternary structures with the general formula M^{II}M^{IV}Q$_4$ (M^{II}=Sr, Eu; M^{IV}=Ge, Si; Q=Se). Both our group and Bernd Mosel [94, 95] simultaneously synthesized Eu$_2$SiSe$_4$ and Eu$_2$GeSe$_4$. Eu$_2$GeSe$_4$ has been shown to undergo a ferroelectric phase transformation between a low-tempera-ture non-centrosymmetric phase and a centrosymmetric high-temperature

a)

b)

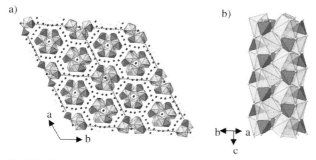

Fig. 14.5 NaEuGeS$_4$ viewed (a) along the c-axis, and (b) perpendicular to EuGeS$_4$ channels. Dark polyhedra are GeS$_4^{4-}$ units, light polyhedra are EuS$_7$ units, dark spheres are Na atoms, light spheres are S atoms.

phase, with a Curie temperature of approximately 600–650 K [95]. We have also recently synthesized Sr$_2$GeSe$_4$ and through X-ray diffraction this structure modeled best in the non-centrosymmetric space group P2$_1$. Studies are under way to determine whether this material undergoes a similar structural phase transformation.

Few examples of complex structures containing homonuclear Ge–Ge or Si–Si bonds have been described in the literature (see Fig. 14.1 b). Isostructural ternary materials with the formula A$_6$M$_2^{III}$Q$_6$ (A = Na, K; M^{III} = Si, Ge; Q = S, Se, Te) have been found with widely varying combinations of elements [96]. Kanatzidis reported a series of isotypic lead and tin structures with the general formula Na$_8$M$_2^{II}$(M$_2^{III}$Q$_6$)$_2$ (M^{II} = Sn, Pb; M^{III} = Si, Ge; Q = S, Se) [97]. These structures contain mixed octahedral Na/M^{II} sites, and the location of these sites within the structure varies with the M^{II} identity. We have recently found that Na$_8$Eu$_2$(Si$_2$Se$_6$)$_2$ is nearly isostructural with Na$_8$Pb$_2$(Si$_2$Se$_6$)$_2$ [10]. We have found another closely related structure, Na$_9$Sm(Ge$_2$Se$_6$)$_2$, which also contains the M$_2^{III}$Q$_6^{6-}$ building block in a layered structure, but the Ge–Ge bonds are rotated relative to the layer planes. Fig. 14.6 displays the structural differences between Na$_8$Eu$_2$(Si$_2$Se$_6$)$_2$ and Na$_9$Sm(Ge$_2$Se$_6$)$_2$. We have also found Na$_8$Eu$_2$(Ge$_2$Se$_6$)$_2$, Na$_8$Eu$_2$(Si$_2$Te$_6$)$_2$, K$_8$Eu$_2$(Si$_2$Te$_6$)$_2$ and K$_8$Eu$_2$(Ge$_2$Te$_6$)$_2$ to be isostructural to Na$_8$Pb$_2$(Si$_2$Se$_6$)$_2$.

A series of quaternary europium structures containing two edge-shared tetrahedra, M$_2^{IV}$Q$_6^{4-}$, see Fig. 14.1 c), have been synthesized with a general formula A$_2$EuM$_2$Se$_6$ (A = Na, K, Rb, Cs; M = Si, Ge). These compounds crystallize in four related structure types, shown in Fig. 14.7. These structures are all monoclinic or orthorhombic, with the β angle generally increasing with increasing cation size for Ge compounds, and decreasing with increasing cation size for Si compounds. Na$_2$EuM$_2$Se$_6$ (M = Si, Ge) was modeled in the Pnma space group, and contains layers of edge-shared (Na/Eu)Se$_7$ bridged by M$_2$Se$_6$ units, with Na cations occupying channel-like cavities. K$_2$EuM$_2$Se$_6$ (M = Si, Ge) was modeled in

a)   b)

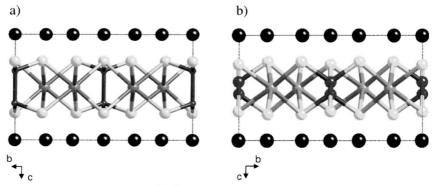

Fig. 14.6 Structures containing $M_2^{III}Se_6^{6-}$ building blocks. In each representation, the alkali metal is black, the rare-earth element is shaded medium grey, the main group element is dark grey, and selenium is light grey. (a) $Na_8Eu_2(Si_2Se_6)_2$, (b) $Na_9Sm(Ge_2Se_6)_2$.

a)   b)   c)   d)

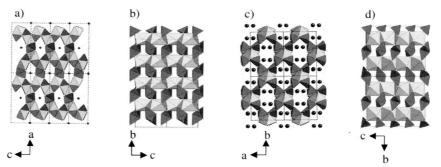

Fig. 14.7 $A_2EuM_2^{IV}Se_6$ (A=Na, K, Rb, Cs; M=Si, Ge) structures. In all structures, the alkali metals are black spheres, Se light spheres, $EuSe_n$ polyhedra are light grey, and M_2Se_6 polyhedra are dark grey. (a) $Na_2EuM_2Se_6$ (M = Si, Ge) viewed along the tunnels. (b) $K_2EuM_2Se_6$ (M=Si, Ge) viewed perpendicular to the tunnels. (c) $A_2EuGe_2Se_6$ (A=Rb, Cs) viewed parallel to the tunnels. (d) $A_2EuSi_2Se_6$ (A=Rb, Cs) viewed perpendicular to the tunnels.

the C2/c space group, and contains chains of edge-shared $EuSe_6$ octahedra interconnected by M_2Se_6 units. $A_2EuGe_2Se_6$ (A=Rb, Cs) are nearly isostructural to $K_2EuM_2Se_6$, and were solved as the pseudo-mirrored (about the a-axis) structure. The structures $A_2EuSi_2Se_6$ (A=Rb, Cs) are also closely related to the K_2Eu-M_2Se_6 structure, but were modeled in the Pccn space group. These structures contain chains of edge-shared $EuSe_6$ octahedra bridged by Si_2Se_6 units, but the relative orientations of these units differ, as shown in Fig. 14.7 b and d.

Although no lanthanide-containing structures containing the bridged $M_4^{III}Q_{10}^{8-}$ building block (Fig. 14.1 d), have been synthesized to date, we have synthesized quaternary structures by substituting some Ag for Na in the ternary compound, resulting in a structure with the formula $Na_{8-x}Ag_xGe_4Se_{10}$ ($x=0$–4). This new structure type is closely related to the ternary $Na_8Ge_4Se_{10}$, but the Ag-containing crystal is centered monoclinic, rather than triclinic, and the $Ge_4Se_{10}^{8-}$ units are aligned in layers separated by mixed Ag/Na sites (Fig. 14.8). Studies are under way to determine the effects of the Na/Ag reaction stoichiometry on the resulting Ag concentration.

Heavy chalcogens are interesting in that Q–Q bonds may be found within a structure. As described above, $Na_{1-2x}SmGeSe_{4+x}$ contains interlayer Se–Se bonds. $Na_2EuGeSe_5$ was also synthesized, containing a tetrahedral Ge^{IV} unit with one corner replaced by a Se_2^{2-} unit (Fig. 14.1e). This structure is isotypical to $K_2EuGeSe_5$ [4], but was modeled as a superstructure with a doubled layer axis, with alternate layers related by a glide plane. Two structures containing the selenium-bridged $M_4^{IV}Q_8^{4-}$ have been synthesized by our group. $Cs_4Si_2Se_8$ and $Rb_4Ge_2Se_8$ containing two MSe_4 tetrahedra linked through Se–Se bonds (Fig. 14.1f), were found to be isostructural to $Cs_4Ge_2Se_8$, although this latter phase was prepared by a solvothermal method versus our molten flux process [98].

As for the chalcopnictates, there are very few quaternary A-Ln-Ge-Q systems [83, 84, 99–105]. Consulting the literature, we found that ternary (II)TdQ$_3$ and (II)TdQ$_4$ phases (II = Sr, Pb, Eu; Td = tetrelide, Si, Ge, Sn) decompose peritectically [101]. In a manner similar to that described for the pnictates above, we proposed a scheme for targeting new quaternary phases of the tetrelates, Table 14.4 using the building blocks introduced in Fig. 14.1a–c. Equation (3) illustrates how this could be accomplished using known phases as building blocks as we have done in Eqs. (1) and (2), where Ln = rare-earth metals. Building blocks from Fig. 14.1a–c are employed in the equation:

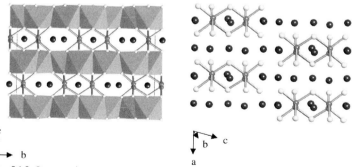

Fig. 14.8 Structural comparison of (a) $Na_{8-x}Ag_xGe_4Se_{10}$ to (b) $Na_8Ge_4Se_{10}$. In each figure, Na is black, Ge dark grey, Se light grey, and $AgSe_6$ octahedra are shaded grey. The view in each figure is along close-packed selenium layers, down the Ge–Ge bonds in the $Ge_2Se_{10}^{8-}$ units.

$$(A_4TdQ_4)_m(ALnTdQ_4)_n(A_3LnTd_2Q_6)_p(ALnTd_2Q_6)_q;$$
$$A = alkali; Td = Ge, Si; Q = S, Se \tag{3}$$

From this equation, we again build up a table of target compounds (Table 14.4).

As with the pnictates, a quasi-ternary phase diagram can be developed to map out possible compounds in this composition phase space using key chalcotetrelate building blocks. We have begun to make use of the peritectic nature of the starting materials, as this has facilitated reactions between phases.

We have developed several examples of fruitful areas in this chemistry that demonstrate feasibility. As one might expect, a list of compounds related to the pnictates proposed in Table 14.4 as the trivalent pnictogens and trielides should yield compounds of similar chemical composition. However, we can think of these systems in a slightly different manner. We can imagine a different formula from that in the previous equations:

$$((II)Q)_m((I)TrQ_2)_n(Tr_2Q_3)_p \text{ where } (I) = Cu, Ag ; (II) = Fe, Pb, Eu; Tr = In, Ga;$$
$$Q = S, Se \tag{4}$$

As an example, $CuInSe_2$ is a known low band-gap material (1.0 eV) that shows promise for use in solar energy conversion [106]. We can imagine preparing rare-earth based materials using the formulation shown in Table 14.5.

Alternatively, we can consider other schemes for structure composition as there is no precedence for the scheme in Eq. (5). We can prepare ternary compounds that peritectically decompose, a feature that may facilitate their reactivity at low temperatures:

$$((II)Tr_2Q_4)_m((II)Ln_2Q_4)_n(LnTrQ_3)_p \text{ where } (II) = Fe, Mn, Pb, Sr; Tr = Ga, In;$$
$$Q = S, Se \tag{5}$$

A table of appropriate m, n, and p values can be generated as in Table 14.5 to yield another list of target compounds (Table 14.6). This may be better suited for the rare earths, Ln, as both $LnInQ_3$ and $(II)In_2Q_4$ peritectically decompose above 800 °C [108]. We can also substitute 2 Cu(I) for one (II) element as another variable in the process.

Table 14.4 Target chalcotetrelate compounds.

mnpq	Compound (empirical)
1100	$A_5Ln(TdQ_4)_2$
0110	$A_4Ln_2(TdQ_4)(Td_2Q_6)$
1010	$A_7Ln(TdQ_4)(Td_2Q_6)$
1001	$A_5Ln(TdQ_4)(Td_2Q_6)$
0210	$A_5Ln_3(TdQ_4)_2(Td_2Q_6)$
0201	$A_4Ln_3(TdQ_4)_2(Td_2Q_6)$

Table 14.5 Target chalcotrielate compounds.

mnp	Compound (empirical)	
110	$(I)(II)TrQ_3$	
210	$(I)(II)_2TrQ_4$	
310	$(I)(II)_3TrQ_5$	
230	$(I)_3(II)_2Tr_3Q_8$	a) $Cu_2^IPb^{II}La^{III}In_3Se_8$
320	$(I)_2(II)_3Tr_2Q_7$	b) $Fe^{II}Nd_2^{III}In_2S_7$
111	$(I)(II)Tr_3Q_6$	
112	$(I)(II)Tr_5Q_9$	
122	$(I)_2(II)Tr_6Q_{11}$	
113	$(I)(II)Tr_7Q_{12}$	

a) $Cu_2^IPb^{II}La^{III}In_3Se_8$ was prepared and has similar lattice constants (but a different XRD pattern) to $CuInSe_2$.

b) $Fe^{II}Nd_2^{III}In_2S_7$ was prepared, and has a structure related to Y_5S_7 [107].

Table 14.6 More target chalcotrielate compounds.

mnp	Compound (empirical)	
110	$(II)LnTrQ_4$	
210	$(II)_3Ln_2Tr_4Q_{12}$	
310	$(II)_4Ln_2Tr_6Q_{19}$	
230	$(II)_5Ln_6Tr_4Q_{20}$	
320	$(II)_5Ln_4Tr_6Q_{20}$	
111	$(II)_2Ln_3Tr_3Q_{11}$	
112	$(II)Ln_2Tr_2Q_7$	
122	$(II)_3Ln_6Tr_4Q_{20}$	
113	$(II)_2Ln_5Tr_5Q_{14}$	
022	$(II)_2Ln_6Tr_2Q_{14}$	a) $Ca_2Ln_6Ga_2Q_{14}$

a) $Ca_2Ln_6Ga_2Q_{14}$ was prepared in our laboratory as Tb, Pr analogs [81].

14.3
Summary

This chapter has provided just a *Streifblick* of the chemistry that our group has pursued over the years since working as a member of Corbett's group and has clearly demonstrated that there are many fertile fields from which we may continue to harvest incredible knowledge about how chemical building blocks behave as they construct interesting structures that we are lucky enough to have imagined. The reduced rare-earth-metal chemistry initiated in his group has enabled us done the right pair of spectacles to view structures with a particular prescription lens that, like so many others before us, has provided the chemical foresight in us to develop logical approaches toward chemical synthesis. The impact that John Corbett has had on structural inorganic chemistry will only be limited by how his chemical offspring share his perception for creative thought and focused action. Finally, this work could not have been accomplished without the financial support of the National Science Foundation.

References

1 C.R. Evenson, P.K. Dorhout, *Inorg. Chem.* **2001**, *40*, 2875.

2 C.R. Evenson, P.K. Dorhout, *Z. Anorg. Allg. Chem.* **2001**, *627*, 2178.

3 C.R. Evenson, P.K. Dorhout, *Inorg. Chem.* **2001**, *40*, 2884.

4 C.R. Evenson, P.K. Dorhout, *Inorg. Chem.* **2001**, *40*, 2409.

5 P.K. Dorhout, C.R.I.V. Evenson, *NATO Science Series, II: Mathematics, Physics and Chemistry* **2001**, *48*, 13.

6 J.L. Burris, I. Orgzall, C.R. Evenson, P.K. Dorhout, H.D. Hochheimer, *J. Physics Chem. Solids* **2002**, *63*, 597.

7 W.W. So, B.R. Martin, P.K. Dorhout, *Z. Krist.* **2002**, *217*, 302.

8 W.-W. So, A. LaCour, V.O. Aliev, P.K. Dorhout, *J. Alloys Compds.* **2004**, *374*, 234.

9 V.O. Aliev, P.K. Dorhout, **2005**, manuscript in preparation.

10 B.R. Martin, L.A. Polyakova, K.P. Dorhout, *Inorg. Chem.* **2005**, in press.

11 J.H. Chen, P.K. Dorhout, *Inorg. Chem.* **1995**, *34*, 5705.

12 J.H. Chen, P.K. Dorhout, J.E. Ostenson, *Inorg. Chem.* **1996**, *35*, 5627.

13 W. Brockner, R. Becker, *Z. Naturforsch.* **1987**, *42a*, 511.

14 C.R. Evenson, Ph.D. thesis, Colorado State University (Fort Collins), **2001**.

15 K. Chondroudis, M.G. Kanatzidis, *Inorg. Chem.* **1995**, *34*, 5401.

16 T.J. McCarthy, M.G. Kanatzidis, *Inorg. Chem.* **1995**, *34*, 1257.

17 T. McCarthy, M.G. Kanatzidis, *J. Alloys Compds.* **1996**, *236*, 70.

18 K. Chondroudis, M.G. Kanatzidis, *Chem. Commun.* **1996**, 1371.

19 K. Chondroudis, M.G. Kanatzidis, *C.R. Acad. Sci. Ser. II B* **1996**, *322*, 887.

20 K. Chondroudis, M.G. Kanatzidis, *J. Am. Chem. Soc.* **1997**, *119*, 2574.

21 K. Chondroudis, M.G. Kanatzidis, *Angew. Chem. Int. Ed.* **1997**, *36*, 1324.

22 K. Chondroudis, J.A. Hanko, M.G. Kanatzidis, *Inorg. Chem.* **1997**, *36*, 2623.

23 M.G. Kanatzidis, *Curr. Opin. Solid State Mat. Sci.* **1997**, *2*, 139.

24 K. Chondroudis, M.G. Kanatzidis, *Inorg. Chem.* **1998**, *37*, 2098.

25 K. Chondroudis, M.G. Kanatzidis, *J. Solid State Chem.* **1998**, *136*, 79.

26 K. Chondroudis, M.G. Kanatzidis, *Inorg. Chem.* **1998**, *37*, 2848.

27 K. Chondroudis, M.G. Kanatzidis, *J. Solid State Chem.* **1998**, *138*, 321.

28 K. Chondroudis, M.G. Kanatzidis, *Inorg. Chem.* **1998**, *37*, 3792.

29 K. Chondroudis, M.G. Kanatzidis, *Inorg. Chem. Commun.* **1998**, *1*, 55.

30 K. Chondroudis, M.G. Kanatzidis, *Inorg. Chem.* **1998**, *37*, 2582.

31 J.A. Hanko, J.H. Chou, M.G. Kanatzidis, *Inorg. Chem.* **1998**, *37*, 1670.

32 J.H. Chen, P.K. Dorhout, *J. Solid State Chem.* **1995**, *117*, 318.

33 K. Chondroudis, T.J. McCarthy, M.G. Kanatzidis, *Inorg. Chem.* **1996**, *35*, 840.

34 M.G. Kanatzidis, *Curr. Opin. Solid State & Mat. Sci.* **1997**, *2*, 139.

35 J.A. Aitken, M. Evain, L. Iordanidis, M.G. Kanatzidis, *Inorg. Chem.* **2002**, *41*, 180.

36 J.A. Aitken, K. Chondroudis, V.G. Young, Jr., M.G. Kanatzidis, *Inorg. Chem.* **2000**, *39*, 1525.

37 K. Chondroudis, M.G. Kanatzidis, *Inorg. Chem.* **1998**, *37*, 3792.

38 B.C. Chan, P.L. Feng, Z. Hulvey, P.K. Dorhout, *Z. Krist.* **2005**, *220*, 11.

39 B.C. Chan, P.L. Feng, P.K. Dorhout, *Z. Krist.* **2005**, *220*, 9.

40 V.A. Efremov, P.P. Mel'nikov, L.N. Komissarova, *Rev. Chim. Miner.* **1985**, *22*, 666.

41 V. A. Efremov, P. P. Mel'nikov, L. N. Komissarova, *Koord. Khim.* **1981**, *17*, 467.

42 O. G. Karpov, D. Y. Pushcharovskii, A. P. Khomyakov, E. A. Pobedimskaya, N. V. Belov, *Kristallografiya* **1980**, *25*, 1135.

43 J. Barbier, J. E. Greedan, T. Asaro, G. J. McCarthy, *Eur. J. Solid State Inorg. Chem.* **1990**, *27*, 855.

44 R. F. Hess, K. D. Abney, J. L. Burris, H. D. Hochheimer, P. K. Dorhout, *Inorg. Chem.* **2001**, *40*, 2851.

45 R. F. Hess, P. L. Gordon, C. D. Tait, K. D. Abney, P. K. Dorhout, *J. Am. Chem. Soc.* **2002**, *124*, 1327.

46 D. M. Smith, C. W. Park, J. A. Ibers, *Inorg. Chem.* **1996**, *35*, 6682.

47 J. A. Ibers, D. M. Smith, M. A. Pell, *Inorg. Chem.* **1998**, *37*, 2340.

48 J. A. Hanko, M. G. Kanatzidis, *Angew. Chem. Int. Ed.* **1998**, *37*, 342.

49 W. Bensch, M. Schur, *Eur. J. Solid State Inorg. Chem.* **1996**, *33*, 1149.

50 W. Bensch, M. Schur, *Z. Naturforsch. [B]* **1997**, *52*, 405.

51 W. Bensch, M. Schur, *Eur. J. Solid State Inorg. Chem.* **1997**, *34*, 457.

52 W. Czado, U. Muller, *Z. Anorg. Allg. Chem.* **1998**, *624*, 239.

53 M. R. Girard, J. Li, D. M. Proserpio, *Main Group Met. Chem.* **1998**, *21*, 231.

54 Y. H. Ko, K. M. Tan, J. B. Parise, A. Darovsky, *Chem. Mater.* **1996**, *8*, 493.

55 J. Li, D. M. Proserpio, M. R. Girard, *Main Group Met. Chem.* **1998**, *21*, 231.

56 T. M. Martin, G. L. Schimek, D. A. Mlsna, J. W. Kolis, *Phosphor Sulfur Silicon* **1994**, *93*, 93.

57 T. M. Martin, G. L. Schimek, W. T. Pennington, J. W. Kolis, *J. Chem. Soc. Dalton Trans.* **1995**, 501.

58 J. B. Parise, Y. H. Ko, *Chem. Mater.* **1992**, *4*, 1446.

59 A. Pfitzner, *Z. Anorg. Allg. Chem.* **1994**, *620*, 1992.

60 A. Pfitzner, *Z. Anorg. Allg. Chem.* **1995**, *621*, 685.

61 H. O. Stephan, M. G. Kanatzidis, *J. Am. Chem. Soc.* **1996**, *118*, 12226.

62 K. M. Tan, Y. H. Ko, J. B. Parise, *Acta Crystallogr. C* **1994**, *50*, 1439.

63 K. M. Tan, Y. H. Ko, J. B. Parise, J. H. Park, A. Darovsky, *Chem. Mater.* **1996**, *8*, 2510.

64 M. Wachhold, M. G. Kanatzidis, *Z. Anorg. Allg. Chem.* **2000**, *626*, 1901.

65 M. Wachhold, M. G. Kanatzidis, *Inorg. Chem.* **2000**, *39*, 2337.

66 R. G. Iyer, M. G. Kanatzidis, *Inorg. Chem.* **2002**, *41*, 3605.

67 R. G. Iyer, J. Do, M. G. Kanatzidis, *Inorg. Chem.* **2003**, *42*, 1475.

68 K.-S. Choi, M. G. Kanatzidis, *Inorg. Chem.* **2000**, *39*, 5655.

69 P. Lemoine, D. Carré, M. Guittard, *Acta Crystallogr. B* **1981**, *37*, 1281.

70 P. Lemoine, D. Carré, M. Guittard, *Acta Crystallogr. B* **1982**, *38*, 727.

71 J. H. Chen, P. K. Dorhout, *Inorg. Chem.*, in preparation.

72 J. H. Chen, P. K. Dorhout, *J. Alloys Compds.* **1997**, *249*, 199.

73 O. M. Aliev, T. F. Maksudova, N. D. Samsonova, L. D. Finkel'shtein, P. G. Rustamov, *Inorg. Mater.* **1986**, *22*, 23.

74 P. G. Rustamov, J. P. Khasaev, O. M. Aliev, *Inorg. Mater.* **1981**, *17*, 1469.

75 J. M. Cowley, in *AIP Conf. Proc., Vol. 53*, Amer. Inst. Phys., **1979**, p. 368.

76 E. Mackovicky, *Fortschr. Mineral.* **1981**, *59*, 137.

77 S.-J. Kim, J. Ireland, C. R. Kannewurf, M. G. Kanatzidis, *Chem. Mater.* **2000**, *12*, 3133.

78 V. A. Gasimov, H. S. Mamedov, *Azerbaijan Chem. Journal* **1979**, 121.

79 V. O. Aliev, K. P. Dorhout **2005**, manuscript in preparation.

80 I. V. Petrova, E. A. Pobedimskaya, N. V. Belov, *Mineral. Zh.* **1980**, *2*, 3.

81 P. M. Van Calcar, P. K. Dorhout, *Rare Earths 98: Proceedings of the International Conference on Rare Earths* **1999**, 322

82 B. R. Martin, K. P. Dorhout, *Inorg. Chem.* **2004**, *43*, 385.

83 R. L. Gitzendanner, F. J. Disalvo, *Inorg. Chem.* **1996**, *35*, 2623.

84 R. L. Gitzendanner, C. M. Spencer, F. J. DiSalvo, M. A. Pell, J. A. Ibers, *J. Solid State Chem.* **1996**, *131*, 399.

85 P. Wu, Y. J. Lu, J. A. Ibers, *J. Solid State Chem.* **1992**, *97*, 383.

86 A. Mar, J. A. Ibers, *J. Am. Chem. Soc.* **1993**, *115*, 3227.

87 P. Wu, J. A. Ibers, *J. Solid State Chem.* **1993**, *107*, 347.

88 C. W. Park, R. J. Salm, J. A. Ibers, *Angew. Chem. Int. Ed.* **1995**, *34*, 1879.

89 M. A. Pell, J. A. Ibers, *Inorg. Chem.* **1996**, *35*, 4559.

90 M. A. Pell, J. A. Ibers, *Chem. Ber. Recl.* **1997**, *130*, 1.

91 C. K. Bucher, S. J. Hwu, *Inorg. Chem.* **1994**, *33*, 5831.

92 J. A. Aitken, G. A. Marking, M. Evain, L. Iordanidis, M. G. Kanatzidis, *J. Solid State Chem.* **2000**, *153*, 158.

93 L. A. Polyakova, A. Choudhury, P. K. Dorhout, *Inorg. Chem.* **2005**, in preparation.

94 M. Tampier, D. Johrendt, R. Pottgen, G. Kotzyba, H. Trill, B. D. Mosel, *Z. Anorg. Allg. Chemie* **2002**, *628*, 1243.

95 M. Tampier, D. Johrendt, R. Pottgen, G. Kotzyba, C. Rosenhahn, B. D. Mosel, *Z. Naturforsch. B.* **2002**, *57*, 133.

96 J. Hansa, Dr.-Ing. thesis, Technischen Hochschule Darmstadt **1987**.

97 G. A. Marking, M. G. Kanatzidis, *J. Alloys Compds.* **1997**, *259*, 122.

98 W. S. Sheldrick, B. Schaaf, *Z. Naturforsch. B* **1994**, *49*, 655.

99 S. Barnier, M. Guittard, J. Flahaut, *Mater. Res. Bull.* **1984**, *19*, 837.

100 S. Barnier, M. Guittard, J. Flahaut, *Mater. Res. Bull.* **1980**, *15*, 689.

101 S. Barnier, M. Guittard, *C. R. Hebd. Seances Acad. Sci. Ser. C* **1978**, *286*, 205.

102 J. Flahaut, P. Laruelle, G. Collin, M. Guittard, J. Etienne, A. Michelet, *Proc. Rare Earth Res. Conf., 9th* **1971**, *2*, 741.

103 G. Bugli, J. Dugué, S. Barnier, *Acta Crystallogr. B* **1979**, *35*, 2690.

104 M. I. Murguzov, R. F. Mamedova, I. O. Nasibov, A. R. Muradov, T. I. Sultanov, *Poluchenie I Issled. Svoistv Soedin. RZM* **1975**, 125.

105 M. I. Murguzov, *Fiz. Tekh. Poluprovodn.* **1978**, *12*, 1825.

106 S. K. Deb, A. Zunger, in "Ternary and Multinary Compounds" *Materials Research Society*, Pittsburgh, **1987**, 39.

107 B. G. Hyde, S. Andersson, *Inorganic Crystal Structures*, Wiley-Interscience, New York, **1989**.

108 M. Guittard, D. Carre, T. S. Kabre, *Mater. Res. Bull.* **1978**, *13*, 279.

15
Cation-deficient Quaternary Thiospinels

Ashok K. Ganguli, Shalabh Gupta and Gunjan Garg

15.1
Introduction

Spinels constitute a large family of structurally simple but robust and stable materials possessing extremely useful properties with the general formula AB_2X_4, where A and B ions may belong to the metals of main group, transition series or to the rare earths and X=O, S [1, 2]. The chalcogenide spinels (including thiospinels) along with their oxides, constitute the largest class of inorganic compounds with close-packed anionic sublattice. A and B cations occupy the tetrahedral (8 a) and octahedral (16 d) sites of a cubic structure crystallizing with Fd$\bar{3}$m space group. Anions occupy the general position (32 e) forming a cubic-close-packed array. Out of the 64 tetrahedral sites available in a typical unit cell of the spinel structure, one-eighth are occupied by the A cations and one half of the 32 octahedral sites are occupied by B cations. The unoccupied 16 c site, which is related to the 16 d site through a symmetry operation is among the vacant sites (interstitial space) available for small ion intercalation in the spinel structure making it an important battery material.

In general, spinel and other chalcogen-based thiospinels with Z=8 leads to the total unit cell content of 8 and 16 (8 a and 16 d sites) cations and 32 anions in the 32 e site corresponding to the formula $A_8B_{16}O_{32}$. In spinels, the AX_4 tetrahedra share corners with BX_6 octahedra and the octahedra are linked together by sharing edges (Fig. 15.1).

In the majority of cases the synthesis of sulphides and other chalcogen-based spinels has been achieved through high-temperature solid-state synthesis carried out under a vacuum ranging from 10^{-3} to 10^{-6} torr. The material handling and other manipulations are generally performed under an inert atmosphere. In a typical reaction, stoichiometric amounts of the required elements are weighed in a nitrogen-filled glove box. The precursors for the reaction in most of the cases are metals with high purity. However, syntheses of these phases have also been reported with binary precursors. The stoichiometric mixture is taken in quartz tubes sealed at one end. The tube is then made air-tight with a stop-cock

Inorganic Chemistry in Focus III.
Edited by G. Meyer, D. Naumann, L. Wesemann
Copyright © 2006 WILEY-VCH Verlag GmbH & Co. KGaA, Weinheim
ISBN: 3-527-31510-1

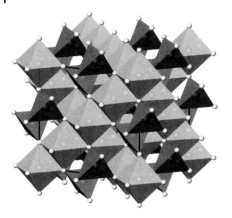

Fig. 15.1 Typical arrangement of octahedra (16d, dark) and tetrahedra (8a, light) in thiospinels structure. Balls represent Sulfur atoms in the 32e site.

and teflon tape, brought out of the glove box and attached to the vacuum line. The evacuated tubes are then sealed off at the other end. The ampoules are then placed in a tube furnace and are heated to the desired temperature. After the reaction is over, the samples are cooled in the furnace. Another method that has proved successful in the growth of chalcogenide spinels is by the chemical vapor transport method. In this process a transport agent such as $AlCl_3$ or $CrCl_3$, which can dissociate at high temperatures to liberate chlorine, is generally employed. For example, single crystals of the Cr thiospinels, MCr_2S_4 (M = Mn, Zn, and Cd) were prepared by using anhydrous $AlCl_3$ as the transport agent for Cr sulfides and thiospinels in transport reactions similar to those reported by Nitsche et al. [3]. Recently some Ni-based thiospinels of the type $M_xM'_{3-x}S_4$, where M = Ni, Co, Fe, Al and M' = Ni, were synthesized by the sol-gel method. Most of the thiospinels were obtained by a pH-controlled homogeneous precipitation method at 70 °C with metal chlorides and thioacetamide in ammonium chloride–ammonia buffer solutions adjusted to pH 10.3 [4].

The synthesis of quaternary thiospinels, like $Ag_2FeSn_3S_8$ [5], $Ag_2MnSn_3S_8$ [6], and $Cu_2NiSn_3S_8$ [7], has been achieved by reacting metals along with sulfur powder in an evacuated silica tube. Pure phases of these compounds were obtained at 750 °C, 670 °C and 750 °C respectively.

Lithium intercalated spinel (AB_2X_4) related compounds are considered as useful reversible cathode materials in an aprotic electrolyte battery system. Due to the presence of a large number of vacant sites, the spinel structure is ideal for ion exchange/intercalation of the monovalent ions suitable for the use as cathode materials. Thiospinels have been studied earlier in an attempt to find suitable electrode materials as Li^+ ions are expected to be more mobile in close-packed arrays of more polarizable anions like S^{2-}. Cation deficiency in the spinel structure may lead to a better diffusion of guest ions through the anionic subarray. Chemical methods have been used earlier to obtain such cation-deficient chalcogenides. For example, copper has been extracted by treatment with mild oxidizing agents from Chevrel phases such as $Cu_xMo_6S_8$ [8–10]. Oxidative extraction of copper has been carried out in $CuZr_2S_4$ by using a large excess of concentrated n-butyllithium. This resulted in the expulsion of almost all the copper ions as metallic copper and subsequent treatment with iodine in acetonitrile allowed the preparation of the defect thiospinel $Cu_{0.05}Zr_2S_4$ [11]. Cation-deficient thiospinels have also been reported for $Cu_xTi_2S_4$ [12, 13].

Apart from the above-mentioned ternary thiospinels, more complex spinels (quaternary thiospinels) belonging to the formula $Cu_8M_4Sn_{12}S_{32}$ or $Cu_2MSn_3S_8$ (M = Ni, Fe, Co, Mn) are also known in the literature for Li-ion intercalation [14].

Recently lithium insertion in quaternary thiospinels like $(Cu_{3.31}Ge\square_{3.69})_{8a}$ $[Fe_4Sn_{12}]_{16d}S_{32}$ and $(Cu_{4-x}Ge\square_{3.0+x})_{8a}[Co_4Sn_{12}]_{16d}S_{32}$ have been reported [15, 16]. It is believed that the vacancies in the tetrahedral 8a sites result from a topotactic substitution of four Cu^I ions by one Ge^{IV} ion.

We have carried out the synthesis and structural characterization of some cation-deficient quaternary thiospinels. No detailed structural studies for cation-deficient quaternary thiospinels have been reported so far. Until recently, only one single-crystal study was reported for quaternary thiospinels [17] which seems to be in error as shown by our recent studies [18]. As part of our ongoing studies on quaternary thiospinels we have investigated the possibility of generating additional vacancies at the 8a site by doping higher-valent ions. We have carried out a single crystal study of a Si-doped quaternary thiospinel, $Cu_{5.5}Si\square_{1.5}Fe_4Sn_{12}S_{32}$ [19]. ^{119}Sn Mössbauer and ^{57}Fe Mössbauer studies of the above compound were carried out. These studies were intended to investigate the presence of Cu-site vacancies in higher-valent substituted thiospinels. In addition, resistivity, magnetic and electrochemical studies of the Si-doped thiospinels were carried out.

Apart from "Si doped" thiospinels, three other cation-deficient thiospinels, $Cu_{5.47}Fe_{2.9}Sn_{13.1}S_{32}$ [20], $Cu_{7.07}Ni_4Sn_{12}S_{32}$ [21] and $Cu_{7.38}Mn_4Sn_{12}S_{32}$ [21] have been reported by us in the recent past. Mössbauer studies (^{119}Sn and ^{57}Fe) of the Fe-containing thiospinels ($Cu_{5.47}Fe_{2.9}Sn_{13.1}S_{32}$ and $Cu_{5.5}SiFe_4Sn_{12}S_{32}$) have been carried out in order to understand the local environment around the M/Sn cations.

15.2
$Cu_{5.5}Si\square_{1.5}Fe_4Sn_{12}S_{32}$

$Cu_{5.5}Si\square_{1.5}Fe_4Sn_{12}S_{32}$ crystallizes in the cubic $Fd\bar{3}m$ space group (Z = 1) with $a = 10.3322(6)$ Å. It may be noted that the pure Fe compound, $Cu_2FeSn_3S_8$, has been reported to crystallize in the $I4_1/a$ tetragonal space group with $a = 7.29$, $c = 10.31$ Å and Z = 2 [17], though our recent studies find that the correct space group is $Fd\bar{3}m$ [18]. The Si-doped compound (present study) shows vacancies in the 8a site (Table 15.1). The final refinement led to the formula $Cu_{5.52(8)}$ $Si_{1.04(8)}\square_{1.44}Fe_4Sn_{12}S_{32}$. All other occupancies were also refined. However, these sites led to full occupancies.

The Si-doped Fe compound, $Cu_{5.5}Si\square_{1.5}Fe_4Sn_{12}S_{32}$, shows a very small decrease in the Cu–S bond length from 2.320 Å in the parent compound (Cu_8

Table 15.1 Atomic positions, occupancies and thermal parameters of $Cu_{5.52(8)}Si_{1.04(8)}\square_{1.44}Fe_4Sn_{12}S_{32}$

Atom	Wyckoff symbol	x	y	z	Occupancy	U (eq.)
Cu	8 a	0.1250	0.1250	0.1250	0.69 (1)	0.0177 (4)
Si	8 a	0.1250	0.1250	0.1250	0.13 (1)	0.0177 (4)
Fe	16 d	0.5000	0.5000	0.5000	0.25	0.0118 (3)
Sn	16 d	0.5000	0.5000	0.5000	0.75	0.0118 (3)
S	32 e	0.74540 (7)	0.74540 (7)	0.74540 (7)	1.00	0.0131 (3)

Table 15.2 Bond distances in $Cu_2FeSn_3S_8$ (Ref. [14]) and $Cu_{5.52(8)}Si_{1.04(8)}\square_{1.44}Fe_4Sn_{12}S_{32}$.

Compound	Cu-4S (Å)	Si-4S (Å)	Fe/Sn-6S (Å)
$Cu_2FeSn_3S_8$	2.320 (1)	–	2.536 (1)
$Cu_{5.5}Si\square_{1.5}Fe_4Sn_{12}S_{32}$	2.3192 (12)	2.3192 (12)	2.5365 (7)

$Fe_4Sn_{12}S_{32}$) to 2.319 Å in the Si-doped compound studied here. This may be due to the doping of Si atoms. The size of Si^{4+} is smaller in comparison to the size of Cu^+ and this would lead to the smaller Cu–S bond length in the Si-doped compound [22]. The Fe/Sn–S distances do not show much change, being 2.536 Å in the pure compound and 2.5365 Å in the Si-doped sample. Selected bond distances are given in Table 15.2.

In general, the substitution of four Cu atoms in $Cu_8Fe_4Sn_{12}S_{32}$ by one Si atom would lead to a formula $Cu_4SiFe_4Sn_{12}S_{32}$. To maintain the electroneutrality this would require that the oxidation state of Fe and Sn be II and IV respectively. The results of our single-crystal X-ray diffraction studies, however, show an excess of Cu atoms in the 8a site resulting in the formula $Cu_{5.5}Si\square_{1.5}Fe_4Sn_{12}S_{32}$. This in turn can take place only if the formal oxidation states of some cations decrease. ^{119}Sn Mössbauer studies show the presence of both Sn^{II} and Sn^{IV}. Approximately 86.4% of Sn is present in the IV oxidation state and 13.6% Sn is present in the II oxidation state. The isomer shift values and quadrupole splitting values are in accordance with those given in the literature [14]. We find a decrease in the isomer shift compared with that of the pure $Cu_8Fe_4Sn_{12}S_{32}$. The increase of the linewidth in the ^{119}Sn Mössbauer spectra of $Cu_{5.5}Si\square_{1.5}Fe_4Sn_{12}S_{32}$ as compared with $Cu_8Fe_4Sn_{12}S_{32}$ is similar in sign and slightly larger than that found in $Cu_8Fe_4Sn_{12}S_{32}$ (Fig. 15.2).

Similar behavior has also been observed earlier for the Ge-doped thiospinel, $Cu_{3.31}Ge\,Fe_4Sn_{12}S_{32}$ [15], where vacancies in the Cu site due to Ge substitution have been observed. In both the above cation-deficient solids, the enhanced broadening is in good agreement with the presence of slightly different Sn^{IV} environments. In $Cu_8Fe_4Sn_{12}S_{32}$, all the tetrahedral neighbors are Cu atoms, while three different neighbors Cu, Si and vacancies are present in our compound $Cu_{5.5}$-

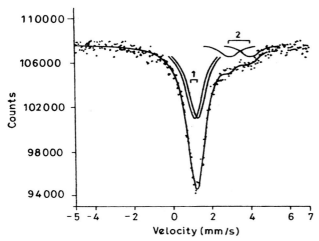

Fig. 15.2 ^{119}Sn Mössbauer spectrum of $Cu_{5.5}Si\square_{1.5}Fe_4Sn_{12}S_{32}$ at 298 K. 1) Sn^{4+}, 2) Sn^{2+}.

$Si\square_{1.5}Fe_4Sn_{12}S_{32}$. The presence of these three neighbors decreases the symmetry around the Sn^{IV} cations leading to an increase in the linewidth of the peak.

We have also studied the ^{57}Fe Mössbauer spectrum of the above compound. A decrease in the isomer shift and quadrupole splitting has been observed in $Cu_{5.5}Si\square_{1.5}Fe_4Sn_{12}S_{32}$ in the same way as in the copper-deficient compound $Cu_{8-x}Fe_4Sn_{12}S_{32}$ [14] compared to that of $Cu_8Fe_4Sn_{12}S_{32}$ (composed of two peaks centered at 0.2 and 1.3 mm s^{-1}). The decrease of the isomer shift and quadrupole splitting after Cu extraction in $Cu_{8-x}Fe_4Sn_{12}S_{32}$ is normally attributed to a slight reduction of the screening effect (less electrons) caused by the partial oxidation of Fe^{II} to Fe^{III}. This correlates well with the stoichiometry and the presence of Sn(II) and Sn(IV) in the compound. Similar behavior has been observed in the Ge-doped compound $Cu_{3.31}Ge\square_{3.69}Fe_4Sn_{12}S_{32}$ [15]. Thus, by comparison of the two cation-deficient compounds, we can conclude that the copper defect in $Cu_{5.5}Si\square_{1.5}Fe_4Sn_{12}S_{32}$ induces part of Fe^{II} to be oxidized to Fe^{III} (Fig. 15.3). However, we could not resolve the separate contributions of Fe^{II} and Fe^{III} to the profile and is probably due to the electron transfer between Fe^{II} and Fe^{III}, which does not allow the two species to be distinguished by Mössbauer spectroscopy.

The electrical resistivity of the Si-doped quaternary thiospinel, $Cu_{5.5}Si\square_{1.5}$ $Fe_4Sn_{12}S_{32}$ has been measured, in the temperature range 100 K to 300 K. It was found that it behaves like a semiconductor from room temperature down to 100 K. From the log ρ vs $1/T$ plot (see inset of Fig. 15.4) the band gap is found to be 0.107 eV in the temperature range (170 –300 K). The room-temperature resistivity is around 3.1×10^2 Ω-cm (Fig. 15.4).

Magnetic studies show that the above compound is paramagnetic in nature in the temperature range 5–300 K. The magnetic susceptibility (χ_M) along with (χ_M^{-1}) vs T plots (1 Tesla) obtained from a DC-magnetization study of the powdered

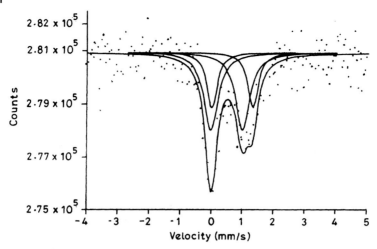

Fig. 15.3 ^{57}Fe Mössbauer spectrum of $Cu_{5.5}Si\square_{1.5}Fe_4Sn_{12}S_{32}$ at 298 K.

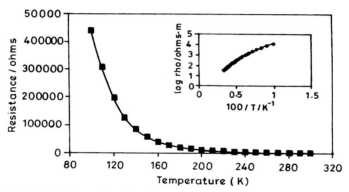

Fig. 15.4 Variation of the electrical resistance of $Cu_{5.5}Si\square_{1.5}Fe_4Sn_{12}S_{32}$ with temperature. Inset shows the log rho vs 100/T plot.

sample of $Cu_{5.5}Si\square_{1.5}Fe_4Sn_{12}S_{32}$ is shown in Fig 15.5. The low-temperature (below 200 K) magnetic susceptibility data behaved according to the Curie-Weiss equation, $1/\chi_M = C/(T-\theta)$. χ_M is the molar magnetic susceptibility, $C =$ the molar Curie constant, θ is the Weiss constant and T is the absolute temperature. However, from the high-temperature (200–300 K) data of the χ_M^{-1} vs T plot, one can see a deviation from the Curie-Weiss behavior. It is possible that there is a magnetic transition at higher temperature ($\sim$400–500 K).

Electrochemical studies were carried out in the above cation-deficient thiospinel. The cyclic voltammogram of $Cu_{5.5}SiFe_4Sn_{12}S_{32}$ in 1 M $LiBF_4$ electrolyte between 1 and 4.6 V versus the Li/Li^+ electrode is given in Fig 15.6. However, no clear peaks could be observed in the CV plot.

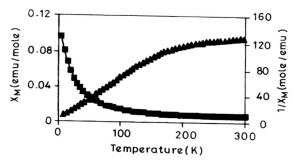

Fig. 15.5 Plot of χ_M and χ_M^{-1} vs. T for $Cu_{5.5}Si\square_{1.5}Fe_4Sn_{12}S_{32}$.

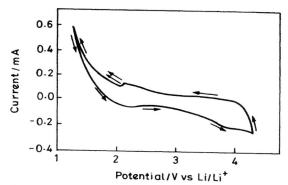

Fig. 15.6 Cyclic voltammogram of $Cu_{5.5}SiFe_4Sn_{12}S_{32}$ in 1 M $LiBF_4$ electrolyte of ethylene carbonate and dimethyl carbonate.

Table 15.3 Charging and discharging capacity of $Cu_{5.5}SiFe_4Sn_{12}S_{32}$ for various cycles.

Cycle No.	Current	Discharging capacity (mAh/g)	Charging capacity (mA h/g)	Cycle efficiency (%)
1.		90.22	102.99	87.6
2.		25.56	56.40	45.2
3.		40.40	90.53	44.6
4.	100 μA	34.76	59.19	58.7
5.		28.18	99.59	28.3
6.		10.04	25.37	39.6
7.		30.06	36.64	82.0
8.		25.36	34.64	69.2

Weight of the electrode: 8.87 mg
Area of the electrode: 1 cm^2

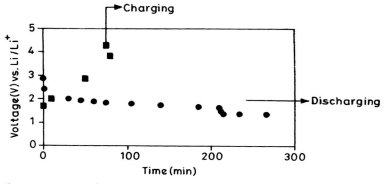

Fig. 15.7 Variation of $Cu_{5.5}SiFe_4Sn_{12}S_{32}$ electrode potential during charge-discharge.

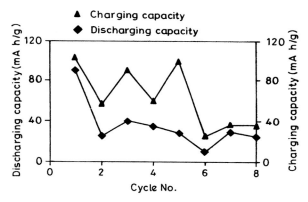

Fig. 15.8 Cycle-life data of $Cu_{5.5}SiFe_4Sn_{12}S_{32}$ electrode

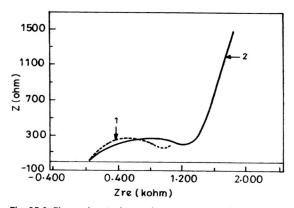

Fig. 15.9 Electrochemical impedance spectrum of $Li/Cu_{5.5}SiFe_4Sn_{12}S_{32}$ cell.

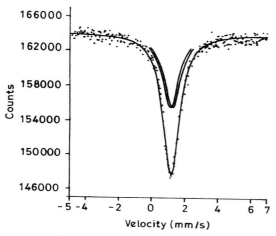

Fig. 15.10 ^{119}Sn Mössbauer spectra of $Cu_{5.47}Fe_{2.9}Sn_{13.1}S_{32}$ at 298 K.

The $Li/Cu_{5.5}SiFe_4Sn_{12}S_{32}$ cells were subjected to charge–discharge cycles. Typical charge–discharge curves showing the variation of $Cu_{5.5}SiFe_4Sn_{12}S_{32}$ electrode potential are presented in Fig. 15.7. The discharging and charging capacity for different cycles are tabulated in Table 15.3. The current density was 0.1 mA cm^{-2} and the weight of the electrode was 8.87 mg. The cycle-life data obtained during repeated cycling of the electrodes are shown in Fig. 15.8.

In Fig. 15.9, the impedance spectra of a freshly assembled $Li/Cu_{5.5}SiFe_4Sn_{12}S_{32}$ cell in discharge state and subsequent to it charging are shown. The spectrum is characterized by a high-frequency small semicircle followed by the low-frequency data in the form of an incomplete large semicircle.

15.3
$Cu_{5.47}Fe_{2.9}Sn_{13.1}S_{32}$

The $Cu_{5.47}Fe_{2.9}Sn_{13.1}S_{32}$ (Z=1) phase crystallizes in the cubic $Fd\bar{3}m$ space group with $a = 10.3362(2)$ Å. Our studies show additional vacancies in the 8a site which is partially occupied by copper. In addition we also find that a significant amount of non-stoichiometry exists between the Fe/Sn sites. This is the first study to report such a behavior. All the earlier studies [23] on quaternary spinels have M^{+2}/Sn^{+4} in the ratio 1:3 and occupy the 16d sites in the $Fd\bar{3}m$ space group. However, most of the earlier studies have been investigated by powder X-ray diffraction.

There is a decrease in (Cu–S) bond length from 2.320 Å in the parent compound, $Cu_8Fe_4Sn_{12}S_{32}$, to 2.312 Å in the cation-deficient compound studied here. On the other hand the Fe/Sn–S distances increase slightly from 2.536 Å in the former to 2.5419 Å in the cation-deficient sample. This effect may be due to the presence of vacancies in 8a sites which increases the covalent character of the Fe/Sn–S bond.

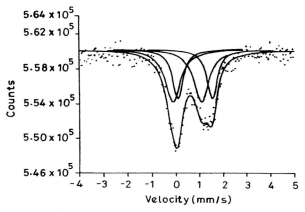

Fig. 15.11 ^{57}Fe Mössbauer spectra of $Cu_{5.47}Fe_{2.9}Sn_{13.1}S_{32}$ at 298 K.

Table 15.4 ^{119}Sn and ^{57}Fe Mössbauer parameters of $Cu_{5.47}Fe_{2.9}Sn_{13.1}S_{32}$.

	δ (mm s^{-1}) ±0.03	Λ (mm s^{-1}) ±0.03	T (mm s^{-1}) ±0.03	Attribution
^{119}Sn	1.173	0.12	1.169	Sn(IV)
^{57}Fe	0.49	1.21	0.77	60% Fe(II) LS
	1.45	1.45	0.53	40% Fe(II) HS

In general, due to the deficiency of cations in the tetrahedral site, the oxidation state of some atoms should increase to maintain the electroneutrality of the solid. However, the results of the single-crystal X-ray diffraction studies show an excess of Sn atoms in the 16d site in $Cu_{5.47}Fe_{2.9}Sn_{13.1}S_{32}$ in comparison with the pure compound $(Cu_8Fe_4Sn_{12}S_{32})$. The refined composition exactly fulfills the conditions of electroneutrality of the compound. Mössbauer data is also in accordance with the single-crystal X-ray studies as it shows the presence of Sn in IV and Fe in II oxidation states in the octahedral environment. ^{119}Sn Mössbauer spectrum of the compound (Fig. 15.10) shows only a single quadrupole-split signal with an isomer shift of 1.173 mm/s (Table 15.4), which may be ascribed to Sn(IV). The linewidth in the ^{119}Sn spectra of $Cu_{5.47}Fe_{2.9}Sn_{13.1}S_{32}$ was found to be 1.169 mm s^{-1}. The statistical distribution of two different neighbors (Cu and vacancies) gives a more asymmetric environment for the Sn(IV) cations. The ^{57}Fe Mössbauer spectra of $Cu_{5.47}Fe_{2.9}Sn_{13.1}S_{32}$ shows the presence of Fe(II) in an octahedral environment. From the analysis of the Mössbauer data it appears that both low-spin and high-spin configurations are present (Fig. 15.11). Details of ^{119}Sn and ^{57}Fe Mössbauer spectra are given in Table 15.4.

15.4
$Cu_{7.38}Mn_4Sn_{12}S_{32}$ (1) and $Cu_{7.07}Ni_4Sn_{12}S_{32}$ (2)

Both $Cu_{7.38}Mn_4Sn_{12}S_{32}$ (Z = 1) and $Cu_{7.07}Ni_4Sn_{12}S_{32}$ (Z = 1) crystallize in the cubic $Fd\bar{3}m$ space group with $a = 10.4145$ and 10.305(1) Å respectively. The occupancies of M and Sn (which occupy the 16d sites) refine to 0.25 and 0.75 and hence were fixed in the final refinement. However in both the structures a small amount of Cu deficiency was observed in the 8a sites. The interatomic distances in both the compounds are in accordance with the earlier known compounds [17]. The ionic radii of Mn^{2+} (VI, HS) is 0.83 and is larger than the size of the Ni^{2+}(VI) which is 0.69 [22]. Accordingly the Mn/Sn–S bond distance is greater than the Ni/Sn–S bond distance.

There is a decrease in the (Cu–S) bond length from 2.3241 Å in the parent compound, $Cu_8Ni_4Sn_{12}S_{32}$, to 2.3171 Å in the cation-deficient compound, $Cu_{7.07}Ni_4Sn_{12}S_{32}$, studied here. On the other hand the Ni/Sn–S distances increase slightly from 2.5227 Å in the former to 2.5276 Å in the latter. In general, due to a deficiency of cations in the tetrahedral site, the oxidation state of some atoms should increase to maintain the electroneutrality of the solid. This can be achieved by considering the presence of II and III oxidation states of Mn and Ni in the corresponding compounds or partial change in the oxidation state of Cu from I to II. It is to be noted that the increase in the oxidation state of the transition metal ion should lead to a smaller M–S distance. However, there would be an increase in the bond distances due to the presence of vacancies in 8a sites which increases the covalent character of the Ni/Sn–S bond. This behavior has also been observed earlier in the case of $Cu_{8-x}Fe_4Sn_{12}S_{32}$ [14]. Note that this compound was obtained by copper extraction from $Cu_8Fe_4Sn_{12}S_{32}$.

We have also synthesized and characterized a silver-based cation-deficient quaternary thiospinel with the formula $Ag_{1.4}Cr_{1.47}Sn_{2.53}S_8$ [24]. The Ag deficiency has been confirmed by solving the structures of crystals prepared in different batches and was observed to vary slightly between crystals. Magnetic studies on a monophasic powder sample indicate antiferromagnetic ordering at low temperature. The high-temperature susceptibility leads to a magnetic moment of 3.45 B.M. suggesting that chromium exists predominantly in a trivalent state.

These studies show that the thiospinel structure is quite flexible with opportunity for cation vacancies at the 8a site. Our investigation on such cation-deficient thiospinels is significant in that it shows that additional vacancies are possible in the 8a site. Most of the cation-deficient compounds known earlier (predominantly copper compounds) were obtained by extraction of Cu by using various oxidizing reagents. These studies show that such cation-deficient quaternary thiospinels can also be obtained by direct solid-state reactions.

15.5
Conclusions

It is clear from the above discussion that cation-deficient quaternary thiospinels have yielded very interesting materials and there is sufficient scope for further work in the field of quaternary thiospinels. Copper and silver-based quaternary thiospinels have been investigated to some extent in the last few years. However, only a few properties have been reported. The discovery of quaternary thiospinels as an active cathode material for lithium batteries has triggered off extensive research towards synthesizing new cation-deficient thiospinels and the results are encouraging. The availability of several cation sites in the thiospinels leads to lot of flexibility as it allows one to tune the structure, the electronic and the electrochemical behavior. As the basic structure is simple and very well understood, the structure–property correlations may be easier to predict. Thus the area of thiospinels is attractive for both theoretical and experimental solid-state scientists.

References

1 F. K. Lotgering, *J. Phys. Chem. Solids*, **29**, 699 (1968).

2 C. N. R. Rao and J. Gopalakrishnan, *New Directions in Solid State Chemistry*, Cambridge University Press, p 488 (1997).

3 R. Nitsche, *Phys. Chem. Solids*, **17**, 163 (1960).

4 T. Ueda, Y. Shimizu, *ITE Letters on Batteries, New Technologies & Medicine*, **5**, 454 (2004).

5 G. Garg, S. Gupta, T. Maddanimath, F. Gascoin, A. K. Ganguli, *Solid State Ionics*, **164**, 205 (2003).

6 G. Garg, K. V. Ramanujachary, S. E. Lofland, M. V. Lobanov, M. Greenblatt, T. Maddanimath, K. Vijayamohanan and A. K. Ganguli, J. Solid State Chem., **174**, 229 (2003).

7 G. Garg, S. Bobev, A. K. Ganguli, *J. Alloys Compds.*, **327**, 113 (2001).

8 W. R. McKinnon and J. R. Dahn, *Solid State Commun.*, **52**, 254 (1984).

9 E. Gocke, R. Schöllhorn, G. Aselmann and E. Muller-Warmuth, *Inorg. Chem.*, **26**, 1805 (1987).

10 R. Kanno, Y. Takeda, Y. Oda, H. Ikeda and O. Yamamoto, *Solid State Ionics*, **18/19**, 1068 (1986).

11 A. C. W. P. James, B. Ellis and J. B. Goodenough, *Solid State Ionics*, **27**, 45 (1988); **24**, 143 (1989).

12 R. Schöllhorn and A. Payer, *Angew. Chem., Int. Ed. Engl.*, **24**, 67 (1985).

13 S. Sinha and D. W. Murphy, *Solid State Ionics*, **20**, 81 (1986).

14 P. Lavela, J. L. Tirado, J. Morales, J. Olivier-Fourcade and J. C. Jumas, *J. Mater. Chem.*, **6**, 41 (1996).

15 C. Bousquet, C. Perez Vicente, A. Kramer, J. Olivier-Fourcade and J. C. Jumas, *J. Mater. Chem.*, **6**, 1399 (1998).

16 C. P. Vicente, J. L. Tirado, C. Bousquet, J. Olivier-Fourcade and J. C. Jumas, *J. Mater. Chem.*, **9**, 2567 (1999).

17 J. C. Jumas, E. Philippot and M. Maurin, *Acta Crystallogr., Sect. B*, **35**, 2195 (1979).

18 S. Gupta and A. K. Ganguli, (unpublished).

19 G. Garg, S. Bobev, A. Roy, J. Ghose, D. Das, A. K. Ganguli, *Journal of Solid State Chemistry*, **161**, 327 (2001).

20 G. Garg, S. Bobev, A. Roy, J. Ghose, D. Das, A. K. Ganguli, *Materials Research Bulletin*, **36**, 2429 (2001).

21 G. Garg, S. Bobev, A. K. Ganguli, *Solid State Ionics*, **46**, 195 (2002).

22 R. D. Shannon, *Acta Crystallogr.*, Sect. **A 32**, 751 (1976).

23 J. Padiou, J. C. Jumas and M. Ribes, *Rev. Chim. Miner.*, **18**, 33 (1981).

24 G. Garg, S. Gupta, K. V. Ramanujachary, S. E. Lofland and A. K. Ganguli, *J. Alloys and Compounds*, **390**, 46 (2005).

16

A New Class of Hybrid Materials via Salt-inclusion Synthesis

Shiou-Jyh Hwu

16.1
Introduction

Mixed-framework compounds, despite their convoluted chemistry, become attractive in advanced materials research because of their integrated structural and physical properties. Inorganic/organic hybrid compounds, for example, are well-known for their unique structural and electronic properties that are otherwise unattainable from their individual components [1–3]. Within a large volume of organically templated micro- and macroporous inorganic solids [1], two classes of compounds are highlighted here to illustrate how the mixed-framework concept is efficiently employed in the bulk synthesis of low-dimensional solids of "controlled" size and geometry. The first example is shown by a family of layered perovskites made of organic-based metal halides [2]. These hybrid materials typify an interesting class of tunable materials in that organic layers can be used to dictate the structure of the inorganic metal halide sheets, in terms of their orientation and thickness, and hence improve the electrical conductivity of the bulk. The second example is given by the formation of crystalline solids containing a periodic array of semiconducting II–VI mixed-metal chalcogenide supertetrahedra, employing organic molecules as a filler of created voids [3]. These supertetrahedra are of broad interest for they can be viewed as semiconducting quantum dots that possess enhanced optical properties due to the uniform size distribution of the corresponding semiconducting components.

 Based on the concept of mixed-framework lattices, we have reported a novel class of hybrid solids that were discovered via salt-inclusion synthesis [4–7]. These new compounds exhibit composite frameworks of covalent and ionic lattices made of transition-metal oxides and alkali and alkaline-earth metal halides, respectively [4]. It has been demonstrated that the covalent frameworks can be tailored by changing the size and concentration of the incorporated salt. The interaction at the interface of these two chemically dissimilar lattices varies depending upon the relative strength of covalent vs. ionic interaction of the corresponding components. In some cases, the weak interaction facilitates an easy

Inorganic Chemistry in Focus III.
Edited by G. Meyer, D. Naumann, L. Wesemann
Copyright © 2006 WILEY-VCH Verlag GmbH & Co. KGaA, Weinheim
ISBN: 3-527-31510-1

removal of the salt by washing at room temperature [5 a]. Studies also show that, like the inorganic/organic hybrid materials, the inclusion of salt, due to its insulating (ionic) nature, seemingly has little effect on expected electronic properties of "host" oxide lattices, but has significant impact on the resulting framework structure. These fascinating salt-inclusion solids have demonstrated potential opportunities across several major fields of advanced materials research, including zeolite-like porous compounds [5], non-centrosymmetric solids [6], and magnetic nanostructures [7].

The chemical systems under investigation have been transition-metal silicates, phosphates and arsenates where the extended covalent frameworks consist of mixed MO_n ($n = 4$–6; M = the first-row transition-metal elements) and $M'O_4$ (M' = Si, P, As) polyhedral linkages. The polyhedra in most cases share common vertex oxygen atoms, as shown in Fig. 16.1 a–e, in order to minimize the repulsion between highly charged cation centers. Occasionally edge sharing does occur, as shown in Fig. 16.1 f, and the resulting polyhedra consequently adopt off-centered distortions due to the cation displacement, which in turn often gives rise to non-centrosymmetric structures and properties. In any event, the inclusion of salt along with the already demonstrated versatility of mixed polyhedral frameworks has enabled new discoveries of a significant number of composite solids containing novel nanostructures.

To demonstrate the utilities of salt inclusion, we review the selected zeolite-like transition-metal-containing open frameworks (TMCOFs) and then describe the structures of non-centrosymmetric solids (NCSs) and, finally, report crystalline solids containing a periodic array of transition metal nanostructures. In particular, we will address the issues concerning the role that molten salt has in

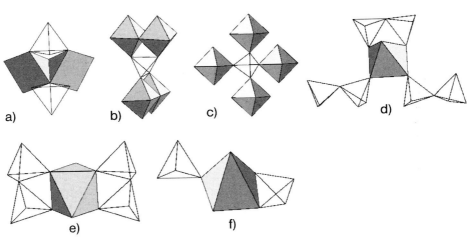

Fig. 16.1 Selected polyhedral representations showing connectivity of mixed MO_nCl_m ($n+m \leq 6$) (shaded polyhedron) and $M'O_4$ (open) units observed in the examples given in this report.

the structure formation and their prospects for continued research development in salt-inclusion synthesis.

16.2
General Approach to Salt-inclusion Synthesis

The new hybrid solids presented in this report were discovered by employing high-temperature, solid-state synthesis using molten salt as a reaction medium. A quick glance at current research shows that hydrothermal reactions ($T < 300\,°C$) have been a popular synthetic strategy for the preparation of mixed-framework solids. This is mostly because the synthesis employs organic solvent as a reaction medium that is subject to decomposition at high temperatures. Using molten salt methods, one can obtain new solids that are stable at high temperatures, as well as acquire a fundamental understanding of structure and bonding complementary to the current state of knowledge in mixed-framework materials.

Salt-inclusion solids described herein were synthesized at high temperature ($> 500\,°C$) in the presence of reactive alkali and alkaline-earth metal halide salt media. For single crystal growth, an extra amount of molten salt is used, typically $3 \sim 5$ times by weight of oxides. The reaction mixtures were placed in a carbon-coated silica ampoule, which was then sealed under vacuum. The reaction temperature was typically set at 100–150°C above the melting point of employed salt. As shown in the schematic drawing in Fig. 16.2, the corresponding metal oxides were first "dissolved" conceivably via decomposition because of cor-

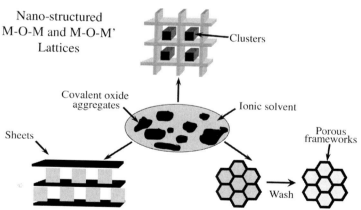

Fig. 16.2 Schematic representation showing the structure features that can be achieved via salt-inclusion synthesis. The covalent oxide is first dissolved (digested) by an ionic solvent in the presence of a reactive molten halide salt. Through slow cooling, crystals containing salt can be grown (see text).

rosive molten salt media. By slow cooling, crystals of new phases, including salt-inclusion phases featuring transition-metal clusters (nanostructures), low-dimensional metal-oxide sheets, and porous frameworks, were grown. Crystalline solids were then retrieved by washing with deionized water using suction filtration methods to remove the water-soluble salt. High-yield synthesis employing a stoichiometric amount of salt was then employed upon acquiring the chemical composition from structure determination. The detailed synthesis, as well as the structure characterization, can be found in the references cited above.

The new phases were discovered by the combination of exploratory synthesis and a phase compatibility study. As commonly practised, the new studies were initially made through the chemical modification of a known phase. Inclusion of salt in some cases is incidental, and the formation of mixed-framework structures can be considered a result of "phase segregation" (for the lack of a better term) between chemically dissimilar covalent oxide lattices and space-filling, charge-compensating salts. Limited-phase compatibility studies were performed around the region where thermodynamically stable phases were discovered. Thus far, we have enjoyed much success in isolating new salt-inclusion solids via exploratory synthesis.

16.3
Examples and Discussion

16.3.1
Zeolite-like Transition Metal Containing Porous Compounds

Open-framework solids that exhibit channel and cage structures like zeolite have drawn much interest for their potential applications in catalysis, ion-exchange, separation, sensors, and molecular recognition [8]. Transition-metal-containing open-framework (TMCOF) solids have attracted particular attention because of their unique properties such as redox catalysis [9], magnetic ordering [10], and cathodic electrolysis [11] attributed to the utilities of TM d electrons. The synthetic endeavors, thus far, have almost exclusively concentrated on the systems consisting of inorganic/organic hybrid frameworks that are largely covalent [12, 13]. We have reported the first example, as far as we know, of open-framework inorganic solids that are composed of covalent/ionic hybrid lattices [4–7]. Based upon the unique salt-inclusion chemistry as well as the importance of ion-exchange properties, we have filed a patent that describes the utilities of salt-inclusion synthesis [14].

By employing alkali metal chloride salts we have thus far isolated several interesting families of TMCOFs whose framework compositions can be written as $M_{2n-1}(M'_2O_7)_n^{2-}$ ($M^{2+} = Mn^{2+}$, Fe^{2+}, Ni^{2+}, Cu^{2+}; $M'^{5+} = P^{5+}$, As^{5+}; $n = 1 \sim 4$). We have designated this class of open-framework compounds as the CU-n series (CU: Clemson University). In CU-2, $A_2M_3(M'_2O_7)_2 \cdot$ (salt) (A = K, Rb, Cs; M = Mn, Fe, Cu; M' = P, As), the framework exhibits pores with diameters of ca. 5.3 Å and 12.7 Å [5b]. Figure 16.3 shows that the transition metals of the

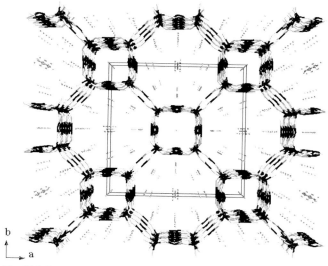

Fig. 16.3 Perspective view of CU-2CuPO showing the channel structure where the salt lattice resides. The black lines highlight the bonding with Cu^{2+} (two-way sharing) and P^{5+} (three-way sharing) cations and the grey are O^{2-}. The grey dots represent K^+, Cs^+ and Cl^-.

MOM' porous framework are distributed in a spatially uniform manner throughout the bulk. The mixed KCl/CsCl salt residing in the channel adopts a new salt lattice showing combined NaCl and CsCl structure features. Because of the weak interaction at the interface of the two chemically dissimilar lattices, the incorporated salt can be removed by washing at room temperature while the resulting covalent frameworks remain intact. This water-filled porous framework is stable up to 220 °C in air and can be re-intercalated with neutral species, such as alkali-metal nitrates. The as-prepared CU-2-CuPO solid demonstrates high-temperature ion-exchange properties in molten salt media. The resulting solid $Na_2Cs_2Cu_3(P_2O_7)_2Cl_2$ (CU-4) [5a] exhibits an elongated pore (Fig. 16.4), which contains smaller alkali metal cations and a smaller amount of corresponding salt than those in $K_2Cs_3Cu_3(P_2O_7)_2Cl_3$ (CU-2). Correlation studies of CU-2 and its structural cousin, CU-4, have revealed that the salt serves as a structure-directing agent. Alternating the identity and amount of the salt can thus vary the shape and geometry of the MOM' covalent framework.

Table 16.1 lists the anionic (host) frameworks that adopt the general formula of $Cu_{2n-1}(P_2O_7)_n^{2-}$ as Cu^{2+} has been the most studied transition-metal cation. In general these frameworks consist of corner-shared CuO_4 and P_2O_7 units, such as the ones shown in Fig. 16.1 d and e. These covalent lattices characteristically form pseudo-one-dimensional channels where the salt lattices reside. Once again, this new family of hybrid materials strongly supports the utilities of salt inclusion as a structural directing agent.

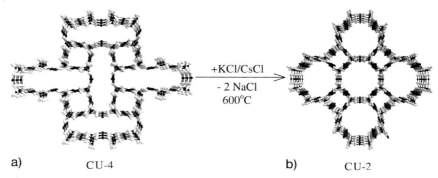

a)   CU-4   b)   CU-2

Fig. 16.4 Microporous structures of CU-4 (a) and CU-2 (b) showing the framework variation upon exchanging the content of salt at high temperature, see text. The frameworks are outlined by interconnecting copper and phosphorus metal cations (solid black circles) through oxygen anions (grey circles).

Table 16.1 The $Cu_{2n-1}(P_2O_7)_n^{2-}$ type host frameworks.

n	P/Cu	Host lattice composition	Ring size
1	2/1	$CuP_2O_7^{2-}$	isolated 8-ring
2	4/3	$Cu_3(P_2O_7)_2^{2-}$	8-, 16-, 24-ring
3	6/4	$Cu_5(P_2O_7)_3^{2-}$	16-ring
4	8/7	$Cu_7(P_2O_7)_4^{2-}$	20-ring

16.3.2
Non-centrosymmetric Solids (NCSs)

Among the dozen or so new frameworks discovered thus far, several mixed-framework structures crystallize in non-centrosymmetric (NCS) space groups. For instance [6c], $Cs_2Cu_7(P_2O_7)_4 \cdot 6\,CsCl$ (CU-9), which is chiral, crystallizes in one of the 11 NCS non-polar crystal classes, 222 (D_2), and $Cs_2Cu_5(P_2O_7)_3 \cdot 3\,CsCl$ (CU-11) in one of the 10 NCS polar crystal classes, $mm2$ (C_{2v}) [15]. As shown by the partial structure of the NCS CU-9 phase (Fig. 16.5), the Cu-O-P lattice (in shaded CuO_4 square planar and P_2O_7 tetrahedral units) wraps around the salt lattice (in fused cubical units), which adopts the NaCl-type core, along the 2_1 screw axis. The extended salt lattice of CU-11 literally adopts one-half of the NaCl-type core in Fig. 16.5. As a result, the pore size in CU-11 is approximately half of that in CU-9. The parallelogram-shaped window in CU-9 (Fig. 16.6a) can thus be described as the result of pore condensation of two inverted heart-shaped windows in CU-11 (Fig. 16.6b).

A close analysis of the chlorine-centered $ClA_{6-n}M_n$ octahedral SBU (secondary building unit) has revealed that the acentricity of the bulk lattice may be introduced via cation substitution. The di-cation substitution (n = 2) in $ClNa_6$

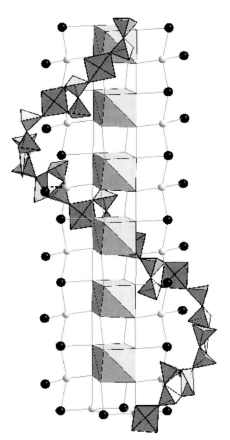

Fig. 16.5 Partial structure of CU-9 showing a polyhedral Cu-O-P chain that wraps around the salt lattice, see text. The former is constructed by sharing corner oxygen atoms of alternating square planar CuO_4 and tetrahedral P_2O_7 units. The salt lattice adopts the NaCl core in which each cubical unit is made of 1/8 of the unit cell structure of the NaCl lattice. Alternating cubical units are highlighted for clarity. The core further extends via the additional chlorine (light grey) and cesium (solid black circle) ions.

(Fig. 16.7 a) of the NaCl structure, for example, can occur either in the *cis* positions (Fig. 16.7 b), as seen in the CU-9 and CU-11 structures (A = Cs, M = Cu), or in the *trans* (Fig. 16.7 c), as shown in $Ba_2MnSi_2O_7Cl$ (CU-13; A = Ba, M = Mn). The latter forms a Fresnoite-type polar framework where the acentric $[Si_2O_7]^{6-}$ poly-anion resides in the anti-ReO$_3$ type $[(Ba_2Mn)Cl]^{6+}$ salt cage [6b]. It should be noted that the incorporation of an acentric SBU is a necessary but not sufficient condition for bulk acentricity. The material can crystallize such that the distortions occur in an antiparallel manner, thus producing bulk centricity. Nevertheless, the formation of CU-13 demonstrates the utilities of salt inclusion with the formation of NCS lattices.

A recent discovery shows that the lone pair electrons of the chlorine atom can also facilitate the formation of NCS frameworks. A novel family of salt-containing, mixed-metal silicates (CU-14), $Ba_6Mn_4Si_{12}O_{34}Cl_3$ and $Ba_6Fe_5Si_{11}O_{34}Cl_3$, was synthesized via the $BaCl_2$ salt-inclusion reaction [6a]. These compounds crystallize in the NCS space group $Pmc2_1$ (No. 26), adopting one of the 10 polar, non-

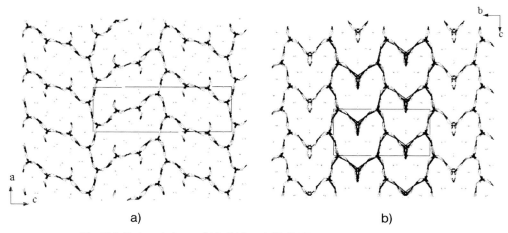

Fig. 16.6 Projected views of (a) CU-9 and (b) CU-11.

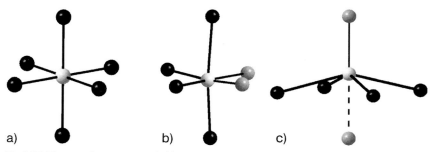

Fig. 16.7 The coordination around chlorine (drawn as light grey solid circles) in (a) NaCl, (b) CU-9 and CU-11, and (c) CU-13. The electropositive cations are shown as solid black circles and transition metal cations as dark grey. The dotted line represents a longer Mn–Cl bond, due to the Jahn–Teller distortion of the Mn^{2+} d^5 cation, see Ref. [6b].

chiral crystal classes, $mm2$ (C_{2v}). The structures form a composite framework made of the $(M_{4+x}Si_{12-x}O_{34})^{9-}$ (M = Mn, $x=0$; M = Fe, $x=1$) covalent oxide and $(Ba_6Cl_3)^{9+}$ ionic chloride sublattices. The covalent framework exhibits a pseudo-one-dimensional channel where the extended barium chloride lattice $(Ba_3Cl_{1.5})_\infty$ resides. Single-crystal structure studies reveal that the $ClBa_4$ unit adopts an interesting seesaw configuration (Fig. 16.8), in which the lone pair electrons of chlorine preferentially face the oxide anions of the transition-metal silicate channel, thus forming the observed polar frameworks. Similar to the synthesis of organic–inorganic hybrid materials, the salt-inclusion method seemingly facilitates a promising approach to introduce "chirality" for the directed synthesis of NCS compounds.

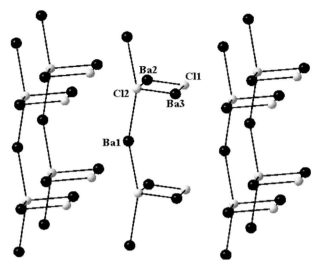

Fig. 16.8 Partial structure of the $(Ba_3Cl_{1.5})_\infty$ chains observed in the CU-14 framework showing the acentric lattice arrangement that facilitates the formation of NCS lattices, see text.

16.3.3
Solids Containing Periodic Arrays of Transition-metal Nanostructures

Recently, we have demonstrated the occurrence of quantum confinement effects in mixed-framework compounds that contain periodic arrays of electronically-confined magnetic nanostructures of transition-metal oxides [16]. The structure of these new phases can be considered as a composite framework where the molecule-like magnetic nanostructures are embedded in the extended salt matrix. As shown in the structure of $Na_5ACu_4(AsO_4)_4Cl_2$ (A=Rb, Cs) (Fig. 16.9a) [7], the parallel slabs of salt lattices made of interconnected Na_6O_8 units (see details below) sandwich a magnetic slab consisting of highly oriented oligomeric μ-oxo $[Cu_4O_{12}]^{16-}$ tetrameric units. The latter contains a *cyclo*-S_8-like Cu_4O_4 magnetic core capping the Cl^- anion (Fig. 16.9b). The tetramers share terminal and bridging oxygen atoms with the AsO_4 units (see Fig. 16.1a) to form a checkerboard pattern (Fig. 16.9c). The ionic slab consists of fascinating mixed alkali metal chloride lattices and rarely seen Na_6O_8 clusters. These compounds are magnetic insulators since the tetrameric units are isolated by closed-shell, non-magnetic ions, and the spin exchange appears to be quasi-two-dimensional due to non-frustrated intra- and inter-cluster spin coupling [7b]. These compounds are subject to thermal decomposition leading to the loss of two moles of chloride salt (NaCl and ACl) per formula unit at 900 °C and beyond, and, in turn, form $NaCuAsO_4$, a new compound containing the $[Cu_4O_{16}]^{24-}$ cluster [17]. The latter exhibits an interesting spin-gap formation as the magnetic clusters settle

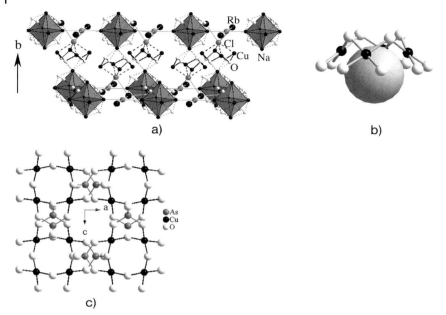

a)

b)

c)

Fig. 16.9 (a) Partial structure of the extended $Na_5RbCu_4(AsO_4)_4Cl_2$ lattice showing the $[Cu_4O_{12}]^{16-}$ clusters embedded in the salt lattice containing, rarely seen, Na_6O_8 (6–8 type) clusters. (b) The crown-like μ-oxo $[Cu_4O_{12}]^{16-}$ tetrameric unit is shown capping the chloride anion. (c) Checkerboard pattern of the $[Cu_4O_{12}]^{16-}$ clusters interlinked by As^{5+} cations.

into the lowest energy magnetic singlet state. In any event, the Na_5ACu_4 $(AsO_4)_4Cl_2$ structure has two crystallographically independent chloride anions, each of which adopts six-coordination (with $2A+4Cu$ or $4Na$) like that in the bulk NaCl structure. It would be interesting to see if the aliovalent salt-exchange reaction with divalent metal chloride (e.g., $BaCl_2$) would alter the structure and charge distribution and, in turn, the geometry and size of the cuprate cluster. Of particular interest, of course, is that these reaction products might give us new insights into the role of salt as a structural directing agent.

16.4
Final Remarks

Over the last few years, we have made a number of novel discoveries using reactive salt fluxes in the crystal growth experiment of mixed-metal oxides. The most important outcome that these salt-inclusion solids have demonstrated is the propensity for structure- directing effects of the employed salt. These hybrid solids have revealed fascinating solid-state structures ranging from nanoclusters to three-dimensional open frameworks of current interest. Solids featuring magnetic nanostructures and large pores can be synthesized through concerted salt inclusion that facilitates the segregation of chemically dissimilar species, i.e., covalent vs. ionic. This synthetic strategy creates a novel approach towards functional materials, and is believed to be complementary to the hydrothermal method commonly employed in exploratory synthesis.

Furthermore, the discovery of new salt-inclusion solids has expanded the utilities of salt-inclusion chemistry beyond zeolite synthesis. While an exhaustive treatment is not the scope of this report, a comprehensive list of the naturally occurring salt-inclusion solids does exist showing feasibility of the salt-inclusion process that originally took place during the cooling event of volcano activities. It should be noted that laboratory synthesis of non-transition-metal-based zeolites, sodalites, for instance, has been proven attainable [18]. In addition, research results concerning salt-inclusion in metal carboxylate compounds, for example, have been recognized following the discovery of the CU series [19]. While a host of such structures has been characterized, the "mechanism" upon which the new salt-inclusion solids are formed needs to be further addressed. Through systematic study, we anticipate that novel hybrid solids with targeted structures and properties can be isolated via salt-inclusion synthesis.

Acknowledgments

First and foremost, the author would like to express his special gratitude to Prof. John D. Corbett for his inspiration in solid state chemistry research. He would also like to thank his former and present graduate students, postdocs and research associates, whose names are listed in the cited references, for their significant research accomplishments. He is equally grateful to his colleagues for their insightful contribution throughout years of collaboration. Last, but not least, he is grateful for the financial support of the National Science Foundation for this research (DMR-0077321, 0322905) and for the purchase of a single-crystal X-ray diffractometer (CHE-9808165).

References

1 (a) J. L. C. Rowsell, O. M. Yaghi, *Microporous and Mesoporous Materials* **2004**, *73*, 3–14, and references cited therein. (b) G. Férey, *Chem. Mater.* **2001**, *13*, 3084–3098.

2 Some earlier examples: (a) D. B. Mitzi, S. Wang, C. A. Feild, C. A. Chess, A. M. Guloy, *Science* **1995**, *267*, 1473–1476. (b) D. B. Mitzi, C. A. Field, W. T. A. Harrison, A. M. Guloy, *Nature* **1994**, *369*, 467–469 and references cited therein.

3 For example: (a) X. Huang, J. Li, Y. Zhang, A. Mascarenhas, *J. Am. Chem. Soc.* **2003**, *125*, 7049–7055. (b) N. Zheng, X. Bu, B. Wang, P. Feng, *Science* **2002**, *198*, 2366–2369. (c) H. Li, J. Kim, T. L. Groy, M. O. O'Keeffe, O. M. Yaghi, *J. Am. Chem. Soc.* **2001**, *123*, 4867–4868. (d) V. I. Klimov, A. A. Mikhailovsky, S. Xu, A. Malko, J. A. Hollingsworth, C. A. Leatherdale, H. J. Eisler, M. G. Bawendi, *Science* **2000**, *290*, 314–317.

4 Q. Huang, M. Ulutagay-Kartin, X. Mo, S.-J. Hwu, *Mat. Res. Soc. Symp. Proc.* **2003**, *755*, DD12.4.

5 (a) CU-4: Q. Huang, S.-J. Hwu, X. Mo, *Angew. Chem. Int. Ed. Engl.* **2001**, *40*, 1690–1693. (b) CU-2: Q. Huang, M.P.A. Ulutagay, S.-J. Hwu, *J. Am. Chem. Soc.* **1999**, *121*, 10323–10326.

6 (a) CU-14: X. Mo, E. Ferguson, S.-J. Hwu, *Inorg. Chem.* **2005**, *44*, 3121–3126. (b) CU-13: X. Mo, S.-J. Hwu, *Inorg. Chem.* **2003**, *42*, 3978–3980. (c) CU-9 and CU-11: Q. Huang, S.-J. Hwu, *Inorg. Chem.* **2003**, *42*, 655–657.

7 For example: (a) S.-J. Hwu, M. Ulutagay-Kartin, J.A. Clayhold, R. Mackay, T.A. Wardojo, C.J. O'Connor, M. Krawiec, *J. Am. Chem. Soc.* **2002**, *124*, 12404–12405. (b) J.A. Clayhold, M. Ulutagay-Kartin, S.-J. Hwu, H.-J. Koo, M.-H. Whangbo, A. Voigt, K. Eaiprasertsak, *Phys. Rev. B* **2002**, *66*, 052403.

8 (a) A.K. Cheetham, G. Férey, T. Loiseau, *Angew. Chem. Int. Ed. Engl.* **1999**, *38*, 3268–3292. (b) F.S. Xiao, S. Qiu, Q.W. Pang, R. Xu, *Adv. Mater.* **1999**, *11*, 1091–2000. (c) M.E. Davis, *Microporous and Mesoporous Materials* **1998**, *21*, 173–182. (d) J.M. Thomas, *Angew. Chem. Int. Ed.* **1988**, *27*, 1673–1691.

9 For example: M. Iwamoto, H. Furukawa, Y. Mine, F. Uemura, S.-I. Mikuriya, S. Kagawa, *J. Chem. Soc., Chem. Commun.* **1986**, 1272–1273.

10 For example: N. Guillou, Q. Gao, M. Nogues, R.E. Morris, M. Hervieu, G. Férey, A.K. Cheetham, *C.R. Acad. Sci. Ser. IIC* **1999**, *2*, 387–392.

11 S.L. Brock, N.G. Duan, Z.R. Tian, O. Giraldo, H. Zhou, S.L. Suib, *Chem. Mater.* **1998**, *10*, 2619–2628.

12 For example: (a) D.E. Akporiaye, *Angew. Chem., Int. Ed. Engl.* **1998**, *37*, 2456–2457. (b) A. Clearfield, *Chem. Mater.* **1998**, *10*, 2801–2810, and references therein. (c) J.Y. Ying, C.P. Mehnert, M.S. Wong, *Angew. Chem., Int. Ed. Engl.* **1999**, *38*, 56–77, and references therein. (d) T. Sun, J.Y. Ying, *Nature* **1997**, *389*, 704–706. (e) M.E. Raimondi, J.M. Seddon, *Liq. Cryst.* **1999**, *26(3)*, 305–339, and references therein. (f) H. Li, A. Laine, M. O'Keeffe, O.M. Yaghi, *Science* **1999**, *283*, 1145–1147. (g) C.C. Freyhardt, M. Tsapatsis, R.F. Lobo, K.J. Balkus Jr, M.E. Davis, *Nature* **1996**, *381*, 295–298. (h) P. Feng, X. Bu, G.D. Stucky, *Nature* **1997**, *388*, 735–741.

13 T.J. Barton, L.M. Bull, W.G. Klemperer, D.A. Loy, B. McEnaney, M. Misono, P.A. Monson, G. Pez, G.W. Scherer, J.C. Vartuli, O.M. Yaghi, *Chem. Mater.* **1999**, *11*, 2633–2656, and references therein.

14 "Salt-Templated Microporous Solids," US Patent 6719955.

15 P. Shiv Halasyamani, K.R. Poeppelmeier, *Chem. Mater.* **1998**, *10*, 2753–2769.

16 Hwu, S.-J. *Chem. Mater.* **1998**, *10*, 2846–2859.

17 M. Ulutagay-Kartin, S.-J. Hwu, J.A. Clayhold, *Inorg. Chem.* **2003**, *42*, 2405–2409.

18 (a) H. Trill, H. Eckert, V.I. Srdanov, *J. Am. Chem. Soc.* **2002**, *124*, 8361–8370. (b) P.A. Cocks, C.G. Pope, *Zeolites* **1995**, *15*, 701–707.

19 C.N.R. Rao, S. Natarajan, R. Vaidhyanathan, *Angew. Chem., Int. Ed. Engl.* **2004**, *43*, 1466–1496 and references cited therein.

17

Layered Perrhenate and Vanadate Hybrid Solids: On the Utility of Structural Relationships

Paul A. Maggard and Bangbo Yan

17.1
Introduction

The discovery of new solid-state compounds and analyses of their structure and bonding is one of the leading driving forces in research, especially as chemists increasingly explore the boundaries between metals, salts, and/or organics. Structures that fall at these boundaries can be indescribably complex and difficult to analyze in terms of counting electrons, structure–property relationships, or in the prediction of future compositions. One approach to this problem, which was shaped by early research with Professor Corbett, focuses on examining structural relationships between similar chemical species expressed in different solid-state structures, such as is found for clusters in metal halides [1], or octahedral chains and layers in metal tellurides [2]. A close analysis of structural relationships among these species has helped to identify the basic structural principles of nature, such as cooperative electronic/steric effects [3], bond-driven metal-lattice distortions [4], and competing modes of chain connectivity [5, 6] or substitutional patterns [6, 7]. Our recent work has been spurred on and guided by a continuing desire to synthesize and uncover new solid-state chemistry at the unexplored boundaries between metal-oxides and organic molecules and to formulate and test new principles of structural design. However, the challenges in the structural chemistry presented below for metal-oxides/organics are different from that in metallic solids, which are not (usually) in understanding electron counts or cluster species, but rather of how to achieve specific structural formations, such as layered or non-centrosymmetric bonding arrangements.

The synthesis of new layered structures is often desired in order to explore the ability of guest molecules to easily intercalate or de-intercalate between their interlamellar regions, such as is useful for applications in heterogeneous catalysis or in battery electrodes. In metal-oxides/organics with layered pillared structures, for example, organic ligands will bridge (or pillar) across metal sites within separated metal-oxide layers, predisposing the structure towards guest intercalation within the gallery or pillared areas. Chemical systems which feature

Inorganic Chemistry in Focus III.
Edited by G. Meyer, D. Naumann, L. Wesemann
Copyright © 2006 WILEY-VCH Verlag GmbH & Co. KGaA, Weinheim
ISBN: 3-527-31510-1

layered pillared structures include many types of phosphonates, sulfates, and molybdates as well as heterometallic solids that incorporate additional transition metals for preferentially bonding to the bridging organic ligands [8–12]. The late transition-metal cations can be of a wide variety (e.g., Cu^{2+}, Ni^{2+}, Co^{2+}, Fe^{2+}) and are incorporated into the layer via coordination to the intralayer anions (e.g., PO_4^{3-}, MoO_4^{2-}). Thus, a significant diversity of layered pillared solids is possible by using a variety of transition metals and/or pillaring ligands, with some recent examples including $[Cu(4,4'\text{-bipyridine})_{0.5}MoO_4)] \cdot 1.5\,H_2O$ [13] $[Cu(\text{tripyridyltriazine})_2Mo_4O_{13}]$ [11], $Cu(\text{pyrazine})(HPO_4)_2F_2]$ [14], and $[Cu(4,4'\text{-dipyridylamine})VO_3]$ [12]. However, investigations of the structural relationships within the basic types of layered solids have been few, and structural principles involving their formation and synthesis, as discussed here, are only slowly emerging.

This highlight examines several new hybrid layered structures in the heterometallic perrhenate and vanadate families, whereby the late transition-metals are incorporated and their roles probed in the structures of layered solids. From these two families, new structural principles have emerged that not only help us understand key structural features and correctly forecast new compositions, but equally, have yielded many surprises (chirality, reduced phases) that show some of the most exciting chemistry is still waiting to be discovered or even imagined!

17.2
Heterometallic Perrhenates

17.2.1
Background: Molecular and Condensed Metal-perrhenates

Rhenium is well known to adopt multiple coordination environments, oxidation states, and many types of chemical bonds ranging from in simple metal-oxides to metal–metal bonded clusters. However, this diverse rhenium chemistry is remarkably absent among the growing family metal-oxides/organics featuring open-frameworks or layered structures. The perrhenate anion (ReO_4^- tetrahedron), for example, is known to exhibit both terminal and bridging bonding modes and therefore can form the basis for the construction of many structures that arise from corner-sharing tetrahedra. However, the short list of known perrhenate-hybrid solids includes several simple metal salts with either urea or thiourea as the organic component [15, 16], as well as a relatively small number of molecular systems [17–20]. These known perrhenate structures do not contain bridging ligands that would be required to connect the perrhenate anion into clusters, chains, or more condensed frameworks.

The known family of heterometallic perrhenate structures provides a basic picture of the local structural units and their connectivity patterns. Shown in Fig. 17.1, these include molecular clusters such as $M(H_2O)_4(ReO_4)_2$ ($M = Fe$, Co, Cu, Zn) [21–23], and $Fe(H_2O)_3(ReO_4)_3$ [24], the chain structure of $Mn(H_2O)_2(ReO_4)_2$ or the

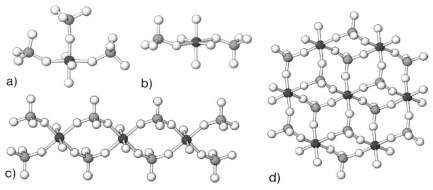

Fig. 17.1 Representative $M(H_2O)_x(ReO_4)_y$ structural units. From molecular species (A, B; $x=3$, 2; $y=3$, 4) to chains (C; $x=2$; $y=2$) and layers (D; $x=0$; $y=2$). Dark spheres = M, Light = Re, and White = O or H_2O.

layered structures of $M(ReO_4)_2$ (M = Mn, Co, Ni, Zn) [25]. As a general trend, higher-dimensional structural units are found with an increasing connectivity of the perrhenate anion (from 0 to 3; molecules to layers), and that is in parallel with a required decrease in the number of terminally-coordinated H_2O molecules to the late transition metal. Not shown are the more condensed perrhenate-containing solids known for $Cd(H_2O)_2(ReO_4)_2$ [26], which has interlayer perrhenate bridges, or $CoReO_4$ [27] and $AgReO_4$ (below) [28]. The structures and bonding features of these solids suggest a rich chemistry of possible building blocks to be found at the unexplored heterometallic–perrhenate/organic boundaries. As many of these simpler solids were synthesized from aqueous solutions, the discovery of hydrothermal synthetic conditions to crystallize perrhenates in new solid-state structures has enabled a significant expansion of their chemistry.

17.2.2
Copper- and Silver-perrhenate Hybrids

Initial research efforts to synthesize heterometallic perrhenate hybrid solids focused on using bridging organic ligands, e.g. pyrazine, to link the chemical species in Fig. 17.1 via the late transition metals. Some of the first basic principles for synthesizing rigid interconnected frameworks containing the perrhenate anion were found in an initial synthetic exploration of the ReO_4/Cu/pyrazine-carboxylate (=pzc) system. The hydrothermal reaction of Cu_2O, Re_2O_7 and pzc at 150 °C yielded both red and blue colored crystals forming together, and that featured some of the first such examples of ReO_4 within a 3D network with channels $[Cu_2(pzc)_2(H_2O)_2ReO_4]$, and also the hydrated layered structure of $[Cu(pzc)(H_2O)ReO_4] \cdot 2 H_2O$ [29]. Both of these structures exhibit two new and different types of $Cu(ReO_4)$ chains that are linked into rigid two- and three-dimensional structures via the pyrazinecarboxylate ligands, as shown in Fig. 17.2.

For the purposes of this discussion the focus here will be primarily on the pillared layered $[Cu_2(pzc)_2(H_2O)_2ReO_4]$ structure, which has served as an illustrative basis for several follow-up syntheses and new structures.

The structure of $[Cu_2(pzc)_2(H_2O)_2ReO_4]$ (Fig. 17.2), is a three-dimensional microporous framework with two types of interconnected chains: namely, infinite Cu(pzc) and Cu(ReO$_4$) chains. In the former, Cu exhibits two different coordination environments, octahedral Cu^{2+} and tetrahedral Cu^{1+}, which alternate and are bridged to each other through the pzc ligand, i.e. $[-Cu^{1+}-pzc-Cu^{2+}-pzc-]$, in Fig. 17.2 b. Each Cu^{2+} is equatorially chelated by two pzc ligands and is also bonded to two H$_2$O molecules, not shown. The Cu^{1+} ions are also bonded to two ReO$_4^-$ anions to form the Cu(ReO$_4$) chains that occur in layers, marked with parentheses in Fig. 17.2, and have a zigzag $[-Cu^{1+}-O-Re-O-]$ repeating pattern. The Cu(ReO$_4$) chain here does not have a structural analog with a simpler copper-perrhenate salt, although this may be from lack of previous synthetic attempts with copper. Thus, the overall structure contains two types of chains, $Cu^{1+}(pzc)/Cu^{2+}(pzc)$ and Cu^{1+}/ReO_4, oriented and interconnected approximately perpendicular to each other via the tetrahedral Cu^{1+}. Highly elongated channels are formed, with dimensions $\sim 3.3 \times 12.5$ Å, that are located roughly between neighboring Cu(ReO$_4$) chains. Hydrogen atoms on the pzc ligand project into and partially fill the upper and lower portions of the ellipsoidal channel wall. However, the

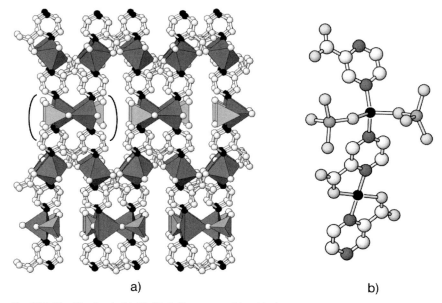

a)   b)

Fig. 17.2 The $[Cu_2(pzc)_2(H_2O)_2(ReO_4)]$ structure (a), with the infinite Cu(ReO$_4$) chain marked. Also, the mixed valence Cu^{1+}/Cu^{2+} chain bonded through the pzc ligand (b). Dark polyhedra = Cu-centered; light = ReO$_4^-$ tetrahedra.

channels remain accessible for H_2O removal, and probably also for guest absorption, as thermogravimetric analysis reveals that up to 64% of its coordinated water is lost starting at 180 °C, before decomposing to an unidentified black material at >225 °C. A constructive way of viewing this structure is as layers of $Cu(ReO_4)$ chains that are bridged by $Cu(pzc)_2(H_2O)_2$ pillars between the chains. This view assigns specific roles for the Cu^{1+} and Cu^{2+} ions, either as part of the inorganic component for the former, or as part of the metal-organic pillar for the latter. The subsequent heterometallic perrhenate structures, seen in the sections below, reveal the utility of this assignment for forming structural relationships.

As no simple copper-perrhenate salts previously existed to serve as a model for structural relationships to $[Cu_2(pzc)_2(H_2O)_2ReO_4]$, our attention turned to the syntheses of the isoelectronic silver-perrhenate systems. Shortly after an analysis of the structure above, a synthesis of the new pyrazine-pillared (=pyz) $AgReO_4(pyz)$ solid was found [30], and is shown in Fig. 17.3 alongside the re-

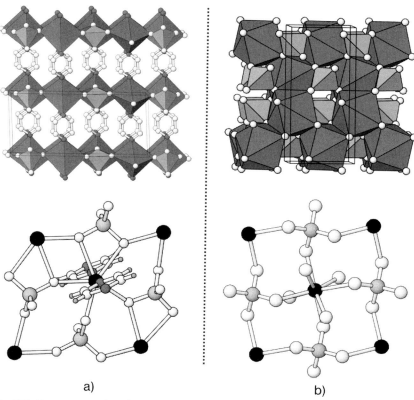

a)   b)

Fig. 17.3 Two structures based on $AgReO_4$ layers: the pillared $AgReO_4(pyz)$ (a) and condensed $AgReO_4$ (b). The bottom figures represent individual $AgReO_4$ layers for comparison. Dark polyhedra = Ag-centered; light = Re-centered tetrahedra.

lated and previously known AgReO$_4$ structure [31]. Both solids are transparent and fluoresce at ~ 550 nm owing to a metal-to-metal charge transfer between Ag and Re. However, while AgReO$_4$ is a condensed structure, AgReO$_4$(pyz) contains isolated AgReO$_4$ layers that are pillared via the bonding of pyrazine ligands to the Ag sites across two layers. The shortest interlayer distance is set by the length of the pyrazine ligand at 2.75 Å, while the ReO$_4$ tetrahedra lining the walls between the pyz pillars are separated at a slightly longer distance of 4.61 Å. The body-centered arrangement of Ag$^+$ ions, with ReO$_4^-$ ions on each edge, is roughly similar within each structure (Fig. 17.3), (bottom), except for subtle differences in the ReO$_4$ orientations. In AgReO$_4$(pyz), it is as if the [110] layers of AgReO$_4$ have been cleaved and pillared by pyrazine ligands. This new relationship provided a unique structural principle among silver-perrhenates, which suggested the subsequent replacement of pyrazine pillars for M(pzc)$_2$(H$_2$O)$_2$ pillars, analogous to that found for the Cu(pzc)$_2$(H$_2$O)-pillared CuReO$_4$ solid.

17.2.3
Metal-coordinated Pillars in Perrhenate Hybrids

One of the first general strategies for synthesizing pillared solids arose from the discovery that metal-coordinated pillars could bridge between CuReO$_4$ or AgReO$_4$ layers [32]. The resultant strategy can also be viewed as the substitution of a simple bridging ligand of a pillared solid (e.g., pyz in AgReO$_4$(pyz)), for one that exhibits two different binding preferences (e.g., pyrazinecarboxylate): one ligand site favors attachment to the metal perrhenate layer and the other site, to the secondary transition metal. In the case of [Cu$_2$(pzc)$_2$(H$_2$O)$_2$ReO$_4$], the chelating COO$^-$ (carboxylate) and N groups bond preferentially to Cu^{2+}, while the lone (*para*) N group favors attachment to Cu^{1+} in the CuReO$_4$ chains. Thus, it should also be possible to utilize the chemistries of two dissimilar transition metals, one that resides in the perrhenate layer (Ag$^+$, Cu$^+$) and the other which is coordinated as part of the organic pillars (Co^{2+}, Ni^{2+}). The strategy of immobilizing the metal sites as part of a pillar, and bridging them to metal-perrhenate layers, will help to stabilize the structure towards ligand removal and guest absorption.

The synthesis of AgReO$_4$(pyz) was modified by using pyrazinecarboxylate in place of pyrazine and adding Co^{2+} and Ni^{2+} for inclusion between the layers [32]. Orange and green colored crystals of M(pzc)$_2$(H$_2$O)$_2$AgReO$_4$ were obtained for M = Co and Ni respectively. As shown in Fig. 17.4, the structures of both solids derive from that of AgReO$_4$(pyz), and have AgReO$_4$ layers that are bridged through pzc to interlayer Co or Ni sites. The [Co(pzc)$_2$] pillars on the left are tilted with respect to the AgReO$_4$ layers, resulting in a relatively smaller interlayer spacing of ~ 13.58 Å (*b*-axis/2), whereas the [Ni(pzc)$_2$] pillars on the right buckle only slightly and have an interlayer spacing of ~ 14.0 Å. The lone (*para*) N atoms of the pzc ligands bond preferentially to the Ag sites of the AgReO$_4$ layer, while the chelating COO$^-$ and N groups of the pzc ligand preferentially

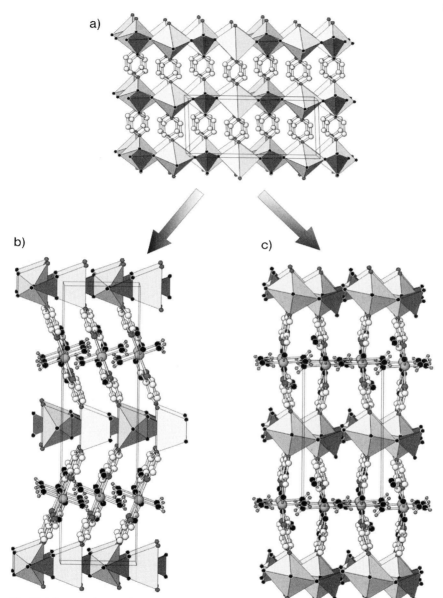

Fig. 17.4 Structural unit-cell views of (a) AgReO$_4$(pyz);
(b) Co(pzc)$_2$(H$_2$O)$_2$AgReO$_4$; and (c) Ni(pzc)$_2$(H$_2$O)$_2$AgReO$_4$.

coordinate to the equatorial positions of Co^{2+}/Ni^{2+}. Water molecules coordinate to the two remaining axial positions on both Co^{2+}/Ni^{2+}. The structures contain a roughly similar two-dimensional packing of the $AgReO_4$ layers, with ReO_4^- in a body-centered arrangement and Ag^+ cations on the edges. The metal-coordinated pillars bond to the $AgReO_4$ layers, which structurally separate the late transition-metal sites which form secondary Co/Ni-pzc layers. Multiple bonds to the carboxylate groups do not occur as in other metal-organic structures.

Another striking feature of both $Co(pzc)_2(H_2O)_2AgReO_4$ and $Ni(pzc)_2(H_2O)_2$ $AgReO_4$ is that coordinated water molecules on the Co^{2+}/Ni^{2+} sites can be easily removed by heating to 120 and 170 °C respectively. Both can be rehydrated with drops of H_2O on a glass slide, with subsequent TGA again revealing the characteristic mass loss for all coordinated H_2O, an indication of porosity. Each solid also maintains its structure and crystallinity through the dehydration and rehydration steps, although with some changes in peak position in the powder X-ray diffraction. Shifts in the diffraction peaks of the dehydrated states correspond to a decrease in the $AgReO_4$ interlayer spacing, probably caused by tilting of the pillars. Both dehydrated structures stabilize at a similar interlayer spacing of ~ 12.8–13.0 Å. The most probable coordination geometries for Co^{2+} and Ni^{2+} are square planar in the dehydrated structures (see Cu example below).

While the full structural details of the dehydrated forms of $M(pzc)_2(H_2O)_2$ $AgReO_4$ (M = Co, Ni) are not yet known, a new dehydrated version was discovered for M = Cu, $Cu(pzc)_2AgReO_4$, which is a surprising and rare example of a chiral structure that is pillared by the same-handed (4_3) helical chains [33]. The blue-colored solid crystallizes in the chiral space group $P4_32_12$, and is active for second harmonic generation (50–75% of SiO_2). A view of the pillared layered structure (Fig. 17.5), shows that the $AgReO_4$ layers are pillared by $Cu(pzc)_2$ bridges that repeat four times per c direction in the unit cell. Interlayer distances are relatively short, at ~ 12.37 Å, owing to the significantly tilted $Cu(pzc)_2$ pillars. This is much shorter than the interlayer distances for the M = Co and Ni versions which have more vertical pillars in their hydrated ~ 13.6–14.0 Å and dehydrated ~ 12.8–13.0 Å forms. Direction of the pillar tilting between oxide layers rotates 90° counterclockwise down the 4_3 screw axis to form the single-handed helical chain, shown in Fig. 17.5 b. Each Cu site is in an axially distorted octahedron, with four short equatorial bonds to two pzc ligands (one COO^- and N each) and much longer axial contacts to carboxylate groups on neighboring pillars. In the hydrated M = Co and Ni structures, the late transition-metal sites are more regularly octahedral, with two bonds to H_2O ligands and farther spaced pillars. The helicity and chirality result from significantly tilted pillars that cause short contacts between the pzc pillars and the apical oxygen atoms on the ReO_4 groups, as described recently [33]. The structure does not absorb water when immersed in solution, but will selectively absorb CO_2 at high pressures, most likely from coordinating to the square-planar copper sites and vertical realignment of the pillars.

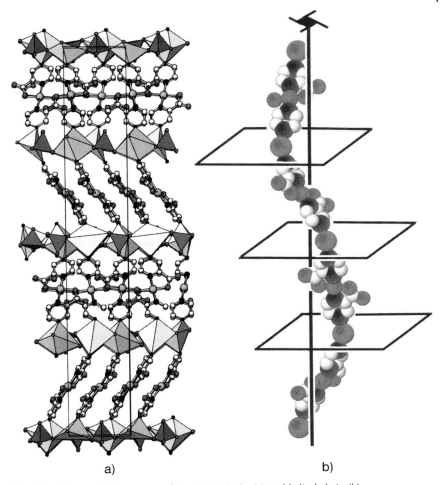

Fig. 17.5 Pillared chiral structure of Cu(pzc)$_2$AgReO$_4$ (a) and helical chain (b).

17.3
Heterometallic Vanadates

17.3.1
Background: Layered Vanadate Species

The flexibility of polyoxovanadate species $[V_xO_y]^{n-}$ to adopt a wide variety of structural configurations, such as rings, chains, and layers, is reflected in their predominance in a wide variety of hybrid solids. For example, approximately sixteen distinct hybrid vanadium oxides based solely on $V_2O_5^{n-}$ layers, with about nine different layer types, have currently been synthesized [34–47]. Layered structures of vanadium oxides which also incorporate late transition-metals are sought for their potential uses as cathodes in lithium secondary batteries, and V_2O_5 in particular has been extensively studied as a guest for the synthesis of intercalated compounds [48, 49]. Three of the approximately nine types of different $V_2O_5^{n-}$ layers are shown in Fig. 17.6. The basic building units of these layers are $\{VO_4\}$ tetrahedra, $\{VO_5\}$ square-pyramids and more rarely $\{VO_6\}$ octahedra, which are condensed via corner- or edge-sharing vertexes. The shared oxygen atoms can bridge between two or three vanadium atoms, with terminal oxygens directed perpendicular to the layers. Generally, these basic building units fuse to form single chains (secondary building units), and the chains then link into double or triple chains (tertiary building units), and then further condense into layers. Among the currently known solids with $V_2O_5^{n-}$ layers, there are five types of secondary building units: (1) straight single chains of corner-sharing square pyramids with all polyhedra in the same direction as in V_2O_5, (2) straight single chains of corner-sharing square pyramids with polyhedra oriented up and down alternately; (3) zigzag single chains of corner-sharing square pyramids with all polyhedra in the same direc-

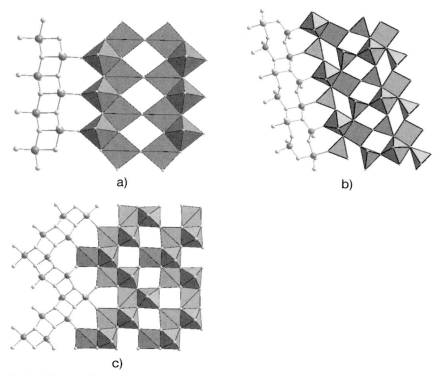

a)   b)

c)

Fig. 17.6 Three different examples of types of $V_2O_5^{n-}$ layers, including in V_2O_5 (a) [50], β-$(H_2pip)V_4O_{10}$ (b) [40], and $(VO_2(terpy))V_4O_{10}$ (c) [39].

tion as in $(VO_2(terpy))V_4O_{10}$ [39]; (4) straight single chains of corner-sharing square pyramids and tetrahedra with all polyhedra in the same direction as in β-$(H_2pip)V_4O_{10}$ and (5) straight single chains of corner-sharing square pyramids and tetrahedra with the polyhedra oriented up and down alternately. The condensation of single chains into double or quadruple chains can involve edge- or corner sharing between polyhedra, while condensation of the adjacent double chains are corner-sharing. Multiple orientations of the neighboring polyhedra between adjacent double chains can also result in many different layer types. This very brief account of the rich and flexible chemistry of polyoxovanadate layers shows a quite fertile ground for examining detailed structural relationships where new structure-guiding principles can be, and are (see below), slowly developing.

17.3.2
Layered Heterometallic Vanadates: Charge Density Matching

To target the synthesis of multilayered heterometallic vanadates, our initial research focused on introducing late transition metals, interconnected via short bridging pyrazine ligands, into the interlamellar regions. However, most of the known vanadate layer types (above) have been synthesized with molecular organic cations as the interlayer species, while two-dimensional networks of metal-organic layers are rarely found between vanadate layers. Thus, some of the first basic secrets were uncovered in our initial synthetic explorations of the copper/pyrazine/ vanadium-oxide system. The hydrothermal reaction of $Cu(NO_3)_2$, V_2O_5 and pyrazine results in dark green crystals of $Cu(pyz)_2V_6O_{16}\cdot(H_2O)_{0.22(1)}$, featuring a structure with a segregated metal distribution into layers of $Cu(pyz)_2^{2+}$ and $V_6O_{16}^{2-}$ [35]. Both layers are unique thus far in layered-vanadate/metal-ligand systems and helped to suggest a new application of the strategy known as charge density matching to synthesize multilayered structures. A discussion is given first for the structure of $Cu(pyz)_2V_6O_{16}\cdot(H_2O)_{0.22(1)}$, which served as a basis for follow-up syntheses of the new $M(pyz)V_4O_{10}$ (M = Co, Ni, and Zn) structures.

The structure of $Cu(pyz)_2V_6O_{16}\cdot(H_2O)_{0.22(1)}$ (Fig. 17.7), contains alternating layers of $Cu(pyz)_2^{2+}$ and $V_6O_{16}^{2-}$ chemical species. The buckled vanadate layer is composed of two different types of vanadium polyhedra, a "3+2" trigonal bipyramidal geometry, with two short vanadyl (V=O) bonds and three intermediate V–O bonds, and a "4+1+1" geometry, with four equatorial bonds, a short vanadyl bond, and one longer axial bond. These polyhedra condense by a complex sharing of edges and vertices to form the vanadate layer (Fig. 17.7b). The $Cu(pyz)_2^{2+}$ square nets are located both above and below each vanadate layer. The octahedral coordination around each Cu^{2+} is Jahn-Teller distorted, with four equatorial bonds to pyrazine ligands and longer axial bonds to two oxygen atoms from the vanadate layers. In addition, H_2O molecules partially occupy the center positions of the squares in the net and hydrogen bond to oxygen atoms on the vanadate layers.

While it is convenient to write the charges on the copper and vanadate layers as 2+/2− per formula unit, this conceals a more accurate and valuable way of describing the formal charge and understanding the structure. To understand why this

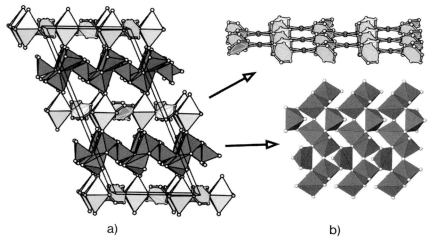

a)   b)

Fig. 17.7 Unit cell of $Cu(pyz)_2V_6O_{16} \cdot (H_2O)_{0.22(1)}$ (a) and the separate layers of $V_6O_{16}^{2-}$ and $Cu(pyz)_2^{2+}$ (b). Dark polyhedra = VO_5 or VO_6, light = $Cu(pyz)_4O_2$.

way of describing the charge is incomplete, consider the following series: $V_{12}O_{32}^{4-}$, $V_{18}O_{48}^{6-}$, and $V_{24}O_{64}^{8-}$, which is the $V_6O_{16}^{2-}$ layer extended out to 2, 3, and 4 unit cells. Thus, the charge (n^-) of a $V_xO_y^{n-}$ layer would increase limitlessly in this way for increasing areas, while the actual charge balance is a constant and for a layer is defined as the charge per area, i.e. (n^-/nm^2). In order to maintain electroneutrality in multilayered structures, the charge per area of the cationic and the anionic layered species must match. This concept, known as charge density matching, has found applications in other chemical systems, such as silica/surfactant composites and zeolitic metal-phosphate solids [51]. The charge densities of known related $V_xO_y^{n-}$ and $Cu(pyz)_z^{m+}$ layers were calculated from simpler (and separate) solids and found to have very similar values around at ~ -0.04 and $+0.04/Å^2$ [35]. Thus, charge density matching was a useful guide for understanding the $Cu(pyz)_2$-$V_6O_{16} \cdot (H_2O)_{0.22(1)}$ structure, and justified further testing as a way to determine the "allowed" combinations of layers in new solids.

17.3.3
Heterometallic Reduced Layered Vanadates

The charge densities and compositions of many known solids with $V_xO_y^{n-}$ layers were screened for possible charge density matches with $M(pyz)_x^{2+/3+}$ (M = Fe, Co, Ni, Cu, Zn) layers [35, 52], as an aid in the synthesis of new heterometallic multilayered vanadates. Close matches were found with $M(pyz)_x^{2+}$ (M = Fe, Co, Ni) coordination polymers, ranging from $\sim 0.022/Å^2 - 0.053/Å^2$, and which might suitably form in the presence of the flexible types of V_2O_5 layers. For example, $Co(pyz)^{2+}$ layers of chains in $Co(pyz)(VO_3)_2$ [53] have a charge density of

+0.046/$\mathring{A}^2$ and several types of known $V_2O_5^{n-}$ layers have charge densities ranging from $\sim$ −0.022/$\mathring{A}^2$ to −0.050/$\mathring{A}^2$. A specific example of the latter is the vanadate layers in β-(H_2pip)V_4O_{10} which have a charge density of −0.050/$\mathring{A}^2$. The $[H_2$pip]$^{2+}$ cations are structurally flexible and can adopt different orientations and geometries, resulting in both a- and β-(H_2pip)V_4O_{10} structures, unlike the more rigid M(pyz)$^{2+}$ chains. Thus, slightly different V_2O_5 layers for M(pyz)$^{2+}$ might be expected. Hydrothermal synthetic conditions were employed to react V_2O_5 with Co(pyz)$^{2+}$ and Ni(pyz)$^{2+}$ in order to test these relationships, resulting in two new reduced layered vanadates, M(pyz)V_4O_{10} (M = Co, Ni, and Zn) [52].

The reduced M(pyz)V_4O_{10} (M = Co, Ni, and Zn) can be prepared in high yield as small black crystals, and is comprised of $V_4O_{10}^{2-}$ layers that are separated and charge balanced by layers of chains of M(pyz)$^{2+}$, shown in Fig. 17.8c. This type of vanadate layer, Fig. 17.8a, was also previously known for (Me$_4$N)V_4O_{10} (Fig. 17.8b), and contains isolated (Me$_4$N)$^+$ cations separating the vanadate layers. The vanadate layer is constructed from edge-sharing ribbons of VO$_5$ tetragonal pyramids that alternate in an up-up-down-down (zigzag) pattern, and which are, in turn, condensed via vertexes to neighboring chains. The formal charge on the $V_4O_{10}^{n-}$ layer is −1 and −2 for the (Me$_4$N)$^+$ and Co(pyz)$^{2+}$ analogs, respectively, with the slightly shorter V–O distances as expected for the former. The late transition metals, either Co or Ni, bond to four apical oxygen atoms of two vertex-shared VO$_5$ units (Fig. 17.9a), two each to layers both above and below, and also bond to two bridging pyrazine ligands in a *trans* fashion. The al-

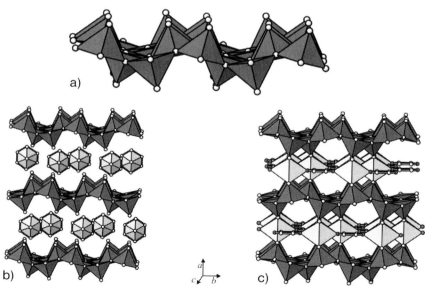

a)

b) c)

Fig. 17.8 An $\sim$[001] view of the structures of V_4O_{10} (a) within (Me$_4$N)V_4O_{10} (b) and Co(pyz)V_4O_{10} (c). Dark polyhedra = VO$_5$, light = Co(pyz)$_2$O$_4$ or (Me$_4$N)$^+$.

ternating up and down pattern of apical oxygen atoms of neighboring VO_5 units matches the repeat distance and pattern of the $Co(pyz)^{2+}$ chain, with Co bonding to two 'towards' oxygens and the pyrazine molecules residing in the empty space provided by two 'away' oxygens. By comparison, the $(Me_4N)^+$ cations in Fig. 17.9b take the same positions as the bridging pyrazine molecules, which is over the bottom faces of two VO_5 tetragonal pyramids. Thus, the replacement of $(Me_4N)^+$ cations in $(Me_4N)V_4O_{10}$ for $Co(pyz)^{2+}$ cations in $Co(pyz)V_4O_{10}$ is possible through a geometrical matching of the apical oxygen atoms in V_4O_{10} with the coordination preferences and repeat distances of the $Co(pyz)^{2+}$ chains.

More significant structural differences are found in the stacking of vanadate layers down the a-axis. A shorter interlayer distance of 6.93 Å (from mid-layer to mid-layer) is found for $Co(pyz)V_4O_{10}$ and a larger interlayer distance of 8.68 Å for $(Me_4N)V_4O_{10}$. The shorter distance is a result of the Co–O bonds which bridge one sheet to another, in contrast to the larger isolated $(Me_4N)^+$ cations. This change in interlayer distance is also reflected in the ~ 3 Å shorter a-axis dimension of 14.311(2) Å versus 17.116(2) Å, for the $Co(pyz)^{2+}$ and $(Me_4N)^+$ analogs, respectively. The b and c lattice dimensions differ much less between the two structures, within ~ 0.3 Å, such that the packing of cations in the bc plane must share more similar repeat distances. Shown in Fig. 17.9 is the surface of the V_4O_{10} layer in the bc plane and the positions of the $Co(pyz)^{2+}$ chains and $(Me_4N)^+$ cations. Representative distances between the apical oxygen atoms down the b and c directions are labeled as **a** and **b** respectively. The **a** (repeat) distance has increased from 6.64 Å to 7.00 Å to accommodate the larger Co-pyz-Co spacing, while the **b** distance has decreased from 3.23 Å to 3.03 Å as a result of Co bridging across these

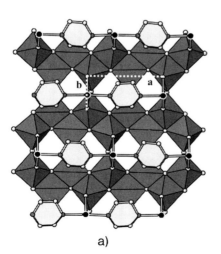
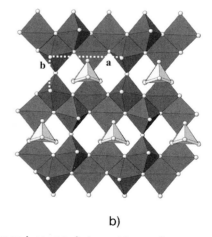

a)　　　　　　　　　　　　　　　　b)

Fig. 17.9 Sections of the $V_4O_{10}^{n-}$ layers in $Co(pyz)V_4O_{10}$ (a) and $(Me_4N)V_4O_{10}$ (b). Repeat distances of the apical oxygen atoms are labeled as dashed lines, which are **a** (7.00 Å (a); 6.64 Å (b)) and **b** (3.03 Å (a); 3.23 Å (b)). Dark polyhedra $= VO_5$, light $= (Me_4N)^+$ or pyrazine, black spheres $=$ Co, white $=$ C or O.

apical oxygen atoms in $Co(pyz)V_4O_{10}$. However, while $Co(pyz)V_4O_{10}$ crystallizes in the centrosymmetric *Cmcm* space group, $(Me_4N)V_4O_{10}$ crystallizes in the non-centrosymmetric subgroup *Cmc2₁*. This structural relationship helps to confirm that the non-centrosymmetric symmetry in the latter arises solely from the arrangement of the $(Me_4N)^+$ cations [36].

17.4
Conclusions

Located at the interfaces of metal-oxides and organics, the heterometallic perrhenate and vanadate families of hybrid solids provide an abundant variety of new structures and compositions that can be understood from the viewpoint of structural relationships. These structural relationships have helped reveal new clues into nature's construction of $MReO_4$ (M = Ag, Cu) layered and pillared phases, and from which it has proven possible to begin to understand and predict new compositions. Although exact structural details, such as in the helical $Cu(pzc)_2AgReO_4$ structure, are intriguing and difficult to foresee. In multi-layered metal-organic/vanadate solids, structural relationships have led to the useful application of charge density matching in order to understand the formation of vanadate layers with new interlamellar $Cu(pyz)_2^{2+}$ and $M(pyz)^{2+}$ (M = Co, Ni, and Zn) networks. Much future solid-state research will be driven by the solid-state structures that are yet to be synthesized, where the structural relationships and structure-guiding principles, as described here, will play a helpful role.

Acknowledgments

P. M. would like to thank all of his previous and current group members who have contributed to the results of this research and is grateful for funding from the American Chemical Society Petroleum Research Fund and the Beckman Foundation.

References

1 M. Lulei, P. A. Maggard, J. D. Corbett, *Angew. Chem. Int. Ed*. **1996**, *35*, 1704.
2 P. A. Maggard, J. D. Corbett, *Angew. Chem. Int. Ed*. **1997**, *36*, 1974.
3 P. A. Maggard, J. D. Corbett, *Inorg. Chem*. **1998**, *37*, 814.
4 P. A. Maggard, J. D. Corbett, *J. Am. Chem. Soc*. **2000**, *122*, 838.
5 P. A. Maggard, J. D. Corbett, *Inorg. Chem*. **1999**, *38*, 1945.
6 P. A. Maggard, J. D. Corbett, *J. Am. Chem. Soc*. **2000**, *122*, 10740.
7 P. A. Maggard, D. A. Knight, J. D. Corbett, *J. Alloys Compds*. **2001**, *315*, 108.
8 A. Clearfield, *Chem. Mater*. **1998**, *10*, 2801.
9 A. P. Cote, G. K. H. Shimizu, *Chem. Comm*. **2001**, *3*, 251.
10 P. J. Hagrman, D. Hagrman, J. Zubieta, *Angew. Chem. Int. Ed*. **1999**, *38*, 2638.

11 D. E. Hagrman, J. Zubieta, *J. Solid. St. Chem.* **2000**, *152*, 141.

12 R. L. Laduca, R. Finn, J. Zubieta, *Chem. Comm.* **1999**, 1669.

13 R. S. Rarig, R. Lam, P. Y. Zavalij, J. K. Ngala, R. L. Laduca, J. E. Greedan, J. Zubieta, *Inorg. Chem.* **2002**, *41*, 2124.

14 W.-K. Chang, R.-K. Chiang, Y.-C. Jiang, S.-L. Wang, S.-F. Lee, K.-H. Lii, *Inorg. Chem.* **2004**, *43*, 2564.

15 J. Macicek, O. Angelova, R. Petrova, *Z. Kristallogr.* **1995**, *210*, 24.

16 J. Macicek, O. Angelova, *Acta Crystallogr. C* **1995**, *51*, 2539.

17 J. Zhao, J. Wang, S. Zang, S. Liu, S. Yinling, *Rare Met.* **2001**, *20*, 28.

18 M. C. Chakravorti, M. B. Sarkar, *J. Ind. Chem. Soc.* **1983**, *60*, 628.

19 A. S. Gardberg, P. E. Doan, B. M. Hoffman, J. A. Ibers, *Angew. Chem., Int. Ed.* **2001**, *40*, 244.

20 A. S. Gardberg, A. E. Sprauve, J. A. Ibers, *Inorg. Chim. Acta* **2002**, *328*, 179.

21 A. Butz, I. Svoboda, H. Paulus, H. Fuess, *J. Solid St. Chem.* **1995**, *115*, 2255.

22 M. B. Varfolomeev, A. N. Zemenkova, V. N. Chrustalev, J. T. Struckov, H.-J. Lunk, B. Ziemer, *J. Alloys Compds* **1994**, *215*, 359.

23 C. Mujica, K. Peters, E.-M. Peters, W. Carillo, H. G. von Schnering, *Z. Kristallogr.* **1998**, *213*, 11.

24 C. Mujica, K. Peters, E.-M. Peters, W. Carillo, H. G. von Schnering, *Bol. Soc. Chil. Quim.* **1999**, *44*, 161.

25 A. Butz, G. Miehe, H. Paulus, P. Strauss, H. Fuess, *J. Solid St. Chem.* **1998**, *138*, 232.

26 O. Angelova, J. Macicek, R. Petrova, T. Todorov, B. Mihailova, *Z. Kristallogr.* **1996**, *211*, 163.

27 W. Baur, W. Joswig, G. Pieper, D. Kassner, *J. Solid St. Chem.* **1992**, *99*, 207.

28 D. Y. Naumov, A. V. Virovets, S. V. Korenev, A. I. Gubanov, *Acta Crystallogr. C* **1999**, *55*, 1.

29 J. Luo, B. Alexander, T. R. Wagner, P. A. Maggard, *Inorg. Chem.* **2004**, *43*, 5537.

30 H. Lin, B. Yan, P. D. Boyle, P. A. Maggard, *J. Solid St. Chem.* **2007**, *179*, 217.

31 F. Buschendorf, *Z. Physik. Chem.* **1993**, *B20*, 237.

32 P. A. Maggard, B. Yan, J. Luo, *Angew. Chem. Int. Ed.* **2005**, *44*, 2553.

33 B. Yan, M. D. Capracotta, P. A. Maggard, *Inorg. Chem.* **2005**, *44*, 6509.

34 A. Bose, P. He, C. Liu, B. D. Ellman, R. J. Twieg, S. D. Huang, *J. Am. Chem. Soc.* **2002**, *124*, 4.

35 P. A. Maggard, P. D. Boyle, *Inorg. Chem.* **2003**, *42*, 4250.

36 D. Riou, O. Roubeau, L. Bouhedja, J. Livage, G. Ferey, *Chem. Mater.* **2000**, *12*, 67.

37 P. Y. Zavalij, M. S. Whittingham, E. A. Boylan, V. K. Pecharsky, R. A. Jacobson, *Zeitschr. Kristall.* **1996**, *211*, 464.

38 Y. K. Shan, R. H. Huang, S. P. D. Huang, *Angew. Chem. Int. Ed.* **1999**, *38*, 1751.

39 P. J. Hagrman, J. Zubieta, *Inorg. Chem.* **2000**, *39*, 3252.

40 Y. P. Zhang, R. C. Haushalter, A. Clearfield, *Inorg. Chem.* **1996**, *35*, 4950.

41 D. Riou, G. Ferey, *Inorg. Chem.* **1995**, *34*, 6520.

42 Y. P. Zhang, C. J. O'Connor, A. Clearfield, R. C. Haushalter, *Chem. Mater.* **1996**, *8*, 595.

43 D. Riou, G. Ferey, *J. Solid St. Chem.* **1995**, *120*, 137.

44 L. R. Zhang, Z. Shi, G. Y. Yang, X. M. Chen, S. H. Feng, *J. Chem. Soc. Dalt. Trans.* **2000**, 275.

45 Y. P. Zhang, J. R. D. Debord, C. J. O'Connor, R. C. Haushalter, A. Clearfield, J. Zubieta, *Angew. Chem. Int. Ed. Engl.* **1996**, *35*, 989.

46 T. Chirayil, P. Y. Zavalij, M. S. Whittingham, *J. Mater. Chem.* **1997**, *7*, 2193.

47 R. J. Chen, P. Y. Zavalij, M. S. Whittingham, J. E. Greedan, N. P. Raju, M. Bieringer, *J. Mater. Chem.* **1999**, *9*, 93.

48 M. Morcrette, P. Rozier, L. Dupont, E. Mugnier, L. Sannier, J. Galy, J.-M. Tarascon, *Nature Materials* **2003**, *2*, 755–761.

49 C. Delmas, H. Cognac-Auradou, J. M. Cocciantelli, M. Menetrier, J. P. Doumerc, *Sol. St. Ionics* **1994**, *69*, 257.

50 R. Enjalbert, J. Galy, *Acta Crystallogr. Sect. C* **1986**, *42*, 1467.

51 X. Bu, P. Feng, G. D. Stucky, *Science* **1997**, 278.

52 B. Yan, P. A. Maggard, *Inorg. Chem.* In press, **2005**.

53 L.-M. Zheng, X. Wang, Y. Wang, A. J. Jacobson, *J. Mater. Chem.* **2001**, *11*, 1100.

18
Hydrogen Bonding in Metal Halides:
Lattice Effects and Electronic Distortions

James D. Martin

18.1
Introduction

In spite of the finite number of space groups that dictate the symmetry of organization in crystals, the number of possible crystalline structures is almost limitless. Various bonding, charge compensation and space-filling principles can generally be employed to understand specific structural organizations at the atomic and molecular levels. Throughout his career, Professor John Corbett has pursued a mission of exploratory discovery of new materials [1]. With new materials in hand, and working with his many coworkers, the Corbett methodology focused synthetic exploration in order to find families of related materials by which the principles of bonding, which are important for given structure types, are deciphered. It was the exploitation of structure/property relationships and the understanding of the underlying principles for both physical and electronic structure, which further enabled the design of materials with specific properties. In this way, changing the interstitial element in clusters, or the electron count in inter-metallics has allowed the Corbett group to tune the structure, magnetic and electronic properties in diverse systems.

Although great advances have been made with regard to understanding local bonding in molecules, clusters and networks, the understanding of the more complex organization into crystalline structures, remains much more elusive. When the specific reasoning as to why a particular compound adopts a specific crystalline structure or why that structure exhibits subtle distortion from an expected geometry is unclear, it is not uncommon to suggest that the structural modification is due to crystal-packing forces or lattice effects. Such terms tend to become "catchall" explanations. However, a detailed understanding of the chemical principles behind such crystal-packing or lattice effects is critical for the advancement of crystal engineering. Building on the tradition of materials discovery that I inherited as a post-doc in the Corbett lab, a component of my research program has focused on understanding and gaining control over such lattice effects to direct the formation of both crystalline and liquid structures [2–9]. Hydrogen bonding is one of the

Inorganic Chemistry in Focus III.
Edited by G. Meyer, D. Naumann, L. Wesemann
Copyright © 2006 WILEY-VCH Verlag GmbH & Co. KGaA, Weinheim
ISBN: 3-527-31510-1

most definable crystal packing forces [10, 11]. In addition to structure direction in crystalline networks, we here describe examples where hydrogen bonding exhibits unique control of the crystallization processes and crystal melting, and causes remarkable perturbations to a material's electronic structure.

18.2
A Hierarchy of Structure-directing Forces

In our efforts towards the rational design of crystalline structures, we frequently consider a hierarchy of structure direction. At the top is the covalent structure of primary building units. Specifically, in an effort to construct metal halide analogs of zeolites and clays, Cu^+, Zn^{2+} and Al^{3+} halides have been selected as tetrahedral building blocks, and Zr^{4+} as an octahedral building block [2–9, 12]. First, because of the relative radius ratio of Cl^- to the M^{n+}, pseudo close-packing of the halide sub-lattice is also a foundational principle of structure organization in these systems. Second, the relative charge density of the inorganic network being formed with respect to that of a templating cation is considered [4, 13]. For example, in the $ACuCl_2$ system, large low-charge-density cations ($A = NPr_4$) result in linear $CuCl_2^-$ anions [14], whereas smaller high-charge-density cations ($A = Hpy$ or H_3NMe) yield complex network structures [4]. Weaker forces such as hydrogen bonding and van der Waals interactions are on the low end of the structure-directing hierarchy but are commonly the underlying basis of crystal-packing forces or lattice effects. Given that in a crystalline lattice there are many such intermolecular contacts, the sum of many weak interactions results in a significant structure-directing influence.

Because hydrogen bonds are the strongest of the weak intermolecular interactions, and can be designed and oriented based on molecular structures, they have been extensively exploited as a tool for crystal engineering [10, 11]. Traditionally viewed as being electrostatic in origin, a result of the dipolar interaction between donors and acceptors, hydrogen bonds have often been classified as non-covalent interactions. Nevertheless, with the observation of low-barrier hydrogen bonds there is an increasing consideration of a covalent component to hydrogen bonds [15–17]. Even in water, evidence suggests that the nature of hydrogen bonding is at least partly covalent [18–20]. We here suggest that the charge transfer (or covalent) component is significantly responsible for hydrogen bonding to species such as metal chlorides, once thought to be too weak an acceptor for hydrogen bonds, yet commonly observed in crystalline structures [21].

18.3
Hydrogen Bond Influence on Melts and Crystallization

Alkylammonium cations are quite commonly used as structure-directing agents, also described as templates, because of their ability to form hydrogen bonds as well as the facility with which cations of various shapes and sizes can be prepared. We, like others [21, 22] have found them to be useful templates for the synthesis of a variety of metal-halide materials. For example, zeolite-type frameworks can be constructed based on a zinc chloride tetrahedral network, instead of a silicate network, by substituting equal molar amounts of CuCl and a templating ACl salt into $ZnCl_2$ to yield $A_n[Cu_nZn_{m-n}Cl_{2m}]$ materials. To date, crystalline structures of such halozeotypes have been observed when the templating cation has a relatively high charge density (e.g., the alkali cation Rb^+) and/or an alkylammonium cation capable of forming hydrogen bonds to the metal-halide network (H_3NMe^+, $H_2NMe_2^+$, $H_2NEt_2^+$ and $HNMe_3^+$) [2, 5]. Interestingly however, no crystalline Cu/Zn/Cl phase has been observed with either NMe_4^+ or NPr_4^+ templates. In fact, halozeotype materials prepared with the latter, yield only amorphous materials, the melts of which solidify to a glass. Neutron-scattering measurements give an indication that zeolite-type network organization is present in the amorphous material [7], but apparently in the absence of the structure-ordering influence of hydrogen bonds or a high-charge-density template, these copper-zinc-chloride materials cannot crystallize.

Not only does hydrogen bonding between the template and the metal halide network appear to be critical for crystal formation, it also appears to be responsible for the rate-determining step in the crystallization process. We have been studying the mechanism of crystallization of the halozeotype CZX-1, which adopts the sodalite structure, by time-resolved synchrotron X-ray diffraction (TRXRD) and differential scanning calorimetry (DSC) [23]. CZX-1 is observed to undergo two crystal-crystal phase transitions within $15°$ prior to melting at $177\,°C$. Our current model, based on X-ray diffraction and inelastic and quasielastic neutron-scattering measurements, implies that these phase transitions correspond to a reordering of the dipoles of the templating $HNMe_3^+$ cation, and the site-to-site hopping of the templating cation within the metal-halide cage, respectively. Nevertheless, the D-Cl pair correlation at $2.8\,Å$ (from the pair distribution function analysis of the neutron structure factor) is observed for both crystalline and molten CZX-1 indicating that hydrogen bonding persists into the liquid state, albeit that the long-range ordering is lost [24]. Interestingly, crystallization of CZX-1 from the melt is not observed to occur until temperatures below that of the phase transition, where it is hypothesized that the cation dipoles are ordered (i.e., below $164\,°C$). This window of no-crystallization is clearly seen in Fig. 18.1 where the crystallization temperature is plotted with respect to the cooling rate. These data are corrected with respect to an indium standard to remove instrumental effects. A plateau in the crystallization temperature for CZX-1 is reached more than $10°$ below its melting point as the cooling rate is slowed. Similar behavior is observed in isothermal DSC experiments where mol-

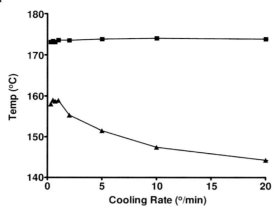

Fig. 18.1 Melting temperature (squares) and temperature of crystallization (triangles) for CZX-1 as a function of the DSC cooling rate. A plateau in the crystallization temperature is observed for cooling rates below 1°/min resulting in a "no-crystallization" region of about 15°.

ten samples of CZX-1 quenched to isotherms between 164 °C and 177 °C were not observed to crystallize in 36 hours. Similarly no crystallization was observed in this temperature regime by X-ray diffraction for samples in sealed capillaries agitated by oscillation on a goniometer. Only when a sample is cooled to a temperature where the hydrogen bonding between template and framework can be locked into an ordered arrangement, does crystallization occur.

A comparison of the series of copper chloride chain compounds ACu_2Cl_3 (A = Cs^+ [25], $H_2NMe_2^+$ [26] and NMe_4^+ [27]) provides a further interesting demonstration of the influence of hydrogen bonding of an alkylammonium cation on a metal-chloride network. The covalent $Cu_2Cl_3^-$ chain common to each of these materials consists of a ribbon chain in which μ^4-chloride ligands bridge alternating faces of the ribbon, and μ^2-chloride ligands bridge edges of the ribbons. As seen in the unit cell views of the crystal structures (Fig. 18.2), the relative size and shape of the respective cations results in an expansion in the unit cell, as well as changes in the relative orientation of the chains to optimize crystal packing. The unit cell volume for each material (Z=4) is 632.5 Å^3, 775.6 Å^3 and 983.4 Å^3, respectively, for the Cs^+, $H_2NMe_2^+$ and NMe_4^+ compounds. The cations in the Cs^+ and NMe_4^+ compounds are reasonably symmetrically located between the three anion chains in their first coordination sphere, with modest contacts of Cs–Cl ~ 3.6 Å and N–Cl ~ 4.3 Å, respectively, to each chain. By contrast, in the $H_2NMe_2^+$ compound, NH–Cl hydrogen bonding results in a strong association between each cation and a single cuprous chloride chain with N–Cl contacts of ~3.3 Å to one chain and much weaker contacts of 3.5–5.0 Å to the other two neighboring chains. Here the NH_2 portion of the ammonium cation sits in the pocket created by one μ^4-Cl at the base and bounded by two μ^2-Cl and two μ^4-Cl ligands around the rim. This structural motif finds the alkylammonium

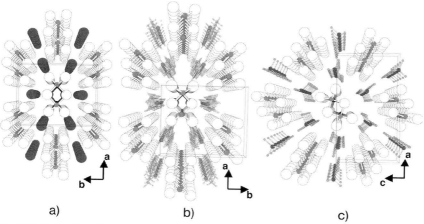

a)   b)   c)

Fig. 18.2 Ball and stick representations of the crystal packing
of (a) CsCu$_2$Cl$_3$, (b) (H$_2$NMe$_2$)Cu$_2$Cl$_3$ and (c) (NMe$_4$)Cu$_2$Cl$_3$.

cation reinforcing the molecular geometry of the metal-halide polymer chain
rather than forming inter-molecular contacts that result in a 3-D network.

In a continuing effort to understand the relationship between crystalline and
molten structures of network materials, these compounds appeared to be reasonable candidates to probe whether the covalent chain structure could persist
into the molten state after the presumably weaker contacts between chains and
cations (so called lattice forces) were lost. Increasing the size, and thus decreasing the charge density of the cations on going from Cs$^+$ to H$_2$NMe$_2^+$ to NMe$_4^+$
(clearly seen in the increase in unit cell volume given above) was anticipated to
result in a systematic decrease in the melting point of the crystalline solids, and
thus an increased probability of retaining the polymeric chain structure into the
melt. Surprisingly, however, the H$_2$NMe$_2^+$ material exhibits a much lower melting point than that of either of the other materials. The melting points are
271 °C for CsCu$_2$Cl$_3$, 226 °C for [NMe$_4$]Cu$_2$Cl$_3$, but only 125 °C for [H$_2$NMe$_2$]-
Cu$_2$Cl$_3$. The modest decrease in the melting points between the Cs$^+$ and NMe$_4^+$
materials is in keeping with expected charge-density-based lattice effects. But
hydrogen bonding is well known to alter such trends. For example, in the well
known case of water, strong intermolecular hydrogen bonds increase lattice effects, resulting in elevated melting and boiling points. In the present case, however, the formation of strong intramolecular hydrogen bonds apparently form a
tight cation-polymer ion-pair that reinforces the polymer structure but decreases
lattice forces and thus dramatically lowers the melting point. We are currently
conducting a series of X-ray and neutron-scattering measurements of the melt
structures of these materials from which preliminary analysis confirms the hypothesis that a significant portion of the polymer chain structure is retained
into the melt.

18.4
Electronic Implications of Hydrogen Bonding

The above examples demonstrate the structure directing and forming role of hydrogen bonding in a lattice, but they give limited insight into the nature of the hydrogen bond itself. Such structure-directing influences are possible irrespective of whether hydrogen bonds are primarily electrostatic, or whether there is a significant covalent component to the bonding. To further probe an understanding of hydrogen bonding, it has been instructive to consider the distortive effects on the metal-halide bonds that serve as the hydrogen bond acceptors. We have previously described significant hydrogen-bond-induced distortions to Zr–Cl bonds in the structure of $[H_2NMe_2]CuZrCl_6$ [9]. There, the Zr–Cl bonds involved in significant hydrogen bonding were elongated by almost 0.1 Å with respect to those not involved in hydrogen bonding. Although complicated by a second-order Jahn-Teller distortion of the Cu(I) center, we suggested that such distortion was a result of mixing between the relatively high-lying Zr–Cl bonding orbitals and the relatively low-lying N–H σ^* orbitals of the hydrogen-bond donor. Similar dramatic distortions to the Zr–Cl bonds that are acceptors of N–H hydrogen bonds are observed in the less complex structure of $[H_2NEt_2]_2Zr_2Cl_{10}$ [28], which allows for a more detailed analysis of these interactions. Furthermore, the $Zr_2Cl_{10}^{2-}$ anion has been structurally characterized with a variety of counter-cations providing comparative data from which to evaluate lattice effects unique to hydrogen bonding [29–40].

The structure of $[H_2NEt_2]_2Zr_2Cl_{10}$, shown in Fig. 18.3, adopts the molecular geometry of an edge-shared bi-octahedron. The well defined ethyl groups of the diethylammonium cation provide confidence for the assignment of the location of the ammonium protons in the crystal structure. Significant hydrogen bonding is observed between the dialkylammonium cations and the axial Cl(2) and equatorial Cl(5) ligands. Interestingly, the Zr–Cl bonds to these chlorides are more than 0.06 Å longer than the corresponding bonds to the other terminal chloride ligands (Zr–Cl(2)=2.461(1) Å, Zr–Cl(3)=2.398(1) Å, Zr–Cl(5)=2.418(1) Å and Zr–Cl(4)=2.351(1) Å). In fact, the equatorial interactions also give rise to a large *trans* influence, such that the bond to the bridging chloride (Zr–Cl(1)=2.567(1) Å), *trans* to the hydrogen-bond-weakened Zr–Cl(5), is more than 0.08 Å shorter than the bridge (Zr–Cl(1')=2.652(1) Å) *trans* to the shorter Zr–Cl(4) bond. These are the largest distortions to the M–X bonds observed for any of the crystallographically characterized $M_2X_{10}^{2-}$ materials (M=Ti, Zr, Hf; X=Cl, Br), with most showing differences in bond distances of no more than a 0.02 Å [29–40]. In no other materials is the counter-cation capable of forming hydrogen bonds to the $M_2X_{10}^{2-}$ anion. Of further note are the angles of the hydrogen-bond interactions. While a simple electrostatic interaction might predict a linear N-H–Cl angle (ϕ), these are observed in the range of 129–149°. Hydrogen-bond angles H–Cl-Zr (θ) are observed to range from 105° to 145°. It has previously been suggested that angular distortions to θ and ϕ result from an interaction between the hydrogen-bond donor and a lone on the acceptor. This has been significantly studied for organic

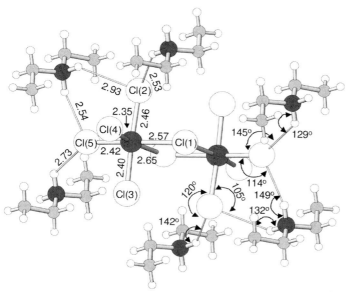

Fig. 18.3 Ball and stick drawing of the centro-symmetric [H$_2$NEt$_2$]$_2$Zr$_2$Cl$_{10}$. Atom labels, bond distances and hydrogen bonds are labeled on the left side of the figure and Zr–Cl–H–N hydrogen-bond angles are labeled on the right side of the figure.

carbonyl molecules where a *trans* arrangement of the D–H–A-X (D = donor, A = acceptor) is described [10, 41]. However, in the present case the N–H–Cl-Zr bonds are oriented in a *syn* confirmation.

As we have previously suggested, this large hydrogen-bond-induced distortion to Zr^{IV}–Cl bonds may be the result of an interaction between the relatively high-lying and filled metal-chloride bonding orbitals and the relatively low-lying and empty N–Hσ* [9]. To further probe this hypothesis, a series of Extended Hückel calculations were performed on the idealized ZrCl$_6^{2-}$ octahedral anion with a single NH$_4^+$ ammonium cation, as described in Scheme 18.1. Because this hypothesis is an orbital-overlap-based argument, the Extended Hückel method was intentionally chosen because its approximation treats orbital overlap but largely ignores electrostatic effects. The calculations were performed using a Zr–Cl distance of 2.4 Å and a H–Cl distance of 2.5 Å. The hydrogen-bond angles were independently varied in two series of calculations. For one series, the angle ϕ was held constant at 120° and the angle θ was varied between 90 and 180°. In the second series, the angle θ was held constant at 142° and the angle ϕ was varied between 90 and 180°. The fixed-angle values for θ and ϕ were taken from the geometry of the strongest hydrogen-bond contact in the crystal structure of [H$_2$NEt$_2$]$_2$Zr$_2$Cl$_{10}$. A calculation on the non-interacting anion and cation, with a Cl–H distance of 5 Å was also performed.

Scheme 18.1

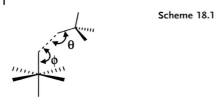

As seen in the orbital-splitting diagram of Fig. 18.4 a, the hydrogen bonding interaction between the ammonium cation and the zirconium chloride anion involves significant mixing of N–H and Zr–Cl orbitals. The N–H hydrogen-bond donor in fact acts as an electron acceptor, or Lewis acid, such that the N–Hσ^* orbital is significantly destabilized. Furthermore, the Zr–Cl bonding character is mixed into the lower energy bonding orbital resulting in a slight destabilization of the formerly Zr–Cl bonding orbitals, and stabilization of the formerly N–H, but now N·H–Cl·Zr bonding orbital. Consideration of the projected overlap populations of the inter-ion Cl–H and N–Cl contacts and the intra-ion N–H and Zr–Cl bonds, shown in Fig. 18.4 b and c provides greater understanding of the nature of the bonding. Whether considering θ or ϕ, the inter-ion contacts are minimized with acute angles approaching $90°$ affording the maximal intra-molecular N–H and Zr–Cl bonding. Of course realistically, steric effects disfavor such acute angles. Increasing the angles θ and ϕ results in a loss of intra-ion bonding and a gain of inter-ion bonding (i.e., hydrogen bonding). Interestingly, these competing factors seek to maximize inter-ion hydrogen bonding with modest cost to the intra-ion Zr–Cl and N–H bonding. These calculations suggest that such balance is achieved with more acute values of ϕ than θ, as experimentally observed. In fact the calculations indicate optimal bonding overlap at a geometry close to that observed in the crystalline structure of $[H_2NEt_2]_2Zr_2Cl_{10}$.

The fact that the geometry of these N·H–Cl·Zr hydrogen bonds is reasonably reproduced with the simple Extended Hückel approximation, provides strong support to an orbital-overlap/charge-transfer component to the hydrogen bond. Such orbital overlap clearly has an impact on the molecular geometry of a hydrogen bond, nevertheless, the electrostatic component is still anticipated to have significant influence on hydrogen bond strength. Extension of this idea suggests that, with a d^{10} metal center, a hydrogen-bond-induced shortening of the metal-chloride bonds might be expected, because the hydrogen-bond donor should withdraw electron density from metal-ligand anti-bonding orbitals. This has been observed for $[Hpy]_3Cu_3Cl_6$ [4]. Most often, however, in the copper(I) chlorides no significant difference is observed in the Cu–Cl distances for bonds that are and are not involved in hydrogen bonding, apparently because the charge-transfer shortening is countered by the electrostatic weakening of Cu-Cl–H-N bonds. In addition, consideration of relative orbital energies provides an understanding as to why hydrocarbon C–Cl bonds do not significantly participate as hydrogen-bond acceptors [42], in contrast to transition-metal chloride bonds [21]. Carbon is much more electronegative than any of the transition me-

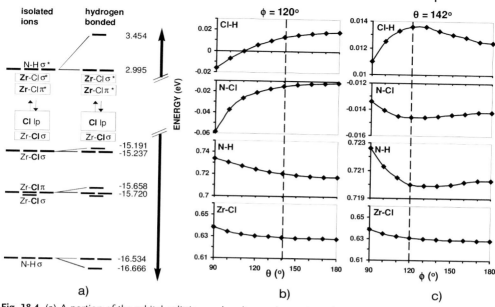

Fig. 18.4 (a) A portion of the orbital splitting for $NH_4^+/ZrCl_6^{2-}$ as isolated ions and as a hydrogen-bonded ion pair. The HOMO is indicated by ↑↓. Projected overlap populations (positive values representing bonding and negative values anti-bonding) for contacts between hydrogen bond donor and acceptor. (b) Keeping ϕ fixed at $120°$ and varying θ, and (c) keeping θ fixed at $142°$ and varying ϕ.

tals, such that its s and p orbitals effectively interact with both the s and p orbitals of Cl. As a result, the C–Cl bonding orbitals are quite low in energy and reasonably delocalized between both carbon and chlorine. By contrast, in a transition-metal chloride, the primary interaction is between the metal d-orbitals and the chlorine p-orbitals. With little participation of the low-energy Cl s-orbitals, the polar covalent M–Cl bond is of relatively high energy and is localized on the Cl p_z-orbital, which can gain additional stabilization by mixing with the empty orbital of a hydrogen-bond donor.

18.5
Conclusions

It is a well known fact that hydrogen bonding is often one of the strongest lattice forces, critical for directing crystal structure organization and exploited for crystal engineering. As described in the brief summary of mechanistic considerations for the crystallization of halozeotypes, and the structural influences on the melts of the ACu_2Cl_3 polymers, hydrogen-bond-derived lattice forces also play a critical role on the mechanism by which crystals form and the structural organization in melts. Such a mechanistic understanding is necessary in order to fully understand crystallization from condensed systems (something not well treated by classical nucleation theory). And development of the ability to design and control structure in the liquid state opens up numerous possibilities in the field of amorphous materials engineering. Nevertheless, the factors that determine the strength and geometry of hydrogen bonds are not thoroughly understood. Carefully examining distortions of hydrogen bond acceptor metal-chloride bonds, it becomes apparent that there is a significant covalent component to metal chloride–ammonium hydrogen bonds. While useful in the understanding of distortions and the nature of hydrogen bonding, it should be possible to exploit this understanding of the charge-transfer component of hydrogen bonding for tuning the energy and width of electronic bands. Such hydrogen bonding to acceptor atoms that contribute to the valence band of MI_4^{2-} perovskites ($M = Sn$ and Pb) has been suggested to be important in stabilizing the top of the valence band, thus increasing the band gap observed as a blue shift in the observed exciton spectrum [43].

Acknowledgments

The work of numerous graduate students and post-docs (particularly A. M. Dattelbaum, M. D. Capracotta and A. H. Josey, and Dr. J. C. W. Folmer) has provided much of the data for this manuscript, and is gratefully acknowledged. Funding to support this work was provided by the National Science Foundation.

References

1 J. D. Corbett, *Inorg. Chem.*, **2000**, *39*, 5178.

2 J. D. Martin and K. B. Greenwood, *Angew. Chem.*, **1997**, *36*, 2072.

3 J. D. Martin and B. R. Leafblad, *Angew. Chem.*, **1998**, *37*, 3318.

4 J. D. Martin, J. Yang, A. M. Dattelbaum, *Chem. Mater.*, **2001**, *13*, 392.

5 J. D. Martin, A. M. Dattelbaum, R. M. Sullivan, T. A. Thornton, J. Wang, and M. T. Peachey, *Chem. Mater.*, **1998**, *10*, 2699.

6 J. D. Martin, *Industrial Applications of Ionic Liquids*, Ed. R. Rogers, ACS Symposium Series, **2002**, *818*, 413.

7 J. D. Martin, S. J. Goettler, N. Fosse, L. Iton, *Nature*, **2002**, *419*, 381.

8 A. M. Dattelbaum and J. D. Martin, *Inorg. Chem.*, **1999**, *38*, 6200.

9 A. M. Dattelbaum and J. D. Martin, *Inorg. Chem.*, **1999**, *38*, 2369.

10 C. B. Aakeröy, D. S. Leinen, in *Crystal Engineering: From Molecules to Materials*, D. Braga et al. (Eds.), Kluwer, Netherlands, **1999**, 89.

11 S. Subramanian, M. J. Zawarotko, *Coord. Chem. Rev.*, **1994**, *137*, 357.

12 A. M. Dattelbaum and James D. Martin, *Polyhedron*, **2005**, in press.

13 C. H. Arnby, S. Jagner, I. Dance, *Cryst. Eng. Comm.*, **2004**, *6*, 257.

14 S. Anderson, S. Jagner, *Acta Chem. Scand. Ser. A*, **1986**, *A40*, 52.

15 G. A. Geffrey, *An Introduction to Hydrogen Bonding*, Oxford University Press, Oxford, **1997**.

16 L. F. Pacios, O. Gálvez, P. C. Gómez, *J. Chem. Phys.*, **2005**, *122*, 214307.

17 B. Schiøtt, B. B. Iversen, G. K. H. Madsen, T. C. Bruice, *J. Am. Chem. Soc.*, **1998**, *120*, 12117.

18 J. Neuefeind, C.J. Benmore, B. Tomberli, and P.A. Egelstaff, *J. Phys.: Condens. Matter.*, **2002**, *14*, L429.

19 T.K. Ghanty, V.N. Staroverov, P.R. Koren, E.R. Davidson, *J. Am. Chem. Soc.*, **2000**, *122*, 1210.

20 W.H. Thompson, J.T. Hynes, *J. Am. Chem. Soc.*, **2000**, *122*, 6278.

21 G. Aullón, D. Bellamy, L. Brammer, E.A. Bruton, A.G. Orpen, *Chem. Commun.*, **1998**, 653.

22 S. Haddad, R.D. Willett, *Inorg. Chem.*, **2001**, *40*, 2457.

23 A. Josey, J.C.W. Folmer, J.D. Martin, manuscript in preparation.

24 A. Josey, J.D. Martin, manuscript in preparation.

25 C. Brink, N.F. Binnendijk, J. van de Linde, *Acta Cryst.*, **1954**, *7*, 176.

26 M.D. Capracotta, J. D. Martin, manuscript in preparation.

27 S. Andersson, S. Jagner, *Acta Chem. Scand.*, **1985**, *A40*, 177.

28 A.M. Dattelbaum thesis, NCSU, **2000**.

29 R. Hart, W. Levason, B. Patel, G. Reid, *J. Chem. Soc. Dalton Trans.*, **2002**, 3153.

30 M.J. Sarsfield, M. Said, M. Thornton-Pett, L.A. Gerrard, M. Bochmann, *J. Chem. Soc. Dalton Trans.*, **2001**, 822.

31 J. Eicher, U. Muller, K. Dehnicke, *Z. Anorg. Allg. Chem.*, **1985**, *521*, 37.

32 D.P. Krutko, M.V. Borzov, E.N. Veksler, A.V. Churakov, J.A.K. Howard, *Polyhedron*, **1998**, *17*, 3889.

33 R. Centore, A. Tuzi, D. Liguori, *Acta Cryst. C*, **2003**, *59*, 241.

34 S.J. Coles, M.B. Hursthouse, D.G. Kelly, N.M. Walker, *Acta Cryst. C*, **1999**, *55*, 1789.

35 A.I. Brusilovets, E.B. Rusanov, A.N. Chernega, *Russ. J. Gen. Ghem.*, **1995**, *65*, 1819.

36 J. Eicher, U. Müller, K. Dehnike, *Z. Anorg. Allg. Chem.*, **1985**, *521*, 37.

37 J. Beck, *Chem. Ber.*, **1991**, *124*, 677.

38 T.J. Kistenmacher, G.D. Stucky, *Inorg. Chem.*, **1971**, *10*, 122.

39 F. Calderazzo, P. Pallavincini, G. Pampaloni, P.F. Zanazzi, *J. Chem. Soc. Dalton Trans.*, **1990**, 2743.

40 B. Neumuller, K. Dehnicke, *Z. Anorg. Allg. Chem.*, **2004**, *630*, 2576.

41 R. Taylor, O. Kennard, W. Versichel, *J. Am. Chem. Soc.*, **1983**, *105*, 5761.

42 J.D. Dunitz, R. Taylor, *Chem. Eur. J.*, **1997**, *3*, 89.

43 J.L. Knutson, J.D. Martin, D.B. Mitzi, *Inorg. Chem.*, **2005**, *44*, 4699.

19

Syntheses and Catalytic Properties of Titanium Nitride Nanoparticles

Stefan Kaskel

19.1
Introduction

In recent years, considerable efforts have been devoted to the design of particles that are only a few nanometers in diameter. They have applications in catalysis [1], sensing [2], coatings [3], and optoelectronic devices [4, 5]. The change of properties with reduced dimensions (quantum size effect) is of fundamental interest. Non-oxide materials such as TiN or Si_3N_4 are used in abrasion-resistant coatings and bulk ceramics due to high hardness and creep resistance [6]. They are typically produced in a dense form. Non-oxide materials with high specific surface area are promising candidates for applications in heterogeneous catalysis [7–11]. High melting points and low sintering tendencies are beneficial for applications in high-temperature catalysis. Additionally, hard ceramics such as TiN are chemically inert, especially under non-oxidizing conditions. Thus, it is necessary to develop chemical methods that allow one to synthesize these materials with high accessible surface area, either as porous materials or nanoparticular powders. In the following, a new approach to the design of titanium nitride (TiN)-based nanoparticles is reported using the ammonolysis of titanium chloride complexes. The particles are active catalysts for the decomposition of complex aluminum hydrides in the solid state. The latter can be used in hydrogen storage applications since acceleration of the alanate decomposition is needed for a rapid release of hydrogen [12, 13]. In addition, the alanate activation can be used in solution to enhance the reactivity of $NaAlH_4$ and to affect the selectivity in alkyne reduction reactions. Acceleration is only achieved with highly dispersed catalyst particles and, therefore, the need for efficient and cheap synthesis routes is evident.

The new method produces TiN powders with surface areas exceeding $200 \ m^2g^{-1}$ that are otherwise only accessible using a forced flow reactor and a microwave plasma activator in which titanium metal is reacted with N_2 in the gas phase [14]. TiN powders with considerably lower specific surface area ($S_g < 60 \ m^2g^{-1}$) were also synthesized using the nitridation of 10–15 nm-sized

Inorganic Chemistry in Focus III.
Edited by G. Meyer, D. Naumann, L. Wesemann
Copyright © 2006 WILEY-VCH Verlag GmbH & Co. KGaA, Weinheim
ISBN: 3-527-31510-1

TiO_2 particles, the benzene-thermal reaction of $TiCl_4$ and NaN_3, chemical vapor reactions, plasma syntheses, or arc-melting [15–20]. The optical properties of low-concentrated TiN nanoparticle systems were studied by Quinten [21]. Highly dispersed TiN was used as an additive to improve the mechanical properties of titanium carbide-based cermets [22].

The ammonolysis of pure $TiCl_4$ has been investigated earlier as a gas-phase reaction, but physisorption measurements were not carried out [23]. In the following, the pyrolysis of solid $TiCl_4L_n$ precursors (L=donor ligand) in flowing ammonia gas is studied at atmospheric pressure for the production of TiN nanoparticles. It is demonstrated that the physisorption properties of the ultrafine powders critically depend on the type of precursor and reaction conditions used in the synthesis. The use of nitrogen and oxygen donor ligands not only allows one to direct the ammonolysis reaction, but also to tune the sintering behavior of the products. The materials are further characterized using nitrogen physisorption isotherms, transmission electron microscopy, CP/MAS NMR, and powder X-ray diffraction.

19.2
Synthesis of TiN Nanoparticles

TiN nanoparticles are produced via ammonolysis of $TiCl_4L_n$ complexes (L = donor ligand, Table 19.1) in the solid state (Eq. 1) [24].

$$TiCl_4L_n \xrightarrow{\ NH_3, 973-1273K\ } TiN \qquad (1)$$

The reaction produces black ultrafine powders with high accessible surface areas. In the course of this reaction, chlorine and the ligand L is successively substituted by nitrogen. Intermediates of the type $Ti(NH)_xCl_y$ are typically X-ray amorphous. TiNCl has been identified as an intermediate in this reaction using longer reaction times and temperatures of 623 to 673 K [25, 26]. Volatility, melting point, and decomposition rate of the precursor $TiCl_4L_n$ vary with the type of ligand. This is why it is possible to affect the crystallization rate and particle sizes chemically by choosing different precursors for the ammonolysis reaction.

Three types of ligands were used: a) aliphatic amines, b) ethers and c) bipyridines. The strength of bonding in the resulting $TiCl_4L_n$ complexes increases from a to c and is one of the factors affecting the reactivity of the precursor towards NH_3. Secondly, within the course of the pyrolysis, the ligands may form a protective shell around the forming particles which may reduce sintering and agglomeration. Thus, the low sintering tendency reported in this work might be caused by deposits on the outer surface of the particles.

Table 19.1 demonstrates how the type of precursor affects the specific surface area (S_g) of the products. Typically, the specific surface area of the powders exceeds $100 \, m^2g^{-1}$ and is higher compared with commercial products [17]. The ef-

Table 19.1 Properties of high surface area TiN.

Sample	Precursor	Ammonolysis temperature [K]	S_g [m^{2}g^{-1}]
(1)	TiCl$_4$(NH$_3$)$_4$	973	100
(2)	TiCl$_4$(n-PrNH$_2$)$_2$	973	117
(3)	TiCl$_4$(n-Pr$_2$NH)$_2$	973	56
(4)	TiCl$_4$(Et$_3$N)$_2$	973	156
(5)	TiCl$_4$(4,4'-bpy)	973	221
(6)	TiCl$_4$(4,4'-bpy)	1273	100
(7)	TiCl$_4$(2,2'-bpy)	973	161
(8)	TiCl$_4$(CH$_3$CN)$_2$	973	37
(9)	TiCl$_4$(Et$_2$O)$_2$	973	108
(10)	TiCl$_4$(THF)$_2$	873	230
(11)	TiCl$_4$(THF)$_2$	973	212
(12)	TiCl$_4$(THF)$_2$	1073	166
(13)	TiCl$_4$(THF)$_2$	1273	96
(14)	TiCl$_4$(diglyme)	973	137

fective particle size d can be estimated from the specific surface area S_g assuming spherical particle morphology [27] (Eq. 2).

$$d = 6/\rho S_g \qquad (2)$$

A specific surface area of 100 m^{2}g^{-1} corresponds to a particle size of 11 nm ($\rho_{TiN} = 5.39$ g cm^{-3}) [28] and p . . . owders with $S_g = 200$ m^2 g^{-1} consist of particles approximately 6 nm in diameter.

Aliphatic amines yield surface areas typically around 100 m^{2}g^{-1} (Table 19.1, 1–4). The physisorption isotherms of samples 1–3 are shown in Fig. 19.1.

For L=NH$_3$ (1) and L=Pr$_2$NH (3), the isotherms are of type II as expected for non-porous materials [27]. Sample 2 shows a significant uptake at $0.6 < P/P_0 < 0.9$ as in type-IV isotherms, indicating the presence of mesopores. They are an indication for a narrow particle-size distribution which results in a more regular packing with interparticle pores of size similar to that of the particles [27]. The latter shows that the ligand-assisted synthesis does not only allow one to affect the total surface area and particle size, but also the size distribution which is an important tool for tailoring the particle properties.

For type-c ligands, the uptake in the mesopore range is even more pronounced (Fig. 19.1, samples 5 and 6). They show typically type-IV isotherms with different hysteresis loops.

Sample 5 is close to an H$_2$-type hysteresis, whereas 6 and 7 can be tentatively assigned to H$_3$- and H$_1$-type hystereses, respectively [27]. The hystereses are caused by capillary condensation in interparticle pores and the shape is an indication of a particular particle morphology. Sample 7 has a more regular narrow mesopore size distribution, whereas sample 5 is more complex with pores of

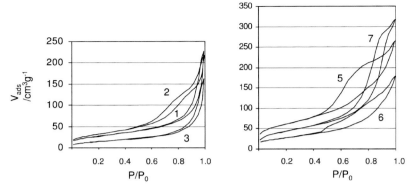

Fig. 19.1 Nitrogen physisorption isotherms at 77 K for TiN nanoparticles (the numbers correspond to the sample code in Table 19.1).

different size and shape. Sample 6 is made from the same precursor as sample 5 but calcined at 1273 K.

Sintering causes an overall decrease in the surface area (from 221 to 100 m^2g^{-1}) and collapse of the interparticle mesopores, as can be seen from the flattening of the hysteresis.

The isotherms for $TiCl_4(THF)_2$-derived samples (Fig. 19.2) are of type II. Sample 11 synthesized at 973 K has a narrow H_4 hysteresis indicating flat particle agglomerates with a broad interparticle mesopore size distribution. At high synthesis temperatures, the hysteresis is closer to an H_3 type. Thus, an irregular size and shape distribution is produced. The surface areas retained at 1073 K (sample 12) and 1273 K (sample 13) are 166 m^2g^{-1} and 96 m^2g^{-1}, still very high compared with commercial materials. This is why the materials are also promising high-temperature supports or catalysts. In particular, metals tend to sinter dramatically under high temperature reaction conditions, lose accessible surface area, and thus become ineffective catalysts, whereas our experiments demonstrate the low sintering tendency of inorganic nitrides.

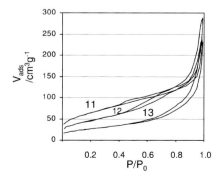

Fig. 19.2 Nitrogen physisorption isotherms at 77 K for TiN nanoparticles (the numbers correspond to the sample code in Table 19.1).

The physisorption isotherms discussed so far clearly demonstrate the ability to affect size and shape of the particles with the ligand-assisted ammonolysis reaction. Whereas high specific surface area is crucial in catalysis, morphology of nanosized particles is also critical for the design of nanocomposites, since the mechanical properties of composites and ceramics are largely governed by the shape of the primary particles and the resulting microstructure. Nanosized TiN particles could also be valuable as abrasives since they are dispersible and have high hardness.

TEM investigations support the interpretation of the nitrogen physisorption isotherms (Fig. 19.3). They show particles 5–20 nm in diameter, whereas the particle size estimated from the specific surface area, assuming spherical particle morphology, is 7 nm. Indeed, the particle morphology for sample 7 is mostly spherical, but for some crystallites edges are discernible.

Thus, the hysteresis loop of sample 7, indicating a relatively narrow pore-size distribution of the interparticle pores, is associated with almost spherical particles. However, the packing is quite irregular, and thus a correlation of particle diameter and average pore size using different idealized types of packing is difficult [27]. Pyrogenic deposits are produced within the course of the heat treatment that can also contribute to the specific surface area since the density of carbon is low. In such a case a specific surface area higher than that of the TiN particles alone would be measured, and the effective particle size, estimated from the specific surface area alone, would be smaller than the particle size determined from TEM pictures.

In X-ray diffraction patterns of high surface area TiN, typically broad reflections are observed due to the small particle size (Fig. 19.4). The peaks can be assigned to cubic TiN [29]. Crystalline side products were not detectable. Particle sizes deduced from "size" broadening of the Bragg reflections come to 12 nm which is close to the particle diameter deduced from TEM studies (5–20 nm) and BET measurements (7 nm). This is, however, within the error limits of the techniques and the assumptions made. As the specific surface area of the materials is reduced with increasing synthesis temperature, the FWHM of the XRD reflection is reduced too due to sintering and grain growth within the nanopowders (Fig. 19.4).

For organic ligands used in the synthesis, carbon impurities of 5–10% are present in the materials. ^{13}C-CP/MAS NMR spectra demonstrate that these impurities are present in the form of pyrogenic carbon (Fig. 19.5).

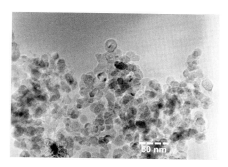

Fig. 19.3 Transmission electron micrograph of TiN nanoparticles (sample 7).

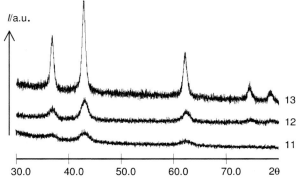

Fig. 19.4 XRD patterns of high surface area TiN synthesized via ammonolysis of $TiCl_4(THF)_2$ at 973 (11), 1073 (12) and 1273 K (13).

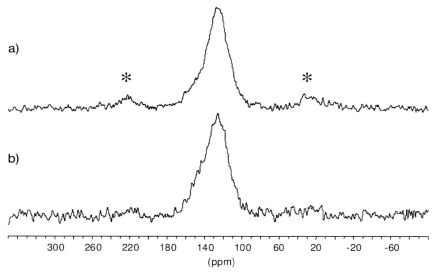

Fig. 19.5 ^{13}C-CP/MAS NMR spectra for sample 5 (a) and 11 (b). The asterisks indicate rotational side bands.

The center of the line is located at 126 ppm. The shape and position of the lines are comparable to those observed in the spectra of pyrogenic deposits (sp^2-hybridized carbon), for example, in coked catalysts where carbon is produced during the deactivation of solid acid catalysts [30, 31].

The peaks at 20 and 220 ppm are rotational side bands, their position depends on the rotation speed. In principle, the presence of carbon could also lead to the formation of TiC. However, the precursor contains no Ti–C bonds and thus TiC formation is not very likely. The peak maximum for the static line of

TiC is observed at 530 ppm and carbonitrides typically have chemical shifts in a range from 330 to 403 ppm [32]. Peaks corresponding to TiC or TiC_xN_y were not observed. However, since the CP technique was used for the measurements, the absence of additional peaks does not necessarily prove the absence of TiC or TiC_xN_y side products with low carbon content.

An important requirement for handling the powders is the stability towards air. In the syntheses described here, contact to air must be avoided, but the final product is relatively inert towards air. For example, heating sample 1 in air to 473 K for 24 h reduces the nitrogen content from 21.26 to 19.25%. The diffraction pattern was identical to that of the as-prepared sample. A small reduction in nitrogen content is tolerable in terms of bulk composition but indicates a passivation layer formation on the outer surface of the particles. The surface chemistry is therefore inevitably changed in air with all the implications for catalytic applications. However, such thin oxide layers may be reconverted into nitride by high-temperature ammonia treatment, in order to achieve reactivation of the materials for catalytic applications. High surface area titanium nitride-based materials are not suitable for long-term catalytic applications under oxidizing conditions, but handling in air and regeneration is possible. Compatible processes are reduction, dehydrogenation, ammonia synthesis, amination, and hydride activation reactions. Complex hydride activation is attractive for hydrogen storage applications [12, 13] as well as in hydrogen transfer reactions [33].

Summarizing, effective tailoring of surface area and morphology of titanium nitride-based particles is achieved using the ammonolysis of solid $TiCl_4$ complexes at high temperature. The method affords TiN powders with high specific surface areas exceeding 200 m^2g^{-1} without using demanding reactor design or plasma-assisted methods. The materials presented here are promising catalysts and catalyst supports for catalytic processes under non-oxidizing conditions, such as hydrogenation and hydrogen storage.

19.3
Titanium Nitride Nanoparticles in Hydrogen Storage Applications [12, 13]

Currently there is an urgent need to develop hydrogen storage materials for mobile and stationary applications. This trend is accelerated by the hydrogen fuel cell technology that could, within the next decades, replace fossil fuel resources such as oil and gas.

Typically, different types of storage are distinguished:
a) pressure tanks
b) liquid hydrogen tanks
c) physisorbed hydrogen in porous materials
d) chemical storage.

In chemical storage, a chemical reaction is used that releases hydrogen, for example, by thermal decomposition of a hydride, but chemical decomposition of hydrides via hydrolysis was also proposed [34–36]. The main developments of intermetallic compounds used as storage materials were carried out earlier [37]. However, transition-metal hydrides are typically of high density and, thus, mass-based capacities of storage materials were low.

In recent years, complex aluminum hydrides have been developed by Bogdanovic for the use as a storage material [12, 13, 38–40]. Hydrogen is herein generated by decomposing $NaAlH_4$ in two separate reactions into NaH and aluminum metal.

Pure $NaAlH_4$ decomposes at relatively high temperatures (Fig. 19.6). The first reaction step is a decomposition into Na_3AlH_6 and aluminum metal at 185 °C (H_2 capacity based on $NaAlH_4$):

$$NaAlH_4 \longrightarrow 1/3 \, Na_3AlH_6 + 2/3 \, Al + H_2 \qquad 3.7 \text{ wt\% H}$$

The second decomposition step is at about 240 °C and produces NaH that is only decomposed above 400 °C (H_2 capacity based on $NaAlH_4$):

$$1/3 \, Na_3AlH_6 \longrightarrow NaH + 1/3 \, Al + 1/2 \, H_2 \qquad 1.8 \text{ wt\% H}$$

The decomposition of pure $NaAlH_4$ produces a NaH/Al mixture that cannot be hydrogenated into $NaAlH_4$ under mild conditions. For mobile applications in hydrogen storage, it is necessary to achieve decomposition at lower temperatures. Also the material should be recyclable, for example, via re-hydrogenation at higher H_2 pressure. Both needs are fulfilled when ultrafine titanium materials are dispersed as a solid catalyst into $NaAlH_4$ [38, 39, 41]. The decomposition temperature is lowered and the decomposed $NaAlH_4$ can be rehydrogenated.

Titanium nitride-based nanoparticles are also efficient dopants for $NaAlH_4$ in hydrogen storage applications [12, 13]. The black solid TiN powder can be dispersed into the hydride via ball milling. Addition of titanium nitride-based na-

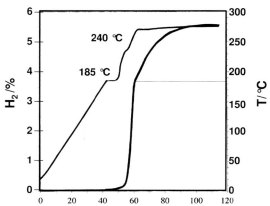

Fig. 19.6 $NaAlH_4$ decomposition.

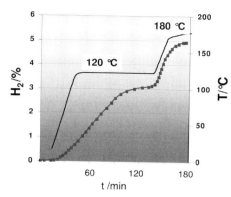

Fig. 19.7 Decomposition of titanium nitride-doped NaAlH₄ (2 mol% TiN).

noparticles lowers the decomposition temperature of NaAlH$_4$ and allows re-charging of alanate batteries.

The decomposition is significantly accelerated and the temperature of the first decomposition reaction is lowered to 120 °C (Fig. 19.7). The decomposition rate is relatively low compared with other titanium-based dopants. The highest activity of a titanium catalyst used in alanate decomposition was observed for ligand-stabilized colloidal titanium metal [42].

Initial activity is important to ensure fast unloading and loading cycles. The second requirement for repeated use of alanate batteries is long-term stability. Whereas titanium colloids show superior performance in terms of decomposition kinetics, titanium nitride-based materials are superior in long-term stability. The latter can be seen comparing both catalytic materials in several runs (Fig. 19.8).

In the first three cycles, the capacity of the alanate is high for both materials, the titanium metal-doped and the nitride-doped material. However, after 15 cycles the hydrogen capacity decreases significantly below 4% for the titanium metal-doped sample, whereas for the nitride-doped sample the capacity remains high at about 5% hydrogen.

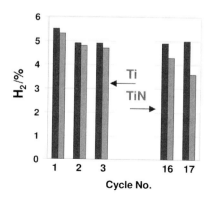

Fig. 19.8 Comparison of titanium metal and titanium nitride-doped NaAlH₄ in hydrogen storage.

The theoretical limit of 5.4% ($NaAlH_4 + 2$ mol% TiN) for the two subsequent decomposition reactions is in both cases only observed in the first cycle. The reason for the decrease in capacity is still unknown and little is known about the mechanism of alanate activation via titanium dopants in the solid state. Certainly, the ease of titanium hydride formation and decomposition plays a key role in this process, but whether titanium substitution in the alanate or the formation of a titanium aluminum alloys, i.e., finely dispersed titanium species in the decomposition products is crucial, is still under debate [41].

However, it is clear that TiN nanoparticles activate $NaAlH_4$ and accelerate the decomposition. Therefore, the activation of $NaAlH_4$ with TiN nanoparticles was also studied in solution.

19.4
Catalytic Properties of TiN Nanoparticles in Solution [33]

For titanium-group nitrides, only few catalytic applications have been reported. In combination with iron, titanium nitride has been studied for ammonia synthesis [43]. The synthesis described above allows one to manufacture titanium nitride materials with very high specific surface areas exceeding 200 m^2g^{-1}. Ultrafine TiN powders catalyze the decomposition of complex aluminum hydrides after mixing via ball-milling at temperatures of about 393 K in the solid state [12, 13]. Thus, the activation of complex hydrides for hydrogen transfer reactions in solution was further studied. The reaction of diphenylethyne (DPE) with $NaAlH_4$ in aprotic solvents using ultrafine titanium nitride as a solid, highly dispersed activator was used as a test reaction. In this reaction, titanium nitride can be used to enhance the reaction rate but, interestingly the selectivity is also significantly affected. In the following we describe the performance of ultrafine titanium nitride materials prepared from various precursors, with specific surface areas ranging from 100 to 200 m^2g^{-1}. The conditions and kinetics of the reaction are analyzed and optimized and a mechanism for the reaction is proposed.

The DPE reduction is used as a test reaction to characterize the materials and optimize the preparation conditions of the catalyst. Since hydroaluminations can also be used for the synthesis of carboxylic acids, deuterated products, or vinyl halides via quenching with CO_2, D_2O or Br_2 [44], the method is also a valuable organic synthesis tool. However, as compared with molecular catalysts like Cp_2TiCl_2 that are known to catalyze hydroaluminations [44], the titanium nitride materials described here are *solid* catalysts and can be separated by centrifugation. Moreover, they can be reused several times, which is an advantage as compared to molecular catalysts.

19.5
Catalytic Properties

The activation of $NaAlH_4$ by titanium nitride for hydrogen storage applications was an accidental discovery and only guided by the fact that titanium compounds catalyze the decomposition of $NaAlH_4$ [12, 13]. Therefore, it should be possible also to use the activation for a transfer of hydrogen from the complex alanate to an organic substrate. Out of several substrates, diphenylethyne (DPE) was found to be suitable to evaluate the catalytic properties of titanium nitride. Whereas the uncatalyzed reduction of carbonyl compounds and nitriles with $NaAlH_4$ proceeds rather fast, the uncatalyzed reduction of DPE is relatively slow at moderate temperatures (328 K). In addition, only two products are obtained after quenching the solution with water, Z- and E-stilbene (Scheme 19.1).

This reduction can also be carried out with molecular hydrogen and as such is probably not of any commercial interest. However, it is suited for the study of the catalytic properties of the ultrafine powders and serves as a characterization and optimization technique for the titanium nitride nanoparticles in this study.

General observations Initially, the reactions of $NaAlH_4$ and diphenylethyne with and without ultrafine TiN were compared. In general, an excess of $NaAlH_4$ with respect to the substrate DPE was used. In the first experiments, a molar $TiN:NaAlH_4$ ratio of 1:3 was used at 328 K in THF using freshly prepared high surface area titanium nitride. Figure 19.9 shows the concentrations of DPE, Z- and E-stilbene for the reaction carried out with (a) and without (b) addition of titanium nitride. The reaction rates are clearly different. For the uncatalyzed reaction, an exponential decrease of the educt concentration is observed and conversion is complete after 24 h. On the other hand, for the reaction mixture containing TiN nanoparticles, the reaction is much faster and approaches 99% conversion after 5 h. Even more surprising are the selectivity differences (Fig. 19.9).

The uncatalyzed reaction (b) gives predominately the thermodynamically more stable E-stilbene and the ratio of E-stilbene to Z-stilbene is typically 70:30. However, a complete reversal of the selectivity is observed if TiN nano-

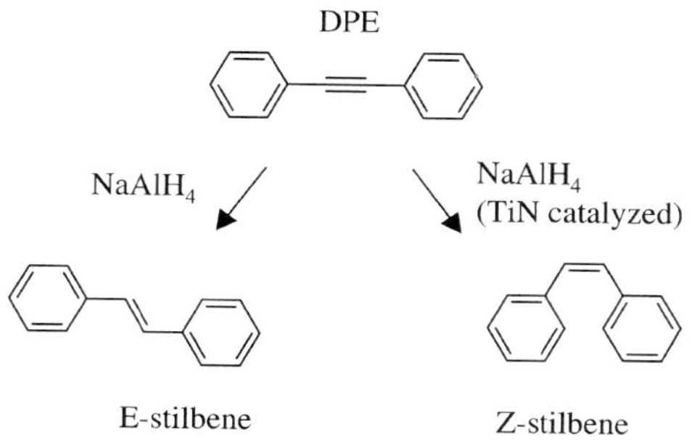

Scheme 19.1 Reduction of diphenylethyne with $NaAlH_4$.

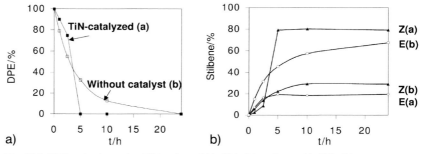

Fig. 19.9 DPE reduction using TiN-activated $NaAlH_4$ (a) and pure $NaAlH_4$ (b).

particles are added to the reaction. Under such conditions, Z-stilbene is the predominant product and typically 80% Z-stilbene and only 20% E-stilbene are obtained (a). The comparison of the uncatalyzed (b) and the catalyzed reaction (a) clearly demonstrate the effect of the addition of titanium nitride. The reaction is significantly accelerated and the product selectivity is reversed.

Re-use of the catalyst The high activity of the solid catalysts also allows one to re-use high surface area titanium nitride several times. By centrifugation, the solid can be separated and used in a new reaction. The catalyst was used up to four times. Figure 19.10 shows the conversion of DPE with time.

The catalyst deactivates, but after four runs the conversion is still significantly higher (>99% after 2 h) as compared with the uncatalyzed reaction. Moreover, the Z-selectivity in all four runs is higher than 80%, whereas in the uncatalyzed reaction, it is typically only 30% (Z). The fact that the solid powder can be used several times furthermore supports the fact that the reaction mechanism is heterogeneous. The reason for the deactivation is unknown. A disadvantage of the nanoparticles is the difficulty of separation. Thus, in some cases the particles form col-

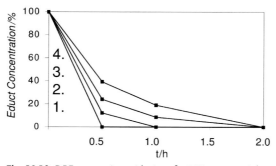

Fig. 19.10 DPE-conversion with time for TiN nanoparticles in four susbsequent runs (after two hours, the catalyst was separated and re-used).

loidal dispersions that are difficult to filtrate and have to be centrifuged. One reason for deactivation could thus simply be the fact that some of the smaller particles are not completely separated and are not available for the following cycle. Another reason is a decrease in the specific surface area. For example, a freshly prepared catalyst with a surface area of $160 \, m^2 g^{-1}$ had a reduced surface area $(65 \, m^2 g^{-1})$ after stirring in a $NaAlH_4$ solution (THF) for two hours. However, the reason for such a reduced surface area can also be agglomeration of the particles during isolation and activation for the BET measurement, or incomplete removal of adsorbed $NaAlH_4$. On the other hand, catalysts isolated after repeated use can still be identified as nanocrystalline TiN in powder X-ray diffraction patterns [33]. The peaks of the fresh and the reused material appear at the same positions. A rearrangement in the bulk structure is not obvious.

Mechanism The understanding of mechanisms in catalytic reactions is sometimes crucial for the creative development of new applications. In a first approach, the main interest was to develop high surface area titanium nitride as a material for catalytic applications and, therefore, evaluation of catalysts prepared under different conditions was performed.

Interestingly, the stereoselective hydrogenation with complex alanates is also feasible using molecular catalysts. This approach has been applied recently for the hydrogenation of alkynols with great success [45]. Initially such reactions were discovered by Ashby [44]. Even though the intermediates in these reactions were never characterized by means of X-ray crystallography, and specific NMR data allowing identification of the addition products are rare, Al–C bond formation was postulated in most of the cases, for catalyzed reactions, as well as for uncatalyzed addition reactions and, thus, the reaction was termed hydroalumination (Scheme 19.2). In that case, evidence for Al–C bond-containing intermediates were based on various quenching reactions which can be used to transform the intermediate into a more valuable product (Scheme 19.2). The latter shows also that hydroaluminations are extremely versatile reactions, which allow one to synthesize a variety of products.

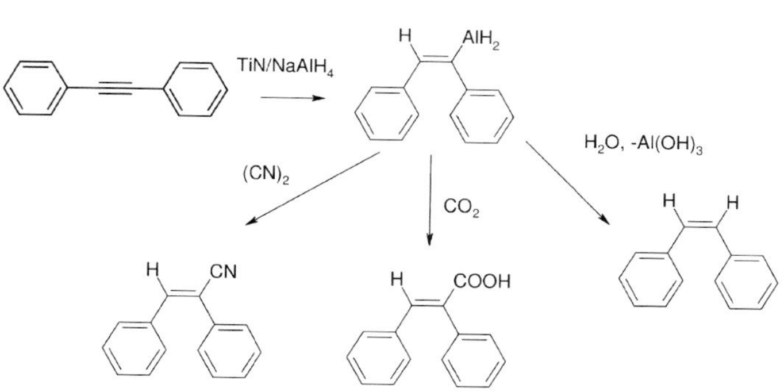

Scheme 19.2 Proposed mechanism for the TiN-catalyzed DPE-reduction.

Identification and characterization of the intermediates was only recently realized by Uhl who reported the structure of several hydroalumination products [46]. In the case of DPE hydroaluminations, structural analyses or NMR investigations have not been carried out. We have therefore separated the intermediates from the catalyst and measured NMR spectra after various reaction times. Identification of the intermediates and assignment of the lines to particular structural fragments is difficult in that case, since the spectra show complicated multiplets which indicate oligomers. However, an important result from NMR data is that neither the lines of DPE nor signals of any of the stilbenes can be recognized in the spectra. From that observation, we conclude that an intermediate is formed in the course of the reaction, probably a hydroalumination product.

Highly dispersed titanium nitride catalyzes the decomposition of $NaAlH_4$. The nanoparticles facilitate subsequent decomposition into Na_3AlH_6 and Al (above). On the other hand, pure AlH_3 is known to add stereoselectively to DPE and Z-stilbene is formed after quenching with water. The role of TiN in the hydrogenation of DPE could thus be to facilitate the dissociation of $NaAlH_4$ into NaH and AlH_3, if a suitable substrate is present. Hereafter, AlH_3, formed on the TiN particle surface, adds to DPE and forms Z-stilbene after hydrolysis of the hydroalumination intermediate.

The catalytic properties discussed in this section are the first example of complex hydride activation for solution phase hydrogenations using high surface area titanium nitride nanoparticles. The activation by TiN results in a significant acceleration of hydrogen transfer and a very high stereoselectivity with over 90% Z-stilbene, whereas the uncatalyzed reaction produces only 30% Z-stilbene. NMR studies suggest that the mechanism of the reaction proceeds via a hydroalumination intermediate, but further characterization of the intermediates using quenching reactions or isolation techniques is necessary. The high stability of the nitrides under extremely reducing conditions in the solid state and in solution is remarkable and shows the potential strength of such materials that should not be seen as competitors of oxide-based catalyst in oxidation reactions, but rather metal-like materials with a high resistance to sintering and high thermal and chemical stability under reducing conditions.

References

1 F. Schüth, K. Unger, in *Handbook of Heterogeneous Catalysis*, G. Ertl, H. Knözinger and J. Weitkamp (Ed.), VCH, Weinheim, 1997, p. 72–86.

2 A. N. Shipway, E. Katz, I. Willner, *ChemPhysChem* **2000**, *1*, 18–52.

3 S. Sepeur, N. Kunze, B. Werner, H. Schmidt, *Thin Solid Films* **1999**, *351*, 216–219.

4 M. Li, H. Schnablegger, S. Mann, *Nature* **1999**, *402*, 393–395.

5 P. V. Kamat, *J. Phys. Chem. B* **2002**, *106*, 7729–7744.

6 A. W. Weimer, *Carbide, Nitride and Boride Materials, Synthesis and Processing*, 1st Ed. Chapman & Hall, London, 1997.

7 D. Farrusseng, K. Schlichte, B. Spliethoff, A. Wingen, S. Kaskel, J. S. Bradley,

F. Schüth, *Angew. Chem., Int. Ed.* **2001**, *40*, 4204–4207.

8 S. Kaskel, D. Farrusseng, K. Schlichte, *Chem. Commun.* **2000**, 2481–2482.

9 S. Kaskel, K. Schlichte, *J. Catal.* **2001**, *201*, 270–274.

10 S. Kaskel, K. Schlichte, B. Zibrowius, *Phys. Chem. Chem. Phys.* **2002**, *4*, 1675–1681.

11 S. Kaskel, in *Handbook of Porous Solids*, F. Schüth, K. S. W. Sing and J. Weitkamp (Ed.), Wiley-VCH, Weinheim, 2002, p. 2063–2086.

12 B. Bogdanovic, M. Felderhoff, S. Kaskel, A. Pommerin, K. Schlichte, F. Schüth, *Reversible Storage of Hydrogen using Doped Alkali Metal Aluminum Hydrides*, **2001**, WO 03/053848 A1.

13 B. Bogdanovic, M. Felderhoff, S. Kaskel, A. Pommerin, K. Schlichte, F. Schüth, *Adv. Mater.* **2003**, *15*, 1012–1015.

14 D. T. Castro, J. Y. Ying, *Nanostruct. Mater.* **1997**, *9*, 67–70.

15 J. Li, L. Gao, J. Sun, Q. Zhang, J. Guo, D. Yan, *J. Am. Ceram. Soc.* **2001**, *84*, 3045–3047.

16 J. Hu, Q. Lu, K. Tang, S. Yu, Y. Qian, G. Zhou, X. Liu, *J. Am. Ceram. Soc.* **2000**, *83*, 430–432.

17 T. Rabe, R. Wäsche, *Nanostruct. Mater.* **1995**, *6*, 357–360.

18 R. A. Andrievski, G. V. Kalinnikov, A. F. Potafeev, V. S. Urbanovich, *Nanostruct. Mater.* **1995**, *6*, 353–356.

19 R. A. Andrievski, *Nanostruct. Mater.* **1997**, *9*, 607–610.

20 I. Narita, T. Oku, *Diamond Relat. Mater.* **2002**, *11*, 949–952.

21 M. Quinten, *Appl. Phys. B-Lasers Opt.* **2001**, *73*, 317–326.

22 N. Liu, Y. D. Xu, H. Li, Z. H. Li, Y. Zhao, G. H. Li, L. D. Zhang, *Mater. Sci. Technol.* **2002**, *18*, 586–590.

23 Y. Saeki, R. Matsuzaki, A. Yajima, M. Akiyama, *Bull. Chem. Soc. Jpn.* **1982**, *55*, 3193–3196.

24 S. Kaskel, K. Schlichte, G. Chaplais, M. Khanna, *J. Mater. Chem.* **2003**, *13*, 1496–1499.

25 R. Juza, J. Heners, *Z. Anorg. Allg. Chem.* **1964**, *332*, 159–172.

26 R. I. Hegde, R. W. Fiordalice, P. J. Tobin, *Appl. Phys. Lett.* **1993**, *62*, 2326–2328.

27 F. Rouquerol, J. Rouquerol, K. Sing, *Adsorption by Powders and Porous Solids*, Academic Press, London, 1999.

28 P. Ettmayer, W. Lengauer, in *Ullmann's Encyclopedia of Industrial Chemistry*, M. Bohnet (Ed.), Wiley Interscience, 2002.

29 A. N. Christensen, *Acta Chem. Scand., Ser. A* **1975**, *29*, 563–564.

30 B. Paweewan, P. J. Barrie, L. F. Gladden, *Appl. Catal., A.* **1998**, *167*, 353–362.

31 J. Weitkamp, S. Maixner, *Zeolites* **1987**, *7*, 6–8.

32 K. J. D. Mackenzie, R. H. Meinhold, D. G. McGavin, J. A. Ripmeester, I. Moudra-kovski, *Solid State Nucl. Magn. Reson.* **1995**, *4*, 193–201.

33 S. Kaskel, K. Schlichte, T. Kratzke, *J. Mol. Catal. A: Chem.* **2004**, *208*, 291–298.

34 V. C. Y. Kong, D. W. Kirk, F. R. Foulkes, J. T. Hinatsu, *Int. J. Hydrogen Energy* **2003**, *28*, 205–214.

35 V. C. Y. Kong, F. R. Foulkes, D. W. Kirk, J. T. Hinatsu, *Int. J. Hydrogen Energy* **1999**, *24*, 665–675.

36 R. Aiello, M. A. Matthews, D. L. Reger, J. E. Collins, *Int. J. Hydrogen Energy* **1998**, *23*, 1103–1108.

37 G. Sandrock, *J. Alloys Compd.* **1999**, *295*, 877–888.

38 B. Bogdanovic, R. A. Brand, A. Marjano-vic, M. Schwickardi, J. Tolle, *J. Alloys Compd.* **2000**, *302*, 36–58.

39 B. Bogdanovic, M. Schwickardi, *Appl. Phys. A: Mater. Sci. Process.* **2001**, *72*, 221–223.

40 F. Schüth, B. Bogdanovic, M. Felderhoff, *Chem. Commun.* **2004**, 2249–2258.

41 B. Bogdanovic, M. Felderhoff, M. Ger-mann, M. Härtel, A. Pommerin, F. Schüth, C. Weidenthaler, B. Zibrowius, *J. Alloys Compd.* **2003**, *350*, 246–255.

42 B. Bogdanovic, 2002, private communi-cation.

43 B. M. Biwer, S. L. Bernasek, *Appl. Surf. Sci.* **1986**, *25*, 41–52.

44 E. C. Ashby, S. R. Noding, *J. Organomet. Chem.* **1979**, *177*, 117–128.

45 A. Parenty, J. M. Campagne, *Tetrahedron Lett.* **2002**, *43*, 1231–1233.

46 W. Uhl, F. Breher, *J. Organomet. Chem.* **2000**, *608*, 54–59.

20

Solventless Thermolysis: A Possible Bridge Between Crystal Structure and Nanosynthesis?

Ling Chen and Li-Ming Wu

20.1
Introduction

Some very interesting properties of nanosized materials have driven the booming development of nanoscience worldwide for the past couple of decades [1]. Examples are the special semiconducting behavior which occurs when an electron is confined to a comparable or smaller length than the Bohr radius (usually a few nanometers) which characterizes electronic motion in bulk semiconductors; the surface plasmon absorption in noble metals, which are tens of nanometers in size, and which results from the collective oscillations of electrons in the conduction band from different particle surfaces; and also the catalytic properties due to the large surface-to-volume ratios. Evidently, the syntheses of new nanomaterials of different sizes and new/special shapes have become the largest activity in this field. Diverse methods, either from bottom-up or top-down techniques, have been explored to gain control over particle size and shape. A variety of shapes such as tubes, wires, belts, rods, cubes, disks, rings, multipods, triangular plates and spheres have already been synthesized. However, there is still a long way to go to fulfill the scientist's dream of being able to make any nanostructure of any shape and in any assembly, and to understand their properties.

20.2
Synthesis Methods

The most important nanomaterial synthesis methods include nanolithography techniques, template-directed syntheses, vapor-phase methods, vapor-liquid-solid (VLS) methods, solution-liquid-solid (SLS) approaches, sol-gel processes, micelle, vapor deposition, solvothermal methods, and pyrolysis methods [1, 2]. For many of these procedures, the control of size and shape, the flexibility in the materials that can be synthesized, and the potential for scaling up, are the main limitations. In general, the understanding of the growth mechanism of any as-

Inorganic Chemistry in Focus III.
Edited by G. Meyer, D. Naumann, L. Wesemann
Copyright © 2006 WILEY-VCH Verlag GmbH & Co. KGaA, Weinheim
ISBN: 3-527-31510-1

synthesized nanostructure is still at a preliminary stage. Many reviews and books have already been written about the synthesis of nanomaterials, including two very recent ones on nanocrystals of different shapes [1] and on one dimensional nanostructures [2] from the point of view of a chemist. In this paper, we will only focus on a newly established solventless thermolysis method.

20.3
Solventless Thermolysis and Some Examples

The solventless thermolysis, first initiated by Korgel in 2003 [3], is rooted in a colloidal precipitation method in which ligands adsorbed at the pre-nanocrystal surface not only provide steric stabilization and eliminate particle collisions-aggregation but also change the growth kinetics and surface energies of different crystal faces. However, such colloidal methods are usually carried out in solution, and the coagulative growth caused by the fast molecular/particle diffusion is, in practice, impossible to eliminate completely. The new approach – solventless thermolysis – avoids these negative effects in solution and takes advantage of the stabilized effects of long-chain alkyl ligands. Generally, such a solventless method involves two steps, the preparation of the metal complex/colloidal precursor in a biphasic organic/inorganic mixed solution and the pyrolysis of the solvent-free thus-synthesized precursor under special conditions. Nanomaterials that have been synthesized by a solventless approach are summarized in Table 20.1.

Table 20.1 Nanomaterial synthesized by solventless thermolysis.

Material	Morphology and description	Reference
Cu_2S (hexagonal)	nanodisk/nanorod dia.: 3–150 nm; thickness: 3–12 nm	4
NiS (trigonal)	nanorod length: 15–50 nm; aspect ratio: 4 nanotriangular plate triangular edge: ~ 20 nm	5
Bi_2S_3 (orthorhombic)	nanowire dia.: 26.0 nm; length: 0.5–70 μm nanorods dia.: 10.7 nm; length: 73.6 μm; aspect ratio: ~ 7 nanofabric	6
$Pb_3O_2Cl_2$ (orthorhombic)	nanobelt width: 29–170 nm; length: 4–20 μm; average thickness: 23.3 nm	7
Cu_2S (hexagonal)	nanowire dia.: 2–6 nm; length: 0.1–several μm	8
Ag (fcc)	nanocoin dia.: 16 nm; thickness: 2.6 nm	9

20.3.1
Cu₂S Nanodisks

Korgel and coworkers utilized a solventless approach to synthesize hexagonal Cu_2S nanocrystals with disk morphologies. The size and shape are controllable in the range of 3–150 nm in diameter and 3–12 nm in thickness and circular to hexagonal prisms depending on the firing conditions from the Cu-dodecane-thiolate precursor [3, 4]. An interesting discovery is, as shown in Fig. 20.1, that monodisperse Cu_2S nanodisks self-assemble into ribbons of stacked platelets, which is probably driven by the dipole–dipole interactions between individual Cu_2S ferroelectric disks. The anisotropic growth of disks appears to be thermodynamically favored because the growth of lower energy {001} planes is limited with respect to the {100} and {110} ones. The morphology may reflect the preferred adsorption and a C–S cleavage of dodecanethiol on different crystal facets.

20.3.2
NiS Nanorods and Nanotrigonal Prisms

The same approach with Ni-thiolate precursor has also successfully produced rhombohedral NiS (millerite) nanorods and triangular nanoplates with a nearly 1:1 ratio (Fig. 20.2) [5]. The lengths of nanorods are controllable through different heating conditions and range from 15 to 50 nm with aspect ratios of approximately 4. The pyrolysis temperature and the reactant concentration, when the precursor was prepared, mainly influence the rod or triangle proportions.

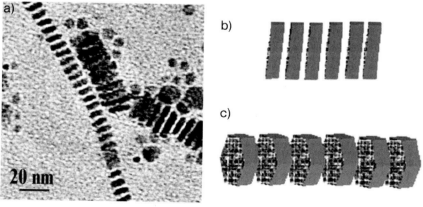

Fig. 20.1 (a) TEM image of Cu_2S ribbons of stacked nanodisks. Schematic of nanodisks oriented (b) perpendicular to the substrate into a rodlike array, and (c) tilted to the substrate, demonstrating the overlap of each disk when imaged with TEM [4].

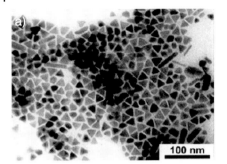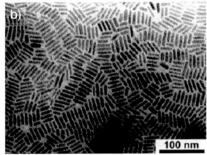

Fig. 20.2 TEM images of NiS nanocrystals synthesized at 190 °C for 1.5 h by a solventless method. The region (a) is enriched in prisms and (b) is enriched in rods [5].

20.3.3
Bi₂S₃ Nanowires, Rods and Fabric

Orthorhombic Bi_2S_3 (bismuthinite) via a solventless method can yield either high aspect ratio (>100) nanowires at about 225 °C (Fig. 20.3 a) or lower aspect ratio (~ 7) (plus sulfur which seems to change the growth kinetics) at around 160 °C. Interestingly, it is possible to fabricate a polymeric matrix from these wires and rods at higher temperatures, ~ 250 °C, as shown in Fig. 20.3 b [6].

20.3.4
Pb₃O₂Cl₂ Nanobelts

A non-thiol precursor, a solid lead chloride-octanoate complex, produces mendipite nanobelts after being heated in air for 1 h at 190 °C (Fig. 20.4). A remark-

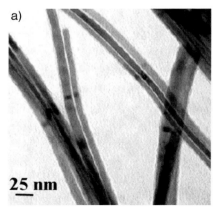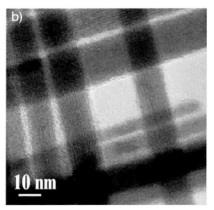

Fig. 20.3 (a) TEM image of Bi_2S_3 nanowires. The wires range from 0.5 to 70 μm, the average diameter was 26.0 nm. (b) TEM image of Bi_2S_3 nanofabric [6].

Fig. 20.4 (a) TEM image of highly crystalline $Pb_3O_2Cl_2$ nanobelts with a width range of 29–170 nm; lengths 4–20 μm and an average thickness of 23.3 nm. (b) SEM image of $Pb_3O_2Cl_2$ nanobelts [7].

able feature of these nanobelts is that they exhibit birefringence enhancement of one order of magnitude as a result of their small size and their preferential elongation in the [010] direction [7].

The foregoing show that the solventless approach has potential for producing a wide range of morphologies in a relatively large quantity, that range from spherical particles, disks, prisms, rods and belts, through to wires. The flexibility of this method also comes from the diverse choice of the ligands for the metal ions, which may provide some new possibilities for accessing the synthesis of new nanoproducts or forms.

20.4
Control of the Nanoproduct Morphology Through the State of the Precursor

We found recently that the viscosity (η_{VAC}) of the colloidal thiolate precursor is a key parameter in controlling the shape of the nanoproducts in the solventless method [8]. Uniform nanowires, rods, or spheres could be made from the corresponding precursors that came from the solutions with different viscosities. The viscosity is a measure of the polymerization of the metal-thiolate complexes. Accordingly, the precursor with the highest viscosity produces nanowires (Fig. 20.5 a), and with decreases in the viscosity, the product morphology changes to rods (Fig. 20.5 b) and then spheres (Fig. 20.5 c).

The significance of this discovery is that control of the degree of polymerization and of coordination modes of the thiolate precursors is likely to affect the shape of the final nanoproduct in a solventless method. Since traditional thiol-metal coordination chemistry is well established, this affords a great opportunity to join the traditional coordination chemistry with an increased variety of nano-

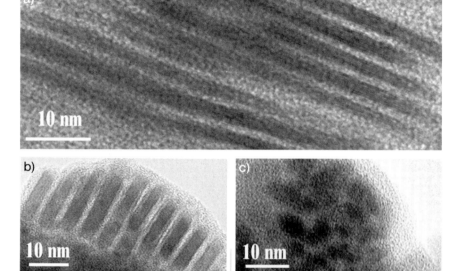

Fig. 20.5 Cu$_2$S nanowires (a), rods (b) and spheres (c) produced at 155 °C for 120 min with Cu-thiolate precursors of different viscosity values: (a) η_{VAC} = 93.5 mPa/s; (b) η_{VAC} = 13.2 mPa/s; (c) η_{VAC} < 10 mPa/s [8].

products on the basis of chemistry methods. And the same strategy may also work for other N-, O-, P- or Se-containing ligand complexes.

20.5
Crystal Structure of the Precursor versus the Morphology and Distribution of the As-synthesized Nanoproduct: A Possible Bridge Between these Two?

The dependence of nanoproduct morphology on the precursor state (viscosity) encourages us to consider whether a precursor with selected crystal structure may generate a nanoproduct with a certain shape. If it is true, this could provide nice access to the control of nanomorphologies. Meanwhile, there are some advantages of solventless method over the solution approaches: (i) crystallization/solidification of the precursor becomes possible and so it becomes practical to determine the structure of the precursor; (ii) mild solid-state thermolysis conditions would be utilized so as to retain the inorganic sub-structure moiety of the precursor, thus the role of the structure on the growth of the nanoproduct might be expected. Considering all these aspects, it seems possible to study the relationship between precursor-structure and nanoproduct-morphology.

With these thoughts in mind, we have recently made $AgSC_{12}H_{25}$ precursor with a layered structure, and determined its structure by X-ray powder diffraction. The XRD pattern (Fig. 20.6a) indicates a very large interlayer distance (34.6 Å), and the intralayer Ag–Ag distance should be ∼ 4 Å [9]. This schematic representation is shown in Fig. 20.6b. This crystalline Ag-thiolate can be reduced to size- and shape-monodispersed-layered silver nanocoins at 180–225 °C under N_2 flow. For example, firing the precursor at 180 °C for 2 hours generates Ag nanocoins with ∼ 16×2.6 nm in size. The diameter increases with the increasing of temperature, e.g. at 200 °C, the average diameter is enlarged to 24.5 nm.

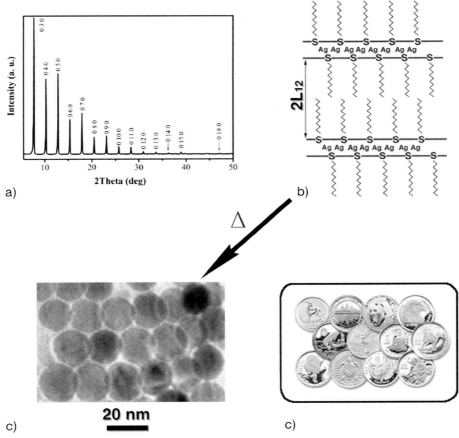

a)

b)

c)

20 nm

c)

Fig. 20.6 (a) XRD of as-synthesized $AgSC_{12}H_{25}$ precursor with indexes of reflections. (b) Cross-sectional representation of the layered crystal structure of $AgSC_{12}H_{25}$. (c) TEM image of Ag nanocoins converted from the layered Ag-thiolate precursor at 180 °C for 2 hours. (d) A picture of some real coins in a pattern similar to that in (c) [9].

A possible layered precursor to the layered nanoproduct conversion mechanism is thus proposed. The silver clusters formed at the initial heating stage by the partial decomposition of AgSR serve as nuclei at further reaction stages, and their distribution naturally inherits the layered pattern of the precursor. The following growth is mainly controlled by the atom concentration and atom diffusion path, which are both constrained by the crystal structure of the precursor [9].

This finding produces some optimism regarding a possible bridge between the nanoproduct morphology and the precursor structure. Such possibilities are certainly worth further exploration to gain fuller understanding.

20.6
Conclusion and Prospects

The newly established solventless thermolysis has been successful in generating nanomaterials over a wide of morphologies (wires, belts, rods, fabric, disks/coins, prisms and spheres). Except for the well understood effects of the capping ligand, pyrolysis temperature/time, and the reactant concentration, it has also been found that the morphology of the nanoproduct is dictated by: (i) the state of the colloidal precursor, such as seen in the viscosity [8]; and (ii) the crystal structure of the precursor [9]. The merits of the solventless method lay in many aspects, such as yields on a gram-scale, the monodispersity in size and shape, and the capability of preparation of a large spectrum of materials with highly crystalline structures and well-controlled composition. While some problems relating to this method need to be clarified, we can be confident that, in the near future, more experimental facts will be available and a deeper understanding of the mechanism will be obtained.

Acknowledgment

This research was supported by the National Natural Science Foundation of China under Projects 20401014 and 20401013, and also by the State Key Laboratory Science Foundation under projects 050086 and 050097.

References

1 C. Burda, X.-B. Chen, R. Narayanan, M. A. El-Sayed, *Chem. Rev.* **2005**, *105*, 1025–1102.

2 Y. Xia, P. Yang, Y. Sun, Y. Wu, B. Mayers, B. Gates, Y. Yin, F. Kim, H. Yan, *Adv. Mater.* **2003**, *15*, 353–389.

3 T. H. Larsen, M. Sigman, A. Ghezelbash, R. C. Doty, B. A. Korgel, *J. Am. Chem. Soc.* **2003**, *125*, 5638–5639.

4 M. B. Sigman, A. Ghezelbash, T. Hanrath, A. E. Saunders, F. Lee, B. A. Korgel, *J. Am. Chem. Soc.* **2003**, *125*, 16050–16057.

5 A. Ghezelbash, M. B. Sigman, B. A. Korgel, *Nano Lett.* **2004**, *4*, 537–542.

6 M. B. Sigman, B. A. Korgel, *Chem. Mater.* **2005**, *17*, 1655–1660.

7 M. B. Sigman, B. A. Korgel, *J. Am. Chem. Soc.* **2005**, *127*, 10089–10095.

8 L. Chen, Y.-B. Chen, L.-M. Wu, *J. Am. Chem. Soc.* **2004**, *126*, 16334–16335.

9 Y.-B. Chen, L. Chen, L.-M. Wu, *Inorg. Chem.* **2005**, *44*, 9817–9822.

21
New Potential Scintillation Materials in Borophosphate Systems

Jing-Tai Zhao and Cheng-Jun Duan

21.1
Introduction

Alkali or alkaline earth borates and phosphates such as BBO (β-BaB$_2$O$_4$) [1], LBO (LiB$_3$O$_5$) [2], and KTP (KTiPO$_4$) [3], have been in research focus for decades for their wide applications in laser and nonlinear optical technologies. With the development of their syntheses and structural characterizations during the last decade, borophosphate, which contains both the borate group and the phosphate group as basic structural units, has also attracted attention recently in the search for new functional materials. Their main structural features are planar BO$_3$ triangles or BO$_4$ tetrahedra sharing corners with PO$_4$ tetrahedra, leading to a great variety of different structures with one-, two- and three-dimensional anion complexes [4].

Therefore, the variety of new compounds in borophosphates, with novel crystal structures, provides a new pool of objects for the exploration of new functional materials. As far as luminescence is concerned, attention has been drawn to alkaline-earth borophosphate lattices doped with rare-earth ions. Blasse et al. [5] reported the UV-excited luminescence of MBPO$_5$:Eu^{2+} (M=Ca, Sr, Ba) prepared in a H$_2$/N$_2$ reducing atmosphere. The luminescence processes of europium in the crystalline and glass modification of calcium borophosphate, have been investigated by Verwey [6]. The X-ray storage phosphor SrBPO$_5$:Eu^{2+} was prepared by Karthikeyani [7]. The luminescence properties of rare-earth-doped alkaline-earth borophosphates in the range of VUV have also been widely investigated [8–10]. However, apart from the advantage of boron containing scintillators for neutron detection, few studies have focused on the scintillation luminescence properties of borophosphates. Scintillation materials have undergone many years of study and have found many applications. Although many of them have been widely used in high-energy physics research and medical imaging systems, new materials with better performance are still to be explored. As a part of our systematic studies on borophosphates and scintillation materials, we have chosen alkaline-earth-based compounds of Ba$_3$BP$_3$O$_{12}$, BaBPO$_5$ and

Inorganic Chemistry in Focus III.
Edited by G. Meyer, D. Naumann, L. Wesemann
Copyright © 2006 WILEY-VCH Verlag GmbH & Co. KGaA, Weinheim
ISBN: 3-527-31510-1

Ba$_3$BPO$_7$ and their rare-earth-doped compounds as starting materials for scintillation studies, with the intention of obtaining new scintillation materials with large, light yield and short decay time.

21.2
Recent Studies on the Scintillation Luminescence Properties of Borophosphates

21.2.1
The Crystal Structures of Ba$_3$BP$_3$O$_{12}$, BaBPO$_5$ and Ba$_3$BPO$_7$

Ba$_3$BP$_3$O$_{12}$ was first prepared by Kniep's group [11] and defined its space group as Ibca. Figure 21.1 shows the crystal structure of Ba$_3$BP$_3$O$_{12}$. The polymeric assembly of tetrahedral anions in crystalline Ba$_3$BP$_3$O$_{12}$ consisted of central "vierer" single chains (BO$_4$ and PO$_4$ alternately) which run parallel to [001]; the two free vertices of each BO$_4$ tetrahedron are connected with terminate PO$_4$ tet-

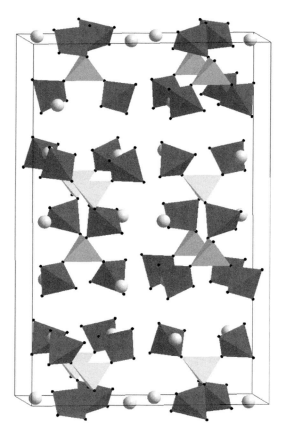

Fig. 21.1 The crystal structure of Ba$_3$BP$_3$O$_{12}$. The dark tetrahedrons represent [PO$_4$]$^{3-}$ groups; the light tetrahedrons represent [BO$_4$]$^{5-}$ groups; the light spheres represent Ba atoms.

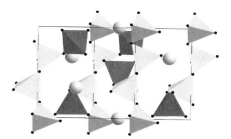

Fig. 21.2 The crystal structure of BaBPO$_5$. The dark tetrahedrons represent [PO$_4$]$^{3-}$ groups; the light tetrahedrons represent [BO$_4$]$^{5-}$ groups; the light spheres represent Ba atoms.

rahedra. Ba$_3$BP$_3$O$_{12}$ contains wide channels in which the barium ions are located.

The BaBPO$_5$ compound was first prepared and structurally characterized by Bauer [12]. Figure 21.2 shows the crystal structure of BaBPO$_5$. Its structure is similar to all stillwellite-like compounds with the space group P3$_2$2. Its main structural elements are spiral tetrahedral chains [001] built of three-membered rings. The contact between the BO$_4$ tetrahedra that form the central part of these chains are reinforced by PO$_4$ tetrahedra and thus [BPO$_5$] heterotetrahedral chain complexes are produced.

Ba$_3$BPO$_7$ was first prepared and characterized by H.W. Ma et al. [13]. Figure 21.3 shows the crystal structure of Ba$_3$BPO$_7$. It adopts a hexagonal space group P6$_3$mc. The basic unit of the structure is the [BaO$_{10}$]-[BO$_3$]-[PO$_4$] polar polyhedra-chain composed of a Ba1-B-P-O cluster. These chains, running along the c-axis, stack in an HCP mode to build the whole structure with triangular prism channels. The channels are also parallel to the c-axis, where Ba2 and Ba3 are located.

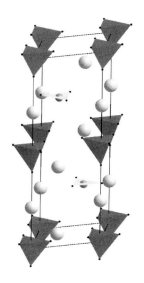

Fig. 21.3 The crystal structure of Ba$_3$BPO$_7$. The dark tetrahedrons represent [PO$_4$]$^{3-}$ groups; the light planar trigons represent [BO$_3$]$^{3-}$ groups; the light spheres represent Ba atoms.

21.2.2

The Preparation and X-ray-excited Intrinsic Scintillation Luminescence Properties of $Ba_3BP_3O_{12}$, $BaBPO_5$ and Ba_3BPO_7 [14, 15]

As described in [14, 15], the samples of $Ba_3BP_3O_{12}$, $BaBPO_5$ and Ba_3BPO_7 used to do scintillation studies were prepared by the solid-state reaction of the components $BaCO_3$ (99.9%), $NH_4H_2PO_4$ (99.9%), H_2BO_3 (99.9%) in molar ratio of $1:1:1$, $3:1:3$ and $3:1:1$, respectively. The reactants were weighed separately on an analytical balance and thoroughly mixed in an agate mortar, and pressed at about 10^8 Pa into 2 cm diameter pellets, heated at 400 °C for 10 h in a covered aluminum oxide crucible, then reground and again pressed into a pellet, heated at 1000 °C for 10 h in a covered aluminum oxide crucible.

Figures 21.4, 21.5 and 21.6 exhibit the X-ray-excited luminescence (XEL) spectrum of $Ba_3BP_3O_{12}$, $BaBPO_5$ and Ba_3BPO_7 powders, respectively. The sequences of the emission intensity of these compounds are $Ba_3BPO_7 > BaBPO_5 > Ba_3BP_3O_{12}$. It can be seen that a broad luminescence band extends from 300 nm to 600 nm with a main emission band at about 400 nm, they all can be defined as a superposition of at least two Gaussian components with the common emission band at about 400 nm. Figure 21.7 shows the comparison between the XEL spectrum of Ba_3BPO_7 powders and that of BGO powders under the same measurement conditions. It is worth noting that the powders of Ba_3BPO_7 have 1/3 of the emission intensity of the $Bi_4Ge_3O_{12}$ powder under X-ray excitation, which may indicate 1/3 light yield of $Bi_4Ge_3O_{12}$ powders. Figures 21.8, 21.9 and 21.10 exhibit the room-temperature fluorescent decay profiles of $Ba_3BP_3O_{12}$, $BaBPO_5$ and Ba_3BPO_7 powders, respectively. All of the studied materials exhibit a two-exponent shape with different decay time, which are in good agreement with the XEL spectra. Compared to the other two compounds, the $Ba_3BP_3O_{12}$ has the fastest decay time with

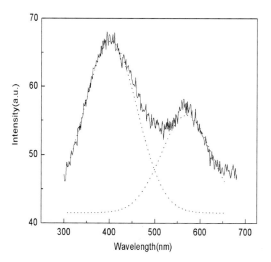

Fig. 21.4 The X-ray-excited luminescence spectrum of $Ba_3BP_3O_{12}$ host. The dotted lines are the Gaussian decomposition.

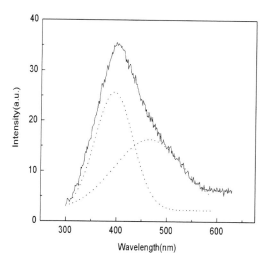

Fig. 21.5 The X-ray-excited luminescence spectrum of the BaBPO$_5$ host. The dotted lines are the Gaussian decomposition.

56.3 ns (99.7%). Table 21.1 lists the emission wavelength and fitted decay components of Ba$_3$BP$_3$O$_{12}$, BaBPO$_5$ and Ba$_3$BPO$_7$.

The defects of the matrix play an important role on luminescent performances in these materials. Taking into consideration the preparation process of these compounds with the solid-state reaction of mixtures of BaCO$_3$, H$_3$BO$_3$, and NH$_4$H$_2$PO$_4$ at different molar ratio, non-equal evaporation during the sintering process of these powders is inevitable and thus results in the formation of intrinsic defects, such as cation and oxygen vacancies. Positional disorder of B and Vacant B (V$_B$)$'''$ have been reported in SrBPO$_5$ crystals on the basis of

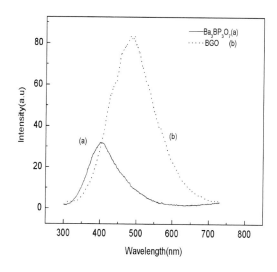

Fig. 21.6 Comparison between the XEL spectrum of Ba$_3$BPO$_7$ powders and that of BGO powders under the same conditions: (a) XEL spectrum of the Ba$_3$BPO$_7$ powders; (b) XEL spectrum of BGO powders.

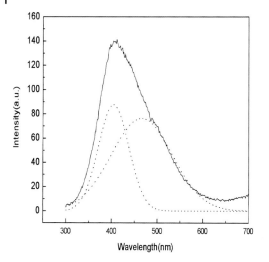

Fig. 21.7 The X-ray-excited luminescence spectrum of the Ba_3BPO_7 host. The dotted lines are the Gaussian decomposition.

thermal displacement parameters for B and XPS data [16]. By analogy, it can be deduced that there would be many more of these defects in the powder crystals of $Ba_3BP_3O_{12}$, $BaBPO_5$ and Ba_3BPO_7 than in single-crystal $SrBPO_5$. As a matter of fact, such centers have been observed in many oxides [17–19]. In addition, taking into account approximate electronegativity (2.04 for B vs 2.19 for P), close ionic radius (0.11 nm for B^{3+} vs 0.17 nm for P^{5+}) [20] and the same coordinating environment (tetrahedral) between B^{3+} and P^{5+} ions, the creation of anti-site defects, B-site P^{5+} ion $(P_B)''$, is possible. It has been noted that analogous phenomena have been found in $LiNbO_3$ and $MgAl_2O_4$ single crystals [21, 22].

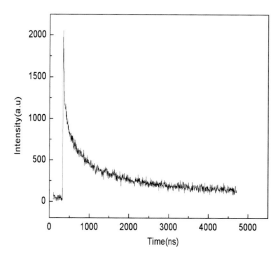

Fig. 21.8 Room-temperature fluorescent decay profile for $Ba_3BP_3O_{12}$.

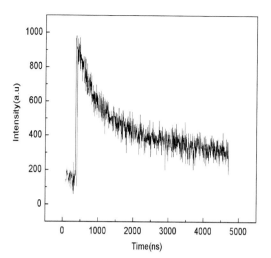

Fig. 21.9 Room-temperature fluorescent decay profile for BaBPO$_5$.

Therefore, there could exist rich defects in Ba$_3$BP$_3$O$_{12}$, BaBPO$_5$ and Ba$_3$BPO$_7$ powders. From the point of energy-band theory, these defects will create defect energy levels in the band gap. It can be suggested that the electrons and holes introduced by X-ray excitation in the host might be mobile and lead to transitions within the conduction band, acceptor levels, donor levels and valence band. Consequently, some X-ray-excited luminescence bands may come into being.

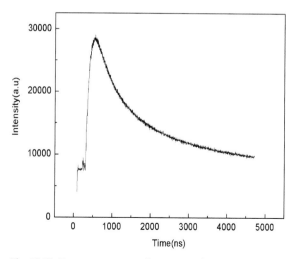

Fig. 21.10 Room-temperature fluorescent decay profile for Ba$_3$BPO$_7$.

Table 21.1 The emission wavelength and fitted decay
components of $Ba_3BP_3O_{12}$, $BaBPO_5$ and Ba_3BPO_7.

Formula	Wavelengths (nm)	Exponential decay times (%)
$Ba_3BP_3O_{12}$	400,569	56.3 ns (99.7%), 835 ns (0.3%)
$BaBPO_5$	400,465	428 ns (73.4%), 2913 ns (26.6%)
Ba_3BPO_7	400,466	592 ns (72.6%), 3279 ns (27.4%)

Considering the emission wavelength, decay time, light output and good mechanical properties of $Ba_3BP_3O_{12}$, $BaBPO_5$ and Ba_3BPO_7 in general, it is thus
significant for the development of $Ba_3BP_3O_{12}$, $BaBPO_5$ and Ba_3BPO_7 crystals to
be used as scintillation material by way of controlling the quality and quantity
of defects in the process of crystal growth.

In order to investigate the luminescence mechanisms of these compounds, a
large $BaBPO_5$ single crystal was grown by the top-seed solution growth (TSSG)
method with $Li_4P_2O_7$ used as the flux [23] and the X-ray-excited luminescence
(XEL) properties were investigated [24]. The grown crystal is transparent and color-
less, free from visible defects such as cracking, scattering centers or growth stria-
tion. The investigated sample was cut into dimensions of about $8\times5\times1.3$ mm^3
with two large facets polished for optical and luminescent measurements. The
sample for TL experiments had the dimensions of $3\times3\times1$ mm^3. Figure 21.11 exhibits the (XEL) spectrum of the $BaBPO_5$ single crystal at room temperature,
which is in good agreement with the XEL spectrum of $BaBPO_5$ powders.

As shown in Fig. 21.12, after X-ray irradiation, the $BaBPO_5$ crystal shows a
few broad, distinct absorption bands from near-infrared to the band gap in the
VUV, which indicates the presence of some trapping sites in the crystal and

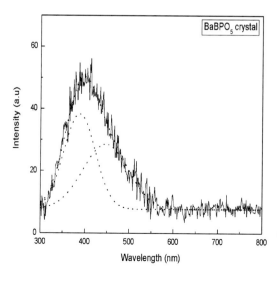

Fig. 21.11 The X-ray-excited
luminescence spectra of the
$BaBPO_5$ crystal. The dotted lines
are the Gaussian decomposition.

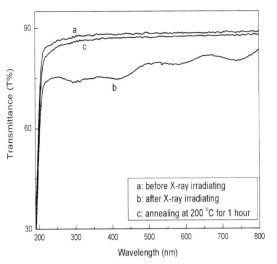

Fig. 21.12 Optical transmittance spectra of the BaBPO$_5$ crystal (a) pristine crystal; (b) after irradiation by X-rays for 15 min; (c) after annealing at 150 °C for 1 hour of the X-ray-irradiated sample.

a: before X-ray irradiating
b: after X-ray irradiating
c: annealing at 200 °C for 1 hour

some local defect energy levels are formed in the energy band gap. Examining Fig. 21.12c, it appears that the optical transmission damage, which originates from X-ray irradiation is totally bleached by post-annealing treatment at 150 °C for 1 hour. On the contrary, UV light irradiation exerts little influence on the OTS of the crystal. To simplify, the OTS under UV irradiation is omitted. Optical transmittance spectra also indicate that irradiation cannot affect the properties of the optical absorption edge of the crystal.

The irradiation-induced absorption coefficient μ may be a better way of evaluating the radiation hardness and it is expressed as:

$$\mu = \frac{1}{d}\ln(t_0/t_1) \tag{1}$$

where d stands for the thickness of the sample in centimeters and t_0 and t_1 are the initial transmission and the transmission after an irradiation procedure, respectively. In Fig. 21.3, induced absorption spectra are given for the X-ray-irradiated sample before and after annealing. Apparently, X-ray irradiation degrades the transmission of BaBPO$_5$ crystal in the UV-VIS spectrum. Moreover, four induced absorption bands are found and they center at about 304, 418, 580 and 740 nm, respectively.

The creation of the induced absorption bands might include the following steps consisting of: (1) the creation of hot electrons and holes by interaction of the lattice with high-energy radiation; (2) their separation during subsequent cooling down and diffusion processes; (3) separate localization of both kinds of charge carriers at suitable lattice sites relating to the defect structure. According to previous discussion, intrinsic defects should be created in the process of crystal growth. They might be in the form of $(V_B)'''$, $(P_B)^{\bullet\bullet}$ and the oxygen vacancy

$(V_O)^{\bullet\bullet}$ in the crystal. $(V_O)^{\bullet\bullet}$ can catch electrons to form F and F^+ centers. $(P_B)^{\bullet\bullet}$ is also able to attract electrons while $(V_B)'''$ can trap holes to give rise to color centers. They will make a contribution to the X-ray irradiation-induced absorption. Of course, the charge balance of the crystal is kept by charge compensation among these defects. Regretfully, the detailed characterization of these defects is too difficult to cover here and further experiments need to be performed.

Although the detailed information on the optical absorption edge of the Ba BPO$_5$ crystal is difficult to estimate because of the limitation of the instrument used, it is a fact that the optical band gap is shorter than 190 nm. For a UV-irradiated sample, the energies of the photons are so low (here, the wavelength of UV light is longer than 254 nm) that few hot electrons and holes can be produced. In fact, UV light hardly influences the optical properties of this material. Contrary to UV light, the higher energy (<190 nm) of the X-ray has the ability to easily excite a great many of hot electrons and holes and then give rise to color centers. The effects of the annealing treatment in air on the induced color centers can be seen in Fig. 21.13 b. All induced absorption bands are completely suppressed.

Figure 21.14 shows the thermo-luminescence (TL) glow curves of BaBPO$_5$ single crystal at temperatures from RT to 300 °C after UV and X-ray irradiation, respectively. The TL curve of ultraviolet light-irradiated crystal presents nearly a straight line, which is in agreement with that of the OTS experiment. In the case of X-ray irradiation, a significant thermo-luminescence glow peak at about 130 °C is observed. The phenomena indicate that there exist electron and/or hole trapping sites in relationship with rechargeable defects in the crystal. They can trap the electron and hole under X-ray irradiation and result in the reconstruction of a local charge balance. For an isolated glow peak, the trap parame-

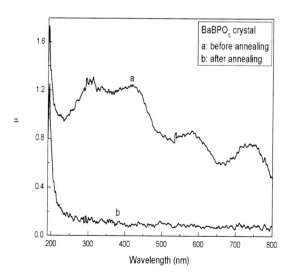

Fig. 21.13 Irradiation-induced absorption spectra of the BaBPO$_5$ crystal (a) after irradiation by X-rays for 15 min; (b) after annealing at 150 °C for 1 hour of the X-ray irradiated sample.

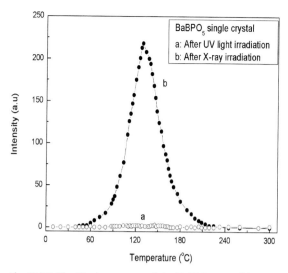

Fig. 21.14 The TL glow curves of the BaBPO₅ crystal between RT and 300 °C after (a) UV and (b) X-ray irradiation for 15 min.

ters such as the thermo-activation energy E and frequency factor S are estimated by the total glow method [25]. The dependence of the intensity of luminescence I on the temperature is expressed as:

$$\beta I(T)/\int_{T}^{T_\infty} IdT = S\exp(-E/kT) \tag{2}$$

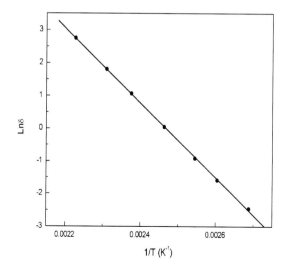

Fig. 21.15 Total glow peak plot of $\ln\delta = \ln(I(T)/\int_{JT}^{T_\infty} IdT)$ versus $1/T$. The points are experimental data and the solid line represents the least-squares fit.

in which β is the heating rate and k is the Boltzmann constant. Giving $\delta = I(T)/\int_T^{T_x} I dT$, Eq. (1) may be written as

$$\ln \delta = -\frac{E}{k} \cdot \frac{1}{T} + \ln \frac{S}{\beta} \tag{3}$$

The plot of $\ln \delta$ vs $1/T$ displayed in Fig. 21.15 is linear, indicating simple first-order kinetics. In this way we determine that E is 0.98 ± 0.02 eV and its S is 3.4×10^{12} s^{-1}.

21.2.3
The X-ray-excited Luminescence Properties of Ce^{3+}-activated Ba$_3$BP$_3$O$_{12}$, BaBPO$_5$ and Ba$_3$BPO$_7$

The Ce^{3+} is an important activator and has been widely used in the preparation of scintillation material. The Ce^{3+} ion has one 4f electron and its lowest excited configuration is 5d^1. It usually gives a band emission in the visible and near-UV regions [26–30]. Furthermore, the decay time of the Ce^{3+} emission is short, namely, a few tens of ns. Besides the VUV characterizations of some rare-earth-activated Ba$_3$BP$_3$O$_{12}$ fluorescence [31], the X-ray-excited luminescence properties of Ce^{3+}-activated Ba$_3$BP$_3$O$_{12}$, BaBPO$_5$ and Ba$_3$BPO$_7$ have also been investigated [31], with the intention of obtaining new kinds of scintillation material with larger, light yield and short decay time.

The samples of Ce^{3+}-activated Ba$_3$BP$_3$O$_{12}$, BaBPO$_5$ and Ba$_3$BPO$_7$ were prepared by a high-temperature solid-state reaction. The following reagent-grade chemicals were used in the synthesis of the samples: BaCO$_3$ (99.99%), NH$_4$H$_2$PO$_4$ (99.99%), H$_3$BO$_3$ (99.99%) and K$_2$CO$_3$ and CeO$_2$ (99.99%). Two firing steps were necessary

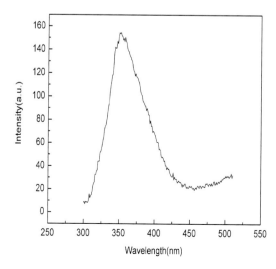

Fig. 21.16 The X-ray-excited luminescence spectrum of Ba$_3$BP$_3$O$_{12}$:Ce^{3+}.

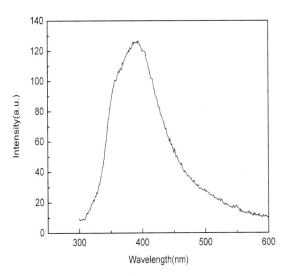

Fig. 21.17 The X-ray-excited luminescence spectrum of $BaBPO_5 : Ce^{3+}$.

to synthesize the pure doped sample. Stoichiometric amounts of the starting materials were firstly thoroughly ground up, mixed and fired in air at 400 °C for 10 h in a covered alumina crucible, then reground and pressed into pellets of 2 cm diameter under 170 Mpa pressure, and subsequently fired in a carbon-reducing atmosphere at 1000 °C for 24 h in a covered alumina crucible. For all samples, the Ce^{3+} activator concentration is 0.5 mol% of $BaCO_3$.

Figures 21.16, 21.17 and 21.18 show the X-ray-excited luminescence properties of Ce^{3+}-activated $Ba_3BP_3O_{12}$, $BaBPO_5$ and Ba_3BPO_7 respectively. They all show a broad emission band with peak center at about 351 nm for $Ba_3BP_3O_{12} : Ce^{3+}$,

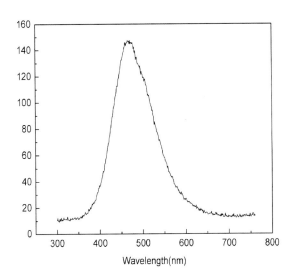

Fig. 21.18 The X-ray-excited luminescence spectrum of $Ba_3BPO_7 : Ce^{3+}$.

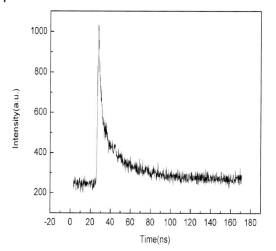

Fig. 21.19 Room-temperature fluorescent decay profile for $Ba_3BP_3O_{12}:Ce^{3+}$.

466 nm for $Ba_3BPO_7:Ce^{3+}$ and 389nm for $BaBPO_5:Ce^{3+}$ respectively. The light yield of these samples is comparable to that of $PbWO_4$ powders. Figure 21.19 exhibits the room-temperature fluorescence decay profiles of $Ba_3BP_3O_{12}a:Ce^{3+}$ powders. The experimental decay curve can be fitted with an equation with three exponential terms corresponding to three decay times of 2 ns (9.7%), 142 ns (55.4%) and 18 ns (34.9%), respectively. Considering the emission wavelength, light yield, decay time, melting point and non-hygroscopic property of $Ba_3BP_3O_{12}:Ce^{3+}$, one has reason to assume that this compound might find an application as a new scintillation material.

21.2.4
Potential Scintillation Material of $Ba_3BP_3O_{12}:Eu^{2+}$ [32]

The divalent rare-earth ion Eu^{2+} has the $4f^7$ electronic configuration at the ground states and the $4f^65d^1$ electronic configuration at the excited states. The broadband absorption and luminescence of Eu^{2+} are due to $4f^7 - 4f^65d^1$ transitions. The emission of Eu^{2+} is very strongly dependent on the host lattice. It can vary from the ultraviolet to the red region of the electromagnetic spectrum. Furthermore, the 4f–5d transition of Eu^{2+} decays relatively fast, less than a few microseconds [33].

The sample of $Ba_3BP_3O_{12}:Eu^{2+}$ was prepared by a high-temperature solid-state reaction. The following reagent-grade chemicals were used in the synthesis of the samples: $BaCO_3$ (99.99%), $NH_4H_2PO_4$ (99.99%), H_3BO_3 (99.99%) and Eu_2O_3 (99.99%). Two firing steps were necessary for synthesizing the pure doped sample. Stoichiometric amounts of the starting materials were first thoroughly ground up, mixed and fired in air at 400 °C for 10 h in a covered alumina crucible, then reground and pressed into pellets of 2cm diameter under

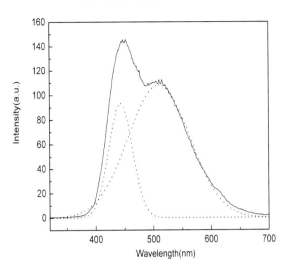

170Mpa pressure, and subsequently fired in air at $1000\,^{\circ}C$ for 24 h in a covered alumina crucible.

Figure 21.20 exhibits the X-ray-excited luminescence (XEL) spectrum of $Ba_3BP_3O_{12}:Eu^{2+}$ at room temperature. It shows a broad emission band extending from 400 nm to 650 nm, which can be defined as a superposition of two Gaussian components with approximate peaking wavelengths of $P_1 \sim 440$ nm and $P_2 \sim 510$ nm, respectively. This is a consequence of having two Eu^{2+} sites in the structure. According to the literature [11], the Ba^{2+} ions are located in two different environments. As shown in Fig. 21.21, the Ba1 is coordinated with eleven, but Ba2 with ten, oxygen atoms. It is believed that the Eu^{2+} ions would replace the sites of Ba^{2+} ions in the host of $Ba_3BP_3O_{12}$. Thus, the Eu^{2+} ions should also be located in two different sites in the host of $Ba_3BP_3O_{12}$.

Figure 21.22 shows the comparison between the XEL spectra of $Ba_3BP_3 O_{12}:Eu^{2+}$ and that of $Bi_4Ge_3O_{12}$ powders with the same measurement conditions. It is worthwhile to note that both spectra have a similar broad emission band. The integral area of the emission bands for $Ba_3BP_3O_{12}:Eu^{2+}$ powders is

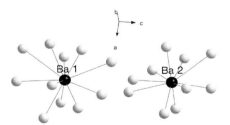

Fig. 21.21 The oxygen atoms coordination of the two different Ba sites in $Ba_3BP_3O_{12}$ (Ba atoms, black spheres; O atoms, grey spheres).

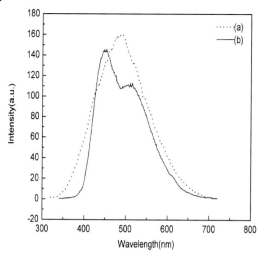

Fig. 21.22 Comparison between the XEL spectrum of $Ba_3BP_3O_{12}:Eu^{2+}$ powders (b) and that of BGO powders (a) under the same conditions.

about 80% as large as that of $Bi_4Ge_3O_{12}$ powders with the same measurement conditions. Thus, the light yield of the two compounds is comparable.

Figure 21.23 exhibits the room-temperature fluorescence decay profiles of $Ba_3BP_3O_{12}:Eu^{2+}$ powders. The experimental decay curve can be fitted by an equation with two exponential terms corresponding to two decay times of 20 ns (98.97%) and 522 ns (1.03%), respectively.

The dependence of the X-ray-excited emission intensities on the amount of doped Eu^{2+} are shown in Fig. 21.24. The intensity of the phosphors increases with the dopant content up to a maximum ($x=0.08$) and then it decreases.

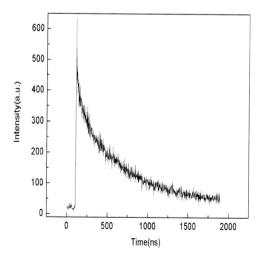

Fig. 21.23 Fluorescence decay profile of $Ba_3BP_3O_{12}:Eu^{2+}$ excited by pulsed X-ray.

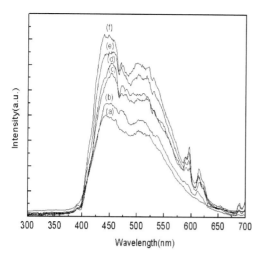

Fig. 21.24 The XEL spectra of $(Ba_{(1-x)}Eu_x)_3BP_3O_{12}$ powders for different replacements of Ba^{2+} by Eu^{2+}. (a) x=0.005; (b) x=0.01; (c) x=0.03; (d) x=0.05; (e) x=0.10; (f) x=0.08.

However, as shown in Fig. 21.24, in highly doped samples, besides the broad emission band of Eu^{2+}, the emission peaks of Eu^{3+} appeared in the XEL spectrum even for samples prepared in a carbon-reducing reaction atmosphere, resulting from incomplete reduction or from impurities of Eu^{3+} under highly doped concentration. In order to obtain the maximum light yield of $Ba_3BP_3O_{12}:Eu^{2+}$, any trace of Eu^{3+} ions should be removed. A stronger reducing atmosphere might be necessary.

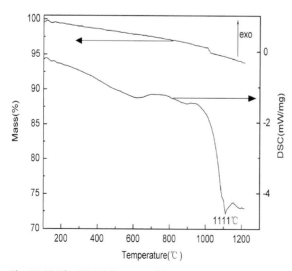

Fig. 21.25 The TG-DSC curves of $Ba_3BP_3O_{12}$.

Table 21.2 Characteristics of typical scintillators and the corresponding characteristics of $Ba_3BP_3O_{12}:Eu^{2+}$.

Quantity	NaI:Tl	PbWO$_4$	Bi$_4$Ge$_3$O$_{12}$	Ba$_3$BP$_3$O$_{12}$:Eu^{2+}
Emission peak (nm)	415	420/450	480	440/510
Relative light yield (%)	100	0.2	9	7.2 [a]
Decay time (ns)	230	6/30	300	20/522
Density (g cm^{-3})	3.67	8.28	7.13	4.17
Melting point (°C)	651	1123	1050	1111
Hygroscopy	Strong	No	No	No

a) Calculated according to comparison with BGO powders.

The TG-DSC studies of $Ba_3BP_3O_{12}:Eu^{2+}$ (Fig. 21.25), show that this compound is chemically stable and melts congruently at about 1111 °C.

In Table 21.2 the characteristics of typical scintillators are summarized and compared with $Ba_3BP_3O_{12}:Eu^{2+}$.

Considering the suitable emission wavelength range, the large light yield, the short decay time, the moderate density, the moderately low melting point and the non-hygroscopic property of $Ba_3BP_3O_{12}:Eu^{2+}$, one has reason to assume that this compound might find an application as a new scintillation material.

21.3
Outlook

Borophosphate research is still a relatively young field although it has already demonstrated rich structural chemistry and physics, and many new compounds with novel structures have been reported. The exploration of this new pool of potential functional materials is still inadequate. From the above, very limited, examples of some borophosphates investigated from the point of view of scintillation usages, one can already see that the potential of this group of compounds is large. The advantages of these compounds as scintillation materials are obvious for the following reasons: they contain neutron-sensitive boron, crystal growth is easy, they have a low melting point, moderate density, and resources are cheap. For the alkaline-earth-containing compounds, relatively easy rare-earth-element doping is another advantage, which makes the fluorescence properties more worthwhile and feasible. It is foreseen that borophosphates, especially when associated with rare-earth elements, will constitute an important part of potential scintillation material systems.

References

1 S. P. Velsko, M. Webb, L. Davis, D. Huang, *IEEE J. Quant. Electron.* 27 (1991) 2182.

2 C. Chen, Y. Wu and R. J. Li, *J. Cryst. Growth.* 99 (1990) 790.

3 T. Y. Fan, C. E. Huang, B. Q. Hu, R. C. Eckhardt, R. C. Fan, *App. Opt.* 26 (1987) 2391.

4 R. Kniep, H. Engelhardt, C. Hauf, *Chem. Mater.* 10 (1998) 2930.

5 G. Blasse, A. Bril, J. de Vries, *J. Inorg. Nucl. Chem.* 31 (1969) 568.

6 J. W. M. Verwey, G. J. Dirken, G. Blasse, *J. Phys. Chem. Solids.* 53 (1992) 367.

7 A. Karthikeyan, R. Jagannathan, *J. Lumin.* 86 (2000) 79.

8 H. B. Liang, Q. Zeng, Y. Tao, S. Wang, Q. Su, *Mater. Sci. Eng. B* 98 (2003) 213.

9 H. B. Liang, Y. Tao, W. Chen, X. Jin, S. Wang, Q. Su, *J. Phys. Chem. Solids.* 65 (2004) 1071.

10 C. J. Duan, X. Y. Wu, H. H. Chen, X. X. Yang, J. T. Zhao, *Mater. Sci. Eng. B.* 121 (2005) 272.

11 R. Kniep, G. Goezel, B. Eisenmann, B. Roehr and C. Asbrand, *Angew. Chem. Ed. Int.* 33 (1994) 749.

12 H. Z. Bauer, *Z. Anorg. Allg. Chem.* 345 (1966) 225.

13 H. W. Ma, J. K. Liang, L. Wu, *J. Solid State Chem.* 177 (2004) 3454.

14 C. J. Duan, X. Y. Wu, H. H. Chen, X. X. Yang, J. T. Zhao, *J. Chin. Inorg. Mater.* (in Chinese), 20 (2005), 1043.

15 C. J. Duan, X. Y. Wu, Y. Liang, H. H. Chen, X. X. Yang, J. T. Zhao, Z. M. Qi, G. B. Zhang, Z. S. Shi, *J. Lumin.*, 117 (2006) 83.

16 B. V. R. Chowdari, G. V. S. Rao, C. J. Leo, *Mater. Res. Bull.* 36 (2001) 727.

17 E. A. Kotomin, A. I. Popov, *Nucl. Instrum. Methods Phys. Res. B* 141(1998) 1.

18 M. Liu, A. Kitai, H. P. Mascher, *J. Lumin.* 54 (1992) 35.

19 G. Corradi, A. Watterich, F. F. Oldvari, R. Voszka, J. R. Niklas, J. M. Spaeth, O. R. Gilliam, L. A. Kappers, *J. Phys: Condens. Matter.* 2 (1990) 4325.

20 R. D. Shannon, *Acta. Crystallogr.* A32 (1976) 751.

21 O. F. Schirmer, O. Thiemann, M. Wohlecke, *J. Phys Chem. Solids* 52 (1991) 185.

22 K. E. Sickafus, A. C. Larson, N. Yu, M. Nastasi, G. W. Hollenberg, F. A. Garner, R. C. Bradt, *J. Nuclear Materials* 219 (1995) 128.

23 S. L. Pan, Y. C. Wu, P. Z. Fu, G. C. Zhang, G. F. Wang, X. G. Guan, C. T. Chen, *J. Cryst. Growth.* 236 (2002) 613.

24 W. F. Li, X. Q. Feng, C. J. Duan, J. T. Zhao, Y. C. Wu, *J. Phys. D: Appl. Phys.* 38 (2005) 385.

25 G. C. Taylor, E. Lilley *J. Phys. D: Appl. Phys.* 11 (1978) 567.

26 A. Bril, H. A. Klasence, *Philips Res. Rep.* 7 (1952) 421.

27 G. Blasse, A. Bril, *Appl. Phys. Lett.* 11 (1967) 53.

28 A. Bril, G. Blasse, J. A. De Poorter, *J. Electrochem. Soc.* 117 (1970) 346.

29 W. J. Miniscalco, J. M. Pellegrino, W. M. Yen, *J. Appl. Phys.* 49 (1978) 6109.

30 R. R. Jacobs, W. F. Krupke, M. J. Weber, *Appl. Phys. Lett.* 33 (1978) 410.

31 C. J. Duan, X. Y. Wu, Y. Liang, H. H. Chen, X. X. Yang, J. T. Zhao, Z. M. Qi, G. B. Zhang, Z. S. Shi, *Mater. Sci. Engineering B* 121 (2005) 272.

32 C. J. Duan, X. Y. Wu, W. F. Li, H. H. Chen, X. X. Yang, J. T. Zhao, *Appl. Phys. Lett.* 87 (2005) 201917.

33 J. S. Kim, P. E. Jeon, J. C. Choi, H. L. Park, *Solid State Commun.* 133 (2005) 187.

Subject Index

Inorganic Chemistry in Focus III.
Edited by G. Meyer, D. Naumann, L. Wesemann
Copyright © 2006 WILEY-VCH Verlag GmbH & Co. KGaA, Weinheim
ISBN: 3-527-31510-1